PATIENT SAFETY CULTURE

Patient Safety Culture
Theory, Methods and Application

Edited by

PATRICK WATERSON
Loughborough University, UK

CRC Press
Taylor & Francis Group
Boca Raton London New York

CRC Press is an imprint of the
Taylor & Francis Group, an **informa** business

CRC Press
Taylor & Francis Group
6000 Broken Sound Parkway NW, Suite 300
Boca Raton, FL 33487-2742

First issued in paperback 2017

© 2014 by Patrick Waterson
CRC Press is an imprint of Taylor & Francis Group, an Informa business

No claim to original U.S. Government works

ISBN-13: 978-1-4094-4814-3 (hbk)
ISBN-13: 978-1-138-74972-6 (pbk)

Visit the Taylor & Francis Web site at
http://www.taylorandfrancis.com

and the CRC Press Web site at
http://www.crcpress.com

Contents

PART IV ADDITIONAL PERSPECTIVES AND FUTURE DIRECTIONS

List of Figures

List of Tables

List of Boxed Text

Author Biographies

Çakil Agnew is a psychologist with a background in training in human factors and organisational psychology. Initially a post-doctoral researcher at the Industrial Psychology Research Centre of the University of Aberdeen, her research interests include the impact of safety culture and leadership at work, mainly in high-risk industries.

Darren M. Ashcroft is a pharmacist and Professor of Pharmacoepidemiology at the University of Manchester, UK. He is the research theme lead for medication safety in the NIHR Greater Manchester Primary Care Patient Safety Translational Research Centre and has led work investigating incident reporting, medication error and safety culture in a range of healthcare settings.

James Battles is a native of Ohio where he did his undergraduate education at Miami University; he received his doctorate in medical education from Ohio State University. In November 2000, Dr. Battles joined the Agency for Healthcare Research and Quality (AHRQ) in Rockville, Maryland as Social Science Analyst for Patient Safety and he is senior content specialist in patient safety for AHRQ's patient safety initiative. He has led AHRQ's efforts in advancing the use of more proactive risk assessment and moving to risk informed design of patient safety interventions. He also leading AHRQ's efforts in the assessment of patient safety culture and improving team work, in collaboration with the Department of Defense.

Peter Bohan has worked in the public sector for over 25 years in a variety of roles and organisational situations. A significant part of his career has involved providing Health and Safety (H&S) advice in Local Authorities covering areas such as Education, Social Services and Construction (Design Consultancy) and he has spent several years within the Environmental Health Department as an H&S Enforcement Officer. He has also worked as a Risk Manager within three NHS organisations which have included a Primary Care Trust and Community Health Services. Peter is currently Head of Safety and Risk at Warrington and Halton NHS Foundation Trust. He is a Chartered Member of the Institute of Occupational Safety and Health; he has a BSc Honours degree in Health, Safety and Environmental Management and has recently completed his Master's degree in Health and Social Care Management.

Henning Boje Anderson is a professor at the Management Engineering Institute at the Technical University of Denmark (DTU). His work examines human performance in safety critical applications, focusing primarily on healthcare and seeking to explore the transfer of experience to and from process industry, aviation and transport sectors. His research themes are centred around human-machine interaction and risk analysis.

Jane Carthey is a Human Factors and Patient Safety Specialist with 17 years' NHS experience. Dr Carthey was formerly Assistant Director of Patient Safety at the National Patient Safety Agency. Her research portfolio includes analysing the influence of human factors on surgical outcomes, identifying system failures that lead to adverse drug events, developing tools to evaluate organisational safety culture, understanding the causes of policy non-compliance and improving team handovers. She has also carried out work on applying human reliability analysis methods to healthcare domains, incident reporting and investigation and facilitating cultural change to develop an open and fair culture.

Ken Catchpole is a research psychologist and human factors practitioner who works with clinicians to develop and scientifically evaluate interventions to improve performance. He has contributed to healthcare research and improvement in the UK, Netherlands, Norway, Australia, New Zealand and the USA and his work has been adopted by hospitals as far away as Dubai and Sao Paulo. After working at Great Ormond Street Hospital and the University of Oxford, he is now based at Cedars-Sinai Medical Centre in Los Angeles.

Jason Etchegaray is an assistant professor at the University of Texas Medical School at Houston and a member of the UT-Memorial Hermann Center for Healthcare Quality and Safety. He is currently working on a K award from the Agency for Healthcare Research and Quality that is focused on linking aspects of high-performance work systems with healthcare quality and patient safety. He is also applying his psychometric and survey methodology background to various patient safety projects.

Rhona Flin is Professor of Applied Psychology and Director of the Industrial Psychology Research Centre at the University of Aberdeen where she leads a team of psychologists conducting research on human performance in high-risk industries and healthcare. Her group's projects include studies of leadership, culture, team skills and decision making in acute medicine, aviation and energy industries. She is currently studying non-technical skills in drilling rig crews, decision making by surgeons and senior managers' safety leadership.

Francesca Giuliani is head of the department for quality management and patient safety at the University Hospital Zurich. She has more than 10 years' practical experience in building up a safety culture at the interplay of corporate strategy, processes and organisational culture, including through coaching and moderating teams. She is a hospital pharmacist by training, with a PhD in environmental risk assessment from the Swiss Federal Institute of Technology, Zurich and further education in quality and risk management.

Mark Griffin is Co-Director of ALL@UWA as well as a Winthrop Professor at UWA School of Psychology. Mark has extensive experience working in organisational psychology throughout Australia, Europe and the USA, investigating areas such as safety, well-being, organisational health, culture and leadership. Mark also has extensive knowledge of statistics and research methods. He currently holds a prestigious Australian Research Council Future Fellowship for his leading edge research in safety leadership.

Frank Guldenmund is a lecturer in the Safety Science Group of Delft University of Technology in the Netherlands. He works primarily in the field of occupational safety, where his research involves modelling and the assessment of (the quality of) safety management systems. In this context he has developed an interest in the topic of safety culture, which he has been pondering since the late 1990s.

Margareta Haelterman is head of the Quality team at the Health Care Department of the Belgian Federal Public Service Health, Food Chain Safety and Environment, Brussels. Trained as a medical doctor and epidemiologist, she now coordinates a team promoting patient safety culture, quality systems, risk management, implementation of innovative processes in care and the use of clinical indicators in different clinical domains.

Antje Hammer is a state-certified nurse and graduate sociologist, working as a research associate at the Institute for Medical Sociology, Health Services Research and Rehabilitation Science (IMVR) of the University of Cologne. Her main research fields are organisational culture, safety culture, patient safety and healthcare quality. She has experience in both qualitative and quantitative methods in empirical social research and completed her dissertation on safety culture in German hospitals.

Johan Hellings is CEO of AZ Delta, a large private-public hospital in the West of Flanders, Belgium. In his PhD research he focused on safety culture and adverse drug events in hospitals. He is a member of the multidisciplinary patient safety research group of Hasselt University, Belgium. He is also president of Flanders Care, an advisory committee of the Flemish Government.

Kenji Itoh is a professor of industrial engineering and management at Tokyo Institute of Technology, Tokyo, Japan. He has been working in the fields of human factors and cognitive engineering and, particularly, in safety research and risk management in several domains including healthcare, transportation – such as rail and maritime transport – and manufacturing for more than 25 years.

Jeanette Jackson joined the Health Quality Council of Alberta (HQCA, Canada) in March 2013 as Health System Data Analyst. Originally from Germany, Jeanette moved to Scotland in 2004 to complete her PhD in Public Health, focusing on improving the measurement of health predictors and outcomes. Before joining the HQCA, Jeanette worked for five years for the Scottish Patient Safety Research Network at the University of Aberdeen (UK) to provide methodological support and advice. She has been involved in various national and international studies in the field of patient safety in collaboration with National Health Services and the World Health Organisation. Jeanette will help the HQCA to elaborate further the survey methodology used to assess Albertans' experiences of several aspects of their healthcare.

Barry Kirwan works for Eurocontrol in France. His principal responsibility is running the European Air Traffic Management Safety Culture Programme, which involves surveys for around 30 European States, helping them evaluate and improve their safety culture. He also chairs a Eurocontrol-FAA group on safety research and periodically works on human factors and human reliability assessment issues with the nuclear power industry in the USA and UK.

Maria Helena Kiss is a project leader in the department of quality management and patient safety at the University Hospital in Zürich. Her background is in psychology and systemic psychotherapy. Her work focuses on patient safety projects, patient surveys and the psychological care of patients with burn injuries and their relatives.

Theresa Kline is Emeritus Professor of the Industrial-Organizational Psychology Program at the University of Calgary Her research programme continues in the areas of psychometrics, organisational effectiveness and work attitudes. She has an active organisational consulting practice with projects ranging from individual and organisational assessment to strategic alignment.

Tanja Manser is Associate Professor of Industrial Psychology and Human Factors at the University of Fribourg. Her research is focused on human performance and patient safety. Her work focuses on acute care settings, studying safety relevant behaviour such as teamwork and communication at organisational interfaces, using clinical and simulated environments.

Mirjam Meier is a project manager in the field of patient safety at the University Hospital Zurich. Her interest focuses on enhancing process reliability within complex systems like healthcare and systems in which the use of technology presents the complex interplay between humans, technology and organisational structures. She studied work and organisational psychology with a focus on human factors and safety psychology at the University of Applied Sciences, Northwestern Switzerland. She also has more than 10 years' experience as a Biomedical Scientist in laboratory diagnostics and clinical research.

Ruth McDonald is Professor of Health Policy and Management at Warwick Business School. Much of her research in recent years has concerned financial incentives for quality in the UK and beyond. Her research brings a sociological perspective to a topic dominated largely by economists.

Kim Lyngby Mikkelsen is a medical doctor with a PhD in public health epidemiology. He worked for 10 years as a senior researcher in occupational health safety research, focusing on safety culture and climate, on the development of a Nordic safety climate questionnaire and on safety intervention research. Since 2009 his work has been centred on patient safety, including learning from adverse events.

Hilde Peleman has a background in psychology. She is a staff member in quality and patient safety at the Health Care Department of the Belgian Federal Public Service Health, Food Chain Safety and Environment, Brussels.

Denham L. Phipps is a research fellow in the NIHR Greater Manchester Primary Care Patient Safety Translational Research Centre, the University of Manchester. With a background in human factors, his research interests include safety management, work design, risk-based regulation and healthcare professionals' fitness to practice.

Elina Pietikäinen works as senior research scientist at the VTT Technical Research Centre in Finland. Her background is in psychology. Her research interests include safety culture evaluation and development, safety management, safety expertise, organisational evaluation and organisational learning. She works in research and development projects in several safety critical industries, such as social and healthcare and the nuclear industry.

Teemu Reiman works as a senior scientist in the organisational psychology research group in VTT Technical Research Centre in Finland. His area of specialty is safety culture and safety management of complex sociotechnical systems and he has acted as a project manager in several national and international research and consultancy projects. He has experience from various safety-critical domains including nuclear power, conventional power, petrochemical, transportation, metal industries and healthcare. In addition to the publication of more than 30 scientific papers and articles, he has written two popular scientific books on human factors and safety critical organisations in Finnish (published by Edita Publishing).

Ward Schrooten has a background in medicine and statistics. He is member of the patient safety research group at Hasselt University, Belgium.

Steven T. Shorrock is Project Leader, Safety Development at EUROCONTROL and Adjunct Senior Lecturer at the University of New South Wales, School of Aviation. Steve has a background in human factors, safety culture and system safety practice. His research encompasses aviation, rail, chemical and energy issues and has been undertaken in higher education and at Federal Government level. He is a Registered Ergonomist and Chartered Psychologist.

Joann Speer Sorra is an organisational psychologist with more than 15 years' experience in organisational and health services research. She has expertise in the areas of organisational culture, medical error and patient safety and patient experience of healthcare, encompassing implementation science, programme evaluation, survey methodology and quantitative data analysis. Dr. Sorra led the development of the AHRQ Hospital, Nursing Home, Medical Office and Pharmacy Surveys on Patient Safety Culture. Her recent work has been in developing surveys to assess value and efficiency in healthcare, patient safety culture in ambulatory surgery and provider and patient experiences in patient-centered medical homes.

Zenobia Talati is an Industrial/Organisational Psychology Masters and PhD candidate at the University of Western Australia. Through her work as a research assistant, she has developed expertise in the areas of leadership, organisational culture/climate, workplace safety, safety leadership and proactivity. She has also co-authored a chapter on how leaders can improve employee safety behaviours and safety culture in high risk industries.

Eric Thomas is Professor of Medicine and Associate Dean for Healthcare Quality at the University of Texas Medical School at Houston. He also directs the UT Houston-Memorial Hermann Center for Healthcare Quality and Safety. Since 1992 he has conducted research on patient safety and his work was heavily cited in the Institute of Medicine's landmark report on medical error. His current research focuses on diagnostic errors, measuring safety culture, measuring and improving teamwork and the use of health information technology to improve quality and safety. He also serves as the Chancellor's Health Fellow for Patient Safety for The University of Texas System. In that role he leads UT System work on disclosure of errors – a quality and safety grants programme – and he serves on the steering committee for the UT System Clinical Safety and Effectiveness Course. As Associate Dean for Healthcare Quality he works with other leaders of the school to develop quality and safety programmes within the education, research and patient care missions of the school. In 2007 he received the John M. Eisenberg Patient Safety and Quality Award for Research from the National Quality Forum and Joint Commission.

Amanda van Vegten-Schmalzl has a background in organisational psychology and management, having obtained a broad experience of safety culture and incident reporting during her PhD research. She has since taken up the post of researcher/lecturer at the Saxion University of Applied Science after working for three years as clinical risk manager at the University Hospital Zurich. She is currently active as a consultant and coach in the field of high reliability organising and safety.

Annemie Vlayen has a background in rehabilitation sciences and healthcare and hospital management. She is a senior researcher in patient safety at Hasselt University, Belgium. In her PhD research she focused on medical record review, prospective risk analysis and safety culture in hospitals. Her research interests include approaches to improving safety culture, leadership and patient safety education.

Justin Waring is Professor of Organisational Sociology at Nottingham University Business School and a Health Foundation Improvement Science Fellow. His research develops theoretical and methodological synergies across social science disciplines to better understand and improve the quality and safety of healthcare. He has researched extensively in the field of health service improvement and contributed to a socio-cultural perspective on patient safety (published by Ashgate).

Patrick Waterson is a Reader in the Human Factors and Complex Systems Group at Loughborough Design School, Loughborough University. His research focuses on applying the systems approach to a variety of topics within healthcare and other domains. He has published widely within the field of human factors and ergonomics, including work on health information technologies, systemic accident analysis and patient safety culture. Dr. Waterson is a Scientific Editor for *Applied Ergonomics* and serves on editorial boards for a number of other journals.

Foreword

Many of us are fortunate to have worked in the last decade with doctors, nurses and other health professionals at the 'sharp end' of healthcare, attempting to understand why and how errors are made and what we can do to stop them. Universally, I am in admiration of every healthcare practitioner[1] I have ever met. It is not just a career, but so all-encompassing as to be a life choice. It is physically, intellectually, philosophically, existentially challenging work. Every day, they must make the right choices for their patients in the face of personal, social and production pressures. They must ask for patients to place trust in them, to trade the uncertainties and higher short-term risks of treatment – it is never a trivial matter to take knife-to-skin – for long term benefits and guide them through that complex, imperfect and sometimes traumatic process. Not a single practitioner I have encountered would ever feel or want to feel that what they did could possibly lead to harm rather than cure. Indeed, one cannot imagine asking for that trust if they did not feel it was warranted.

Yet the statistics say otherwise; that medical error is highly prevalent and occasionally catastrophic. Every single practitioner at work has made an error. It might be an instantaneous mis-perception, immediately identified; a wrong diagnostic choice easily corrected; mistaking one patient for another but quickly caught; calculating the wrong drug dosage, rectified by a colleague; or going home exhausted without informing another team member about a test result. Mostly, these errors are insignificant, quickly forgotten and perhaps not even noticed. However, sometimes events conspire in the ways we now know to allow tragedies to occur. When the 'holes in the Swiss cheese' line up, not only is the patient a victim, but frequently the clinician too – the legal system, and sometimes colleagues, seeking to vilify and isolate, coupled with personal guilt and confusion about how such a thing could have happened. Indeed, no matter how perfect or error-free the care that is delivered, many clinicians will inevitably, albeit occasionally, lose their patients – and when they do, they must wake the next day, go to work, face their colleagues, perhaps a board of enquiry, and their own feelings and once again ask their patients to trust them to deliver the care that is needed.

The last decade has seen a huge rise in interventions, training and new processes related to safety. Now most hospitals worldwide have a set of professionals dedicated to gathering, analysing and reporting the many safety

---

1 I use this term to refer collectively to those who care for patients or contribute directly to the healthcare process.

incidents that occur every year. In the rush to improve safety, the spread and use of some ideas has outstripped evidence of effectiveness. Looking back, it is easy to see how naïve many of us were. There are no 'low hanging fruit'. Healthcare is person-driven and extremely complex, with only peripheral technological mediation. Unlike other high-risk industries, it was never engineered to achieve a specific goal, but grew organically over many centuries to ease as many threats to the human condition as practice and science could deliver; yet of course we can never ultimately prevent death. Many studies individually report improvements but sustainability and the spread of good ideas are both challenges. For a 'simple checklist' to be functional and effective is neither simple nor purely related to the contents of a sheet of paper. To be adopted initially, a checklist needs to deliver something of what is needed to the people who are using it, while to be permanently effective, the use needs to be continually reinforced in the face of competing pressures such as throughput or the attitude that 'we are safe around here'.

In the UK in the 1970s, 1980s and 1990s considerable effort was put into two aspects of driving safety; the seatbelt, and the reduction in drunk driving. For the former, new laws were introduced; for the latter, the penalties became higher. However, in both cases, it was not the 'stick' approach that was effective, but something more deeply rooted. Those of us growing up in that era were highly influenced by the campaigns for behavioural change. Many people now feel extremely uncomfortable about riding without a seatbelt – putting it on is habitual behaviour upon entering a car – and most find it socially unacceptable to drink and drive; they will go to great lengths to ensure they and their friends do not do this, ostracising those who break those social norms. Some people, however, still feel that such rules do not apply to them. Thus, having processes, behaviours, safety systems, or regulation is not sufficient to ensure safety. At a fundamental level, it is humans who deliver safety, and thus any component of safety needs to be understood, accepted, habitually used and socially embedded to be effective.

It is certainly true that many clinicians feel that safety processes and checks can be irritating, make their work more difficult, require them to work longer hours without extra pay, or reduce their ability to care for the largest number of patients possible. Some feel that safety is solely their choice – that diligence, discipline and awareness alone will ensure safety. Both views are flawed. Working harder or being more vigilant cannot necessarily prevent errors. Indeed, this view of individual choice is insidious and destructive in many ways and can be difficult to change in the face of medical schools and professional colleges that continuously emphasise individual self-determination over collective delivery of care. However, it is also clear that many interventions can add checks, bureaucracy or other forms of non-value adding behaviour that lead to higher pressures on those involved. In a tightly resource-constrained world, the costs of safety interventions can reduce the ability to care for those patients who really need it. It is therefore right that there is resistance to change that is not warranted and does not benefit patients. But this is not so when rules are broken or change resisted because the view is that this isn't part of 'what we do'. Thus, accidents are mitigated or avoided through a complex social, technical and procedural milieu that encourages people to be the 'elastic glue' that holds the system together. This culture is not just about prevention but about understanding and learning after something terrible has happened. We know that no system can be perfect and that every effort should be expended to deliver the highest possible care to the largest number of people, for the least cost; but we also know that there will always be trade-offs with these other life and work pressures; even given every effort to make the right decisions, we know that, sometimes, bad things happen.

As a clinician, you are provided with many years of training, much of which becomes secondary to the processes of working in hospitals – dealing with broken equipment, difficult patients or colleagues, limited resources, complex communications, distributed multi-skilled teams, unclear information and frequent delays at every level. Sometimes it feels like the real work of a clinician is in navigating patients through a series of disparate and indifferent systems or processes that are frequently dysfunctional. Visiting many hospitals and speaking with many practitioners, across Europe, Australia and the USA, it is clear that while some hospitals appear 'better' than others (better staffed, better funded, busier, with more modern buildings, better facilities, a good reputation and competition for staff positions) I have never visited a hospital that did not have organisational, professional and financial challenges. As with any industry it is easily possible to find disillusionment and disengagement running side-by-side with enthusiasm and motivation. We can never make the perfect system for everyone but we can make a system that is good enough. It is perhaps the idea of what is 'good enough round here' that describes the collective views of those who strive to find balance in these universally beneficial organisations, and thus in the culture of an organisation. From Bristol in the 1990s, more recently in Mid Staffordshire, and in a swathe of other events around the world, we know that those perceptions can go awry. Thus, being able to identify where our balance point is – what is acceptable and what is not – the way we do things here – is a vital component of our safety awareness and the application of our safety systems.

The idea of 'culture' is perhaps similar to that of 'intelligence' – everyone thinks they know what it is but conceptual clarity is more elusive. It is 'the way we do things around here'; the interactions of work units and the behavioural norms that promote safety; a product of individual and group values, attitudes, perceptions, competencies and behaviour that determine the proficiency of an organisation's health and safety management, situated in national, and professional contexts. Definitions are not universal and are sometimes not even explicit. It is challenging to measure. Safety is multi-dimensional, but scales attempt to simplify this complexity by aggregating or reducing to a single or smaller number of dimensions. Different professionals may have very different perceptions of safety. Calibration may be difficult as representative measures are difficult to collect and are themselves multidimensional. It is these complexities that the present book seeks to address; and as such it forms a significant advance in aiding our clinicians to understand their own attitudes to safety, and to deliver ever better care for our patients.

Ken Catchpole,
Los Angeles

Acknowledgements

I owe a huge debt of gratitude to the authors who contributed chapters to this book and the support and encouragement many of them gave to me during its preparation. I should also like to thank Guy Loft at Ashgate for initially believing in the idea and spurring me into action on the many occasions we bumped into each other at conferences in Europe and the USA. Jude Chillman, Lianne Sherlock and Barbara Spender also did a sterling job of nursing the book into production. Finally and most importantly, I would like to thank Angelika, Orla and Hannes for putting up with an often absentee (and grumpy) husband and father and for not relentlessly asking me how it was going. I hope the book makes a contribution towards improving patient safety, whilst at the same time demonstrating the potential of human factors in helping to find ways to reduce the number of accidents, errors and near misses which occur worldwide each year in healthcare.

Patrick Waterson,
Loughborough

Chapter 1

Patient Safety Culture – Setting the Scene

Patrick Waterson

> Few phrases occur more frequently in discussions about hazardous technologies
> than safety culture. Few things are so sought after and yet so little understood.
>
> (Reason 1997)

Introduction

Twenty-five years ago a series of explosions on the Piper Alpha oil and gas
platform resulted in the deaths of 167 men on board. The subsequent inquiry
report (Cullen 1990) identified a wide range of factors contributing to the
accident. Chief amongst these were the relaxed attitudes of managers towards
safety and lack of adequate training provision alongside a widespread 'culture
of complacency' (McGinty 2008: 259) within the operating company. The Piper
Alpha accident certainly wasn't a 'one-off event'. During the period 1970–90
a number of other high profile accidents occurred. These included the Tenerife
Airport disaster (1977), the explosion of the Challenger space shuttle (1986), the
Ladbroke Grove train accident (1996), the underground rail fire at Kings Cross
(1987) and the Three Mile Island and Chernobyl nuclear accidents (1979 and
1986). These and other large scale accidents/disasters resulted in widespread
public debate about the most appropriate ways in which to manage safety and
the role of external bodies such as regulators within many of the safety critical
industries (e.g. aviation, nuclear power and rail transportation).

A second outcome was a flurry of activity and interest amongst the scientific
research community in understanding accident causation. What characterised
much of this work was the application of the systems approach to understanding
the contributory factors underlying accident causation (e.g. Vaughan 1996;
Turner and Pidgeon 1997). A central idea of the systems approach is that complex
systems, (e.g. organisations, teams and technologies) are composed of interrelated
components, the properties of which are changed if the system is disassembled
in any way. The approach also emphasises two specific aspects of social and
organisational behaviour: (1) their systems character, such that movement in
one part leads in a predictable fashion to movement in other parts and (2) their
openness to environmental inputs, so that they are continually in a state of flux
(Katz and Kahn 1966: 3). Adopting a systems point of view often affords insights
into how actions or occurrences at one level (e.g. an error made by a process
operator) collectively interact with phenomena at team and organisational levels
of analysis (Waterson 2009; Leveson 2012; Karsh et al. 2013; Wilson 2013).

Early Work on Safety Culture

One of the most important elements which draws on the systems approach is the concept of safety culture. The Chernobyl accident investigation report by the International Atomic Energy Agency (IAEA) for example, described the accident as partly arising through a 'poor safety culture' at the plant and within the wider Soviet society (INSAG 1986; Antonsen 2009). Since that time there has been an enormous amount of research on the topic and a wide variety of measurement tools and frameworks exist across a range of application domains (e.g. rail – RSSB 2011; aviation – Isaac et al. 2002; the nuclear industry – Lee 1998 and Lee and Harrison 2000; offshore installations – Mearns et al. 2003; and construction – Fang and Wu 2013).

Almost from the outset the concept of safety culture has tended to be something which has elicited strong opinions. The word 'culture' is enough in itself to generate widely differing opinions. The literary theorist Raymond Williams (1983) listed several hundred definitions in his book *Keywords*. The quotation from James Reason at the beginning of this chapter and the title of a paper by Cox and Flin (1998), 'Safety culture: philosopher's stone or man of straw?' reflect the character of some of the debates which have taken place over the years. Subsequent chapters of this book (e.g. Chapter 2 by Guldenmund; Chapter 6 by McDonald and Waring) take up discussion of this theme and put forward the case for alternative concepts of safety culture and behaviour in organisations.

Culture and Climate

Defining what we mean by safety culture has taken up many of the pages of scientific articles and books in the last few decades. A recent round table involving experts, organised by the Healthcare Foundation in March 2013, touched upon one of the thornier issues which was raised by a number of these articles, namely the 'culture' vs 'climate' debate (e.g. Schein 1984; Mearns and Flin 1999). The definitions provided by the round table (Healthcare Foundation 2013: 3) attempted to distinguish between the two, whilst noting that definitions vary within the research literature:

> *Climate* emerges through a social process, where staff attach meaning to the policy and practice they experience and the behaviours they observe. *Culture* concerns the values, beliefs and assumptions that staff infer through story, myth and socialisation, and the behaviours they observe that promote success. In other words, culture is more interpretative.

For the purposes of simplicity, and partly because it seems to be the most widely used term within industries including healthcare, we use the term 'culture' in this book.

Characteristics and Components of a 'Safe Culture'

In contrast to the debates which surround the differences between 'culture' and climate, the characteristics and components of what constitutes safety culture tend to have elicited more agreement amongst researchers and practitioners. The UK Health and Safety Executive (HSE) for example lists the following 'markers' of what constitutes a 'good' company safety culture (HSE 2002):

- Managers regularly visit the workplace and discuss safety matters with the workforce.
- The company gives regular, clear information on safety matters.
- We can raise a safety concern, knowing the company take it seriously and they will tell us what they are doing about it.
- Safety is always the company's top priority, we can stop a job if we don't feel safe.
- The company investigates all accidents and near misses, does something about it and gives Feedback.
- The company keeps up to date with new ideas on safety.
- We can get safety equipment and training if needed – the budget for this seems about right.
- Everyone is included in decisions affecting safety and is regularly asked for input.
- It's rare for anyone here to take shortcuts or unnecessary risks.
- We can be open and honest about safety: the company doesn't simply find someone to blame.
- Morale is generally high.

Typically, these types of components are further broken down and decomposed into sets of factors and survey items which are used quantitatively to assess and measure safety culture in organisations. Examples of these factors include: levels of staffing and workload; supervisor support; trust in management decision-making; levels of organisational commitment and employee communication (see Itoh et al., 2012 for a review of patient safety culture assessment methods; Chapter 3 by Talati and Griffin and Chapter 4 by Itoh, Boje and Mikkelson). In other cases, the use of qualitative methods such as interviews, observational and ethnographic studies and participatory workshops is common in assessing safety culture. These include the use of a combination of both quantitative and qualitative measures, in parallel with maturity frameworks, which can be used to assess the extent to which an organisation is progressing with its efforts to improve its safety culture. One of the most well-known of these is the 'Hearts and Minds™' safety programme which has been used extensively by a number of petrochemical companies including Shell (2006) to measure and benchmark progress from what is termed a 'pathological' attitude to safety (characterised by a 'who cares as long as we are not caught' attitude) to a generative approach (characterised as 'health and safety is how we do our business round here'). These types of approaches have also been influential within healthcare and include the Manchester Patient Safety Framework (MaPSaF) which was adopted widely within the UK by the National Patient Safety Agency (NPSA) – see Chapter 7 by Carthey and Chapter 16 by Bohan.

Patient Safety and 'Patient Safety Culture'

Medical error has been described and studied for the best part of a century. However, the extent and seriousness of the problem was either not recognised or not acknowledged within the medical profession (Vincent 2010). Part of the reason for this 'denial' was that the patterns of socialisation and training within the medical profession ill-equipped them to deal with situations which acknowledged fallibility or error. As Charles Bosk showed in his detailed ethnography of the work of surgeons in the USA, *Forgive and Remember* (1979, 2003), the ability to stand back and take an objective view of error was not normal practice for members of the medical profession. Likewise, the full extent and high rates of error within medicine were not as well documented or understood, as compared to today.

In 1994 the Harvard-based surgeon Lucian Leape published a paper which summarised evidence showing that error rates in medicine were very high (Leape 1994). The publication of two reports in 1999 and 2000 on either side of the Atlantic (the UK Department of Health, 2000, *An Organisation with a Memory* and the US Institute of Medicine (IOM) *To Err is Human* – Kohn et al. 1999) provided further backing for Leape's arguments and almost overnight resulted in the issue of medical harm and patient safety grabbing the newspaper headlines and reaching the attention of the politicians and the general public. A second outcome was the start of a programme of research which has developed and expanded dramatically over the last few decades. Both reports focused on statistics which showed that medical error was often the cause of unnecessary deaths amongst patients undergoing treatment in a variety of healthcare settings (e.g. hospitals). Perhaps the most striking of these statistics is from the IOM report which showed that between 44,000 and 98,000 people die in US hospitals each year as a result of medical errors (Leape 2000). Within the UK the investigations into the causes of high numbers of paediatric deaths following cardiac surgery at the Bristol Royal Infirmary during the 1980s (Department of Health 2001; Walshe and Offen 2001) raised the spectre of widespread systemic failings amongst hospital staff and greatly added to the debate surrounding ways of preventing error in healthcare.

Human Factors and Patient Safety

Hale and Hovden (1998) described management and culture as the third age of safety. The first age was about technical measures, the second about human factors and individual behaviour (Hale and Glendon 1987) and these merged with the technological approaches. Catchpole et al. (2011) in their summary of the development of patient safety research in healthcare, characterised early human factors work in patient safety as focusing on attempts to locate the source of error within medicine. Other exploratory work, took a number of forms, including review of patient's case notes (Vincent et al. 2001), observational studies (e.g. de Leval et al. 2000) and the implementation of quality improvement programmes such as incident reporting systems (Webster and Anderson 2002). Much of this work was conducted by retrospective review of documentation or using direct observation of medical practice and without the involvement of clinical practitioners. What was missing at the time was a vocabulary or framework for understanding error. At about this time some of the human error models which had been used in other safety domains (e.g., the aviation, nuclear and rail industries) started to be used within healthcare. James Reason's work on active and latent failure modes in organisations and his well-known 'Swiss Cheese' model of accident causation (Reason 1990) influenced the development of generic models of errors and accidents in healthcare (e.g., the London Protocol – Vincent et al. 1998; Rogers 2002), but was also applied to specific areas within healthcare (e.g. surgery – Catchpole et al. 2005).

More recently, models drawing on establishing work within human factors and ergonomics, as well as sociotechnical systems theory (Trist and Bamforth 1951; Waterson 2013) have begun to appear. It might be argued that human factors and ergonomics, alongside socio-technical systems, represent one of the unifying strands in the emerging field of research on patient safety (Norris 2012; Flin et al. 2013). One of the most well-known examples of systems of sociotechnical models is the Systems Engineering Initiative for Patient Safety (SEIPS) model (Carayon et al. 2006 – Figure 1.1).

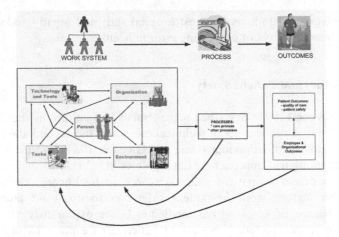

Figure 1.1   The SEIPS model
Source: Carayon et al. 2006.

The Systems Engineering Initiative for Patient Safety (SEIPS) model contends that patient safety hazards can emerge from a variety of work system factors, including the technology and tools being used by medical professionals, the particular way in which work is organised and allocated (e.g. team working arrangements, leadership), situational and individual factors (e.g. individual competences, extent of training and skills), as well as the degree to which work tasks match or mismatch with other elements. A critical factor determining patient safety outcomes and the quality of care processes is the degree to which environmental factors support work tasks. These environmental factors might take the form of traditional human factors and ergonomics variables (e.g. noise, temperature, lighting), but more critically, they might also involve aspects of the culture within the work systems (e.g. surgical unit) and the wider organisation (e.g. hospital). In common with the SEIPS model, many of the chapters in this book draw on sociotechnical and systems theory developed over the last 50 years within human factors and socio-technical systems (e.g. Chapter 5 – Phipps and Ashcroft; Chapter 7 – Carthey; and Chapter 17 – Kirwan and Shorrock).

The First Patient Safety Culture Measurement Tools

Around 2004 the first safety culture tools designed for healthcare began to appear. Many of these tools are in the form of survey instruments or questionnaires, the two most well-known being the Hospital Survey on Patient Safety Culture (HSPSC) developed by the US Agency for Healthcare Research and Quality (AHRQ) and the Safety Attitudes Questionnaire (SAQ – Sexton et al. 2006). Both instruments are described in detail within the book and illustrated with examples of their latest application in hospitals and other healthcare settings (Chapter 10 – Agnew and Flin; Chapter 11 – Hammer and Manser; Chapter 12 – Sorra and Battles; Chapter 13 – Etchegaray and Thomas; Chapter 14 – Vlayen, Schrooten, Hellings, Haelterman and Peleman). A number of other tools exist, some of which aim to target specific aspects of safety culture (e.g., leadership behaviours, communication during surgical handover – World Health Organisation 2013). These tools have been applied across a wide arrange of healthcare contexts and healthcare systems around the world (e.g. Chapter 11 on the use of the HSPSC in Europe). Despite their popularity, it is fair to say these tools are still very much under development and assessments and improvements to their content and psychometric properties continue at a pace (see Chapter 8 – Jackson and Kline). Taken as a whole, these are exciting times with respect to the development of patient safety culture tools and instruments. Much is likely to change in the future and many of the chapters in this book reflect this optimism and recognition that patient safety culture is still a relatively new area in which there remains much to be done (Chapter 18 – Waterson).

Origins and Organisation of the Book

The goal of this book was to bring together some of the most recent work within the area of patient safety culture and to reflect some of current debates and discussions which dominate the field of safety culture within patient safety. The book is aimed at not only other researchers, but also practitioners working in hospitals and other healthcare settings. In order to meet the needs of both audiences, the book is organised along the lines of three main themes (theory, methods and application). The following five chapters (Part I) provide in-depth coverage of safety culture, as well as patient safety culture. These chapters review some of the research which has focused on operationalising and deconstructing the meaning of safety culture in healthcare. In addition, alternative perspectives on safety culture are presented. Part II of the book contains three chapters which provide accounts of the range of methods and tools used to measure patient safety culture, as well as some of the techniques (e.g. psychometrics) which are used to assess their efficacy. Part III focuses on examples the application of patient safety culture and includes accounts from healthcare professionals in the 'front line' (Chapter 15 – Giuliani, Meier, Kiss and Van Vegten; Chapter 16 – Bohan). The penultimate chapter of the book (Chapter 17 – Kirwan and Shorrock) provides a set of 'lessons learnt' from a more established and developed application domain of safety culture, aviation. Further discussion of these 'lessons' is taken up in the final chapter (Chapter 18 – Waterson).

Figure 1.2 summarises some of the themes described in this chapter in the form of a timeline, the concluding point being the UK Francis report published in 2013 which described some of the consequences and multiple problems brought about over time by a poor hospital safety culture (Francis 2013).

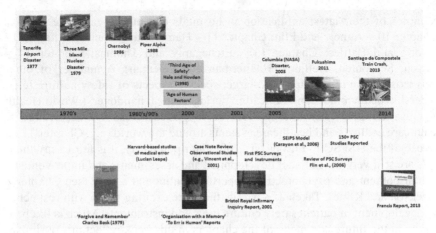

Figure 1.2   Patient safety and safety culture timeline

References

Antonsen, S. 2009. *Safety Culture: Theory, Method and Improvement*. Farnham: Ashgate.

Bosk, C. 1979, 2003. *Forgive and Remember – Managing Medical Failure*. Chicago: Chicago University Press (1st edition 1979; 2nd edition 2003).

Carayon, P., Hundt, A.S., Karsh, B.T., Gurses, A.P., Alvarado, C.J., Smith, M. and Brennan, P.F. 2006. Work system design for patient safety: The SEIPS model. *Quality and Safety in Health Care*, 15 (Suppl. I), 50–58.

Catchpole, K., Godden, P.J., Giddings, A.E.B., Hirst, G., Dale, T., Utley, M., Gallivan, S. and de Leval, M.R. 2005. *Identifying and Reducing Errors in the Operating Theatre*. Patient Safety Research Programme Final Report PS012. Available at: http://www.pcpoh.bham.ac.uk/publichealth/psrp/PS012_Project_ Summary.shtml (last accessed on 14 July 2013).

Catchpole, K., Hadi, M., Morgan, L., Pickering, S., Robertson, E., New, S. and McCulloch, P. 2011. Human factors in surgical safety: That was then, this is now. In Albolino, S., Bagnara, S., Bellandi, T., Llanexa, J., Rosal, G. and Tartaglia, R. (eds), *Healthcare Systems Ergonomics and Patient Safety*. London: CRC Press.

Cox, S. and Flin, R. 1998. Safety culture: Philosopher's stone or man of straw? *Work & Stress*, 12(3), 189–201.

Cullen, The Hon. Lord W.D. 1990. *The Public Inquiry into the Piper Alpha Disaster*. London: H.M. Stationery Office.

de Leval, M.R., Carthey, J., Wright, D.J. and Reason, J.T. 2000. Human factors and cardiac surgery: A multicentre study. *Journal of Thoracic and Cardiovascular Surgery*, 119(4), 661–72.

Department of Health 2000. *An Organisation with a Memory*. London: The Stationery Office.

Department of Health. 2001. *Learning from Bristol. The Report of the Public Inquiry into Children's Heart Surgery at the Bristol Royal Infirmary, 1984–95.* London: The Stationery Office.

Fang, D. and Wu, H. 2013. Development of a Safety Culture Interaction (SCI) model for construction projects. *Safety Science*, 57, 138–49.

Flin, R., Bromiley, M., Buckle P. and Reid, J. 2013. Mid Staffs Inquiry: Changing behaviour with a human factors approach. *British Medical Journal*, 5 March 2013, 346.

Francis Report 2013. *The Mid Staffordshire NHS Foundation Trust Public Inquiry.* Available at: http://www.midstaffspublicinquiry.com/ (last accessed on 14 April 2014).

Hale, A.R. and Glendon, A.I. 1987. *Individual Behaviour in the Control of Danger.* Amsterdam: Elsevier.

Hale, A.R. and Hovden, J. 1998. Management and culture: The third age of safety. In A.-M. Feyer and A. Williamson (eds), *Occupational Injury: Risk, Prevention and Intervention.* London: Taylor and Francis.

Health and Safety Executive (HSE) 2002. *Health and Safety Laboratory, Human Factors Group. 'Safety Culture: A Review of the Literature'.* HSL/2002/25.

Healthcare Foundation 2013. *Safety Culture: What is It and How Do We Monitor and Measure It?* Event Report. Available at: http://www.health.org.uk/publications/safety-culture-what-is-it-and-how-do-we-monitor-and-measure-it/ (last accessed on 11 July 2013).

INSAG (International Nuclear Safety Advisory Group) 1986. *Summary Report on the Post-accident Review Meeting on the Chernobyl Accident.* Vienna (August–September).

Isaac, A., Shorrock, S.T. and Kirwan, B. 2002. Human error in European air traffic management: The HERA project. *Reliability Engineering and System Safety*, 75, 257–72.

Itoh, K., Anderson, H.B. and Madsen, M.D. 2012. Safety culture in health. In P. Carayon (ed.), *Human Factors and Ergonomics in Health Care and Patient Safety.* Boca Raton: CRC Press.

Karsh, B.-T., Waterson, P.E. and Holden, R. 2013. Crossing levels in systems ergonomics: A framework to support 'mesoergonomic' inquiry. *Applied Ergonomics* (Invited paper for special issue on systems ergonomics), 45, 45–54.

Katz, D. and Kahn, R. 1966. *The Social Psychology of Organizations.* Wiley: New York.

Kohn, L.T., Corrigan, J.M. and Donaldson, M.S. 1999. *To Err is Human.* Washington, DC: National Academy Press.

Leape, L. 1994. Error in medicine. *Journal of the American Medical Association (JAMA)*, 272, 23.

Leape, L. 2000. Institute of Medicine figures are not exaggerated. *Journal of the American Medical Association*, 272, 1851–7.

Lee, T. 1998. Assessment of safety culture at a nuclear reprocessing plant. *Work and Stress*, 12(3), 217–37.

Lee, T. and Harrison, K. 2000. Assessing safety culture in nuclear power stations. *Safety Science*, 30, 61–97.

Leveson, N. 2012. *Engineering a Safer World*. Cambridge, MA: MIT Press.

McGinty, S. 2008. *Fire in the Night – The Piper Alpha Disaster*. London: Macmillan.

Mearns, K.J. and Flin, R. 1999. Assessing the state of organizational safety culture or climate? *Current Psychology*, 18(1), 5–17.

Mearns, K., Whitaker, S. and Flin, R. 2003. Safety climate, safety management practice and safety performance in offshore environments. *Safety Science*, 41, 641–80.

Norris, B. 2012. Systems human factors – how far have we come? *BMJ: Quality and Safety*, doi:10.1136/bmjqs-2011-000476.

Rail Safety and Standards Board (RSSB) 2011. *Safety Culture Toolkit*. Available at: http://safetyculturetoolkit.rssb.co.uk/home.aspx (last accessed on 14 July 2013).

Reason, J. 1990. *Human Error*. Cambridge: Cambridge University Press.

Reason, J. 1997. *Managing the Risks of Organizational Accidents*. Cambridge: Cambridge University Press.

Rogers, S. 2002. A structured approach for the investigation of clinical incidents in health care: application in a general practice setting. *British Journal of Medical Practice*, 52(Suppl), S30–S32.

Schein, E. 1984. Coming to a new awareness of organisational culture. *Sloan Management Review*, 25(2), 3–6.

Sexton J.B., Helmreich, R.L., Neilands, T.B., Rowan, K., Vella, K., Boyden, J., Roberts, P.R. and Thomas. E.J. 2006. The Safety Attitudes Questionnaire: Psychometric properties, benchmarking data, and emerging research. *BMC Health Serv Res.*, 6, 44.

Shell International Exploration and Production B.V. 2006. *Winning Hearts and Minds: The Road Map*. Available at: http://www.eimicrosites.org/heartsandminds/userfiles/file/Homepage/HP%20PDF%20roadmap.pdf (last accessed on 11 July 2013).

Trist, E.L. and Bamforth, K.W. 1951. Some social and psychological consequences of the long-wall method for coal-getting. *Human Relations*, 4, 3–38.

Turner, B.A. and Pidgeon, N.F. 1997. *Man-made Disasters*, 2nd edition. London: Butterworth-Heinemann.

Vaughan, D. 1996. *The Challenger Launch Decision: Risky Technology, Culture, and Deviance at NASA*. Chicago: The University of Chicago Press.

Vincent, C. 2010. *Patient Safety* (2nd edition). New York: Wiley.

Vincent, C., Neale, G. and Woloshynowych, M. 2001. Adverse events in British hospitals: Preliminary retrospective record review. *BMJ*, 322, 517–19.

Vincent, C., Taylor-Adams, S. and Stanhope, N. 1998. Framework for analysing risk and safety in clinical medicine. *British Medical Journal* (316)(7138), 1154–7.

Walshe, K. and Offen, N. 2001. A very public failure: Lessons for quality improvement in healthcare organisations from the Bristol Royal Infirmary. *Quality in Health Care*, 10(3), 250–56.

Waterson, P.E. 2009. A critical review of the systems approach within patient safety research. *Ergonomics*, 52(10), 1185–95.

Waterson, P.E. 2015 (in press). Sociotechnical design of work systems. To appear in J.R. Wilson and S. Sharples (eds), *Evaluation of Human Work* (fourth edition). London: Taylor and Francis.

Webster, C.S. and Anderson, D.J. 2002. A practical guide to the implementation of an effective incident reporting scheme to reduce medication error on the hospital wards. *International Journal of Nursing Practice*, 8(4), 176–83.

Williams, R. 1983. *Keywords: A Vocabulary of Culture and Society*. London: Fontana.

Wilson, J.R. 2014. Fundamentals of systems ergonomics/human factors. *Applied Ergonomics*, 45, 5–13.

World Health Organization 2013. *Patient Safety – Organizational Tools*. Available at: http://www.who.int/patientsafety/research/methods_measures/human_factors/organizational_tools/en/ (last accessed on 16 April 2014).

PART I
Background and Theory

Chapter 2

Organisational Safety Culture Principles

Frank Guldenmund

What is This Thing Called Culture?

Man and Culture

The scientific study of culture reveals great variety and various disputes and scholars often differ on what culture actually 'is' (Keesing 1981). Unlike animals, humans develop a culture. Whereas the meaning of an animal's behaviour at one end of the world will be comparable to the meaning of a similar animal's behaviour at the other end, the ideational systems and convictions of humans from both these ends are often quite dissimilar. According to Geertz there is no culture without humans but also, 'more significantly, without culture no men' (Geertz 1973: 49).

An early notion, still echoing imperialistic times, placed cultures on a single continuum ranging from savage (low) to civilised (high), with high cultures obviously enjoyed by the colonialists and low cultures held by the conquered natives (Avruch 1998). This view could be labelled 'colonialist', being both ethnocentric and evaluative and putting much emphasis on refinement and (evolutionary) development. This notion of social evolution was later dismissed by many in favour of a descriptive stance, emphasising the uniqueness and variety of cultures, none of them superior over or more developed than the others (ibid.).

An important function of culture is related to the reduction of uncertainty (Van Hoewijk 1988) or even anxiety (Schein 2004) which, consequently, leads to more continuity, because less time is spent on various mutual adjustments within a group. The fact that people know what to expect in a variety of situations – e.g. with regard to particular rituals (like celebrations, meetings, appointments and so on), the expression of emotions, dress codes, behaviours, *et cetera* – makes life more predictable and hence more fluent. Culture has also been linked to adaptation (Schein 1992) and habituation. Habituation is well-developed in all organisms that have a nervous system – the working of this mechanism has been described in as primitive a life form as the marine snail (Kandel and Schwartz 1985: 817 ff.).[1] Adaptation is important for learning, for continuity and therefore for survival.

---

1   For instance, Castellucci et al. (1978) have shown that repeated stimulation of a single nerve cell results in this cell not responding to that stimulus anymore.

Forces from outside the organism that demand its adaptation will initiate change;[2] in this view, cultures are considered both functional and well adapted to their environment. However, while adaptation and learning are both necessary aspects of culture, they define neither its essence nor its working mechanisms.

As early as 1952 Kroeber and Kluckhohn had already compiled a list of 164 definitions of culture (Kroeber and Kluckhohn 1952) so it does not appear useful to embark on a personal definitional cruise. Hofstede defines culture briefly as 'the collective programming of the mind, which distinguishes the members of one group or category of people from another' (Hofstede 2001: 9) and considers culture 'mental software'.[3] He distinguishes three levels of such mental programming (Hofstede 1991: 6, 2001: 3):

1. Human nature: universal
2. Culture: collective
3. Personality: individual.

Human nature corresponds to the programs all humans around the world are instilled with, but this 'software' can be influenced by both culture and personality. For instance, the way an individual expresses his or her anger will be determined both by this person's personality and by their culture (and by situational conditions, but these are kept out of the discussion for the time being). Applying the psychoanalytical idiom to this three-way split, human nature would represent the *Id*, personality would be considered the *Ego* and culture, also encompassing various assumptions about ethics and behaviour, would represent the *Superego*. Considered in this way the attention given to (organisational) culture from a managerial point of view is certainly not surprising.

Hence, culture is distinguished from human nature and personality in that it is shared by a defined group of people, whereas human nature and personality are not. Culture is often considered the 'collective memory'[4] of a group and is therefore thoroughly intertwined with the history of that group. Moreover, the term 'memory' implies that culture is learned, not inherited. Importantly, one person can belong to many groups and can therefore share several cultures with different people. This particular characteristic makes the study of culture extremely difficult, because to what particular culture should any observed or otherwise assessed regularities of groups be attributed? This issue will be taken up more extensively later on in the chapter.

---

2   Please note that Schein (Schein 1992: 298 ff.) follows a similar reasoning about culture change.

3   Following Geertz, who refers to 'plans, recipes, rules, instructions, [...] programs' (1973: 44).

4   Human nature is shared by everybody and a personality is held by only one person. Additionally, Hofstede considers culture the 'personality' of a group (2001: 10).

(National) cultures should not be compared normatively. However, within its bounds a culture provides norms for thoughts and action, perceptions and behaviour. Therefore, within a (national) culture actions and justifications for these actions can be compared to the norms that have developed within that culture (Hofstede 2001: 15). Indeed, such norms can become part of the culture and define its core, alongside its values. Consequently, culture provides one of the anchors for behaviour. This behavioural aspect is actually not captured in the definition supplied by Hofstede. Anthropologists Spradley and McCurdy (1975: 4) define culture as 'the acquired knowledge people use to interpret experience and generate behavior'. Combining this definition with Hofstede's produces the following:

> Culture is the acquired and collective knowledge groups or categories of people use to interpret experience and generate behaviour, which distinguishes them from other groups or categories of people.

In this definition the learned and shared aspects of culture as well as its sense making and action components are captured. As satisfying a definition as it might seem, it still misses the fuzziness of the concept, which is captured in Spencer-Oatey's (2000) definition:

> Culture is a fuzzy set of attitudes, beliefs, behavioural conventions, and basic assumptions and values that are shared by a group of people, and that influence each member's behaviour and each member's interpretations of the 'meaning' of other people's behaviour.

Attempting to reveal the essence of a culture raises an important question; i.e. to what extent are cultures comparable and to what extent are they unique (Hofstede 2001: 24 ff.)? This distinction is discussed in various (social) sciences, e.g. sociology, anthropology, cross-cultural psychology, and brings along its own vocabulary (ibid.). Basically, it pertains to the issue of generality and specificity; *Gestalt* (unique holistic configurations) versus *Gesetze* (general laws); *idiography* versus *nomothetic*; and *emic* (as in phonemic, i.e. unique) versus *etic* (as in phonetic, i.e. general). Evidently, this discussion also throws some light on the issue of safety culture and it will be taken up further below. An argument of the generalists could be that each group (collective, category, society) has to face similar problems during its lifetime. However – as the specifists would retort – each group will develop solutions based on its unique personal situation. It would be too much of a simplification to narrow this discussion down to a 'basic problems focus' versus 'unique solutions focus' dichotomy although the aspect of survival is quite important in this discussion. Survival of the organisation is also the primary incentive for change in Schein's conception of organisational culture, resulting in (external) adaptation and (internal) integration (1992: 51 ff.).

An outcome of a generalist approach is that cultures can be described with a limited number of aspects, e.g. dimensions, facets or factors. A unique culture approach does not have this common underlying framework and its descriptions are limited to single cultures. However, either approach can ultimately lead to a third approach, that is a typology of cultures. All three approaches are well represented in the organisational culture literature and can be discerned also in literature on safety culture. Again, this topic will be discussed more extensively below.

Summarising, humans develop a culture when they interact and try to achieve something. This culture is primarily locally defined. Having acquired this culture not only means that an influence is exerted on behaviour, but also that other people's behaviour is interpreted in this culture's way.

Layers of Culture

Next to the levels of mental programming present in humans – i.e. universal, collective and individual – and the various levels of aggregation at which culture can be studied – e.g. societal, regional, occupational, organisational – most scholars consider culture as something consisting of a core surrounded by one or more layers, not unlike the anatomy of an onion. Whereas the core is something (deeply) hidden, the culture projects itself gradually through and onto the outer layers. The more remotely a layer is located from the core, the more easily it can be observed but also the more indirect, or interpretive, its relation with the core becomes. This simply implies that it is not straightforward to understand a culture from its outer layer(s). With regard to changing a culture a similar rule is sometimes put forward: the more deeply a layer is located, the more difficult it becomes actually to change it (Meijer 1999; Sanders and Neuijen 1987). Hofstede, citing Bem, argues that a particular culture can be more effectively changed by starting with the practices of the outer layers, not the values of the core (Bem 1970; Hofstede 2001: 12). The latter change only gradually, with different time estimates for different levels of culture. For instance, a substantial change in national culture might take no less than a millennium (Hofstede 2001), whereas an organisational culture might take around 25 years (Schein 1992). Various conceptions of the layers of culture are presented in Table 2.1.

Table 2.1   The layers of culture according to various authors

Author(s)	Central core	Layer 1	Layer 2	Layer 3
Deal and Kennedy (1982)	Values	Heroes	Rites and rituals	Communication network
Van Hoewijk (1988)	Fixed convictions	Norms and values	Myths, heroes, symbols, stories	Codes of conduct, rituals, procedures
Hofstede (2001)	Values	Rituals	Heroes	Symbols
Meijer (1999)	Fundamentals	Practices	–	–
Rousseau (1990)	Fundamental assumptions	Values	Behavioural norms	Patterns of behaviour; and artefacts (= 4th layer)
Sanders and Nuijen (1987)	Values and principles	Rituals	Heroes	Symbols
Schein (2004)	Basic underlying assumptions	Espoused values	Artefacts	–
Spencer-Oatey (2000)	Basic assumptions and values	Beliefs, attitudes and conventions	Systems and institutions	Artefacts and products; rituals and behaviour
Trompenaars and Hampden-Turner (1997)	Basic assumptions	Norms and values	Explicit culture (e.g. behaviour, clothes, food, language, housing)	–

All authors have something quite deep and profound positioned at the core – values, convictions, principles, fundamental or basic assumptions – but beyond that there are differences, not so much concerning the nature of the layers, but regarding their position in the onion. Importantly, of the authors mentioned in Table 2.1 the scholars Hofstede, Spencer-Oatey and Trompenaars and Hampden-Turner focus mostly on national culture, whereas the others have primarily organisational culture in mind. Regarding organisational culture, Hofstede argues that the core – i.e. the values – is less relevant for the study of organisations, although it offers a reflection of the organisation's national values, i.e. the values of the country where the organisation is situated. Hofstede therefore maintains that the notion of (national) culture does not apply so much to differences between organisations within a country. They only differ in what he calls 'practices', i.e. the outer three layers of his onion: rituals, heroes and symbols (Hofstede 1991: 182–3).

Schein does not differentiate between the more visible aspects of culture, i.e. between rituals, heroes and symbols, all of which he sweeps up under the heading of 'artefacts', along with all visible behaviour.[5] However, he divides the core into 'espoused values' and 'basic assumptions', thereby indicating that he does not take for granted the values that members of an organisation express when asked about these. Schein also makes a point of calling his core 'basic assumptions' and not 'values'. To him, values are still negotiable whereas basic assumptions are not (Schein 1992: 16). As can be seen in Table 2.1, more authors use this distinction between (basic) assumptions on the one hand and values on the other; this way values (and attitudes and beliefs) are modelled to change still more radically, whereas the (basic) assumptions will not.

Spencer-Oatey introduces the notion of institutions, a topic that will be discussed later when the process of culture development is discussed. Institutions either teach or otherwise develop and disseminate some of the values of a culture. As is clear from the table, at this stage these values are not yet internalised, to the extent that they are cultural values.

This rather extensive discussion should make another point clear, namely that the labels given to the layers are typically assigned from an analyst's point of view. For a member of a particular culture these aspects are thoroughly intertwined and their meaning is obvious. It is therefore the researcher who labels these activities as such and in many cases their differences are not clear-cut.

Regarding research on culture it is possible to distinguish two contrasting approaches; one approach considers culture a sociocultural (i.e. behavioural) system, whereas the other considers it an ideational system, i.e. a system comprising ideas, concepts, rules and meanings (Keesing 1981: 68). Whether it is sufficient to observe the practices and not understand their underlying rationale seems much more a matter of preference for a particular paradigm than something that can be resolved through scientific inquiry. On the one hand, researchers observing only practices might sometimes be bothered by their inconsistency, their irrationality or their incongruence and might end up relying on basic, behavioural psychology (cf. Avruch 1998: 19). On the other hand, researchers focussing on the core have a hard time untangling it.

It is, however, important to look a little deeper into what is inside the core. Several authors refer to the core as 'deep' (Schein 1990: 109). This immediately triggers the question as to what deep exactly is, or entails. Deep appears to refer to something fundamental and pre-conscious. People become emotional when their fundamentals are questioned or under attack (Avruch 1998; Hofstede 1991), often without being aware of why this is so important to them. Moreover, '[t]he more deeply internalised and affectively loaded, the more certain images or schemas are able to motivate action' (Avruch 1998: 19).

---

5   Pedersen and Sorensen, taking Schein's research model as a starting point, bring some diversity to his rather amorphous artefacts, distinguishing (1) physical symbols, (2) language, (3) traditions and (4) stories amongst them, all of which they consider important for a cultural analysis (Pederson and Sorensen 1989: 29).

It is quite illuminating to bring up the reason why Schein considered organisational culture as something that goes beyond the notion of 'practices'. After the Korean War, Schein and his colleagues worked closely with prisoners of war (POWs) who had been brainwashed by the Chinese.[6] Whereas some of them simply distanced themselves from the ideas being forced upon them, others had adopted a communist worldview and had even confessed to 'crimes' they did not commit, that is, not from a Western point of view. Rather later, Schein began to see parallels between the beliefs of these POWs and the beliefs schools, both private and public organisations try to establish in their pupils and personnel, albeit through a much milder process (ibid.). According to Schein, it is possible to provide people with such strong tacit beliefs, which are indeed much deeper than the more superficial 'practices' Hofstede has in mind regarding his distinction between organisations. This is not to say that Schein's basic assumptions and Hofstede's values coincide. Hofstede's values are indeed acquired at a much earlier stage – Hofstede claims before the age of 10 – and are therefore quite static and rather fixed. Schein's basic assumptions are more dynamic and subject to change, but changing these requires much effort and unleashes 'large quantities of basic anxiety' (Schein 1992: 22) because members of the organisation lose many of their certainties for a period of time. It is therefore not surprising that this organisational change process has been likened to the process of mourning (Kets de Vries 1999).

Yet culture is not only deep because it is so fundamental and covert, it is also immensely patterned and therefore related to everything we think, perceive and do. When attempting to change one belief, we have to change many related ones and much that has been built upon these. The 'large quantities of basic anxiety' and the process of mourning mentioned here are quite understandable when such basic belief networks are taken apart.

Trying to formulate such deeply seated assumptions, these 'webs of significance' as Geertz calls them (1973: 5), will be particularly difficult because they are so taken for granted (Schein 1992) that, within the boundaries of a culture, they are never challenged and, consequently, never have to be verbalised. Because of its fundamental nature, culture can be blinded by itself to itself. Schein's distinction between basic assumptions and their verbal counterpart, i.e. espoused values, seems therefore quite valid and sensible. Comparable reasoning can be found with Bloch (1998) who proposes that much (conceptual) knowledge – and hence also cultural 'knowledge' – is essentially non-linguistic and acquired primarily through experience, not through explanation, i.e. communication. When such knowledge is 'rendered into language', its character is also changed (ibid.: 7). Hence, what is considered 'deep' can also be considered non-linguistic and implicit. Making this deep knowledge explicit also changes its overall character.

---

6   Afterwards, the process through which these POWs had been converted by the Chinese was named 'coercive persuasion' (Schein 1992: 327–9, 1999).

Summarising, the whole idea behind the onion model seems to depict the essence of culture as something hidden rather deeply under a layered set of more or less visible manifestations upon which it exerts its influence. These layers can function as a key to the nature of the underlying culture.[7]

The Development of Culture

A straightforward account of the development of culture comes from Hardin (2009); simply stated – (behavioural) experience leads to knowledge leads to culture. This is not to say that all experience leads to knowledge and that all knowledge leads to culture. First of all, because culture is shared, this knowledge should be shared too. Moreover, for experience to become shared knowledge it should be shared between (some of) the members of a group and an agreement should be reached on what the experience is about and what the knowledge should entail.

Halfway through the 1960s, Berger and Luckmann published *The Social Construction of Reality*, in which they put forward a process model along which societies develop their version of reality (Berger and Luckmann 1966). Organisations too, can be regarded as social communities that also share a particular version of reality, on which they act and respond. Berger and Luckmann's model has been taken as a starting point for the model outlined below. This model describes the process of organisational culture formation and its internalisation over time.

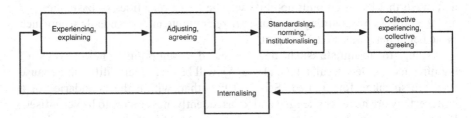

**Figure 2.1   The development of organisational culture based on a model
             by Berger and Luckmann (1966)**

---

7   In using the culture concept this way, one should beware of the reification of culture with a thing that can act, almost on its own (Avruch 1998: 14). Such notions often lead to quite simplistic linear models of influence and modification.

In Figure 2.1, a model of culture development is put forward, which describes the formation of cultures in groups, like organisations. The first box of the model pictures the situation a member of the group finds himself in; the individual is trying to make sense of the experiences he encounters. With regard to safety and risk these are particular perceptions of both constructs that will partly determine this individual's behaviour, e.g. what is safe and risky behaviour. The result of this process is an individual's understanding of reality. This particular understanding is brought into the second box, the process of interaction with group members. This interaction is often based on communication, i.e. dialogue, discussion, correction, and results in mutual adjustments, agreements and various expectations of each other's behaviours. The outcome of this box is shared understandings, e.g. standards of behaviour, roles and norms. In the third box the formal processing of standards and norms is pictured, i.e. the establishment of norms and the institutionalisation of behaviour and expectations. The fourth box pictures the situation in which norms, standards and expectations are accepted to the extent that they are considered the 'best' or, perhaps, the 'only' way of doing things. Members of the group share a comparable understanding of reality, at least with regard to the part of reality the group acts on. This understanding is internalised by the members of the group and forms the 'basic assumptions' with which individuals within the group understand reality.

This model makes a few things clear with regard to culture. Firstly, this process takes some time to complete. Secondly, it is not easy to predict the outcome of this process, as it is dependent on, for instance, the composition of the group, the communication the group members have, the distribution of power within the group and the particular context the group operates in. Thirdly, the outcome is more arbitrary than intentional, although some members will deliberately try to influence the process. However, the result may be that particular standards are established which are not the result of consensus and are therefore not internalised as 'basic assumptions' but rather as 'obligations', i.e. 'the way we *have* to do things around here' instead of 'the way we do things around here'.

Obviously, when individuals enter a group, this process has been going on for some time and therefore many assumptions are already widely established. The individual is then either trained or otherwise socialised into the group. It may be that the individual does not agree with the various assumptions of the group and he can either pretend that he does or leave the group. Going against the assumptions is yet another option but, depending on the age of the group and various other conditions, this is often a futile quest.

Importantly, this is not the only way a culture develops itself. For instance, Schein describes a process of culture formation based on the reduction of anxiety all members have when facing a new group that has to work together from some time. Initially, the group has to resolve the issue of power and has to develop routines that work for it. After that, the group can start working within the boundaries it has developed for itself (Schein 2004: 63 ff.). Nevertheless, a process of adjustment and agreement is also at work here, leading to a shared understanding of what is going on.

In the next section, the concept of culture applied to organisations will be explored in more detail.

Culture and Organisations

The Organisational Context of Culture

According to Schein, an organisational culture develops in organisations that have existed for some time and that have experienced significant external or internal difficulties or changes. Alongside the influence of founder(s) of a company or of significant leaders (heroes), the solution for problems that are effectively resolved or overcome might become part of the leading but tacit assumptions a company entertains (Schein 1992). Such internal difficulties could very well be major safety problems, like fatal accidents, explosions or releases of dangerous chemicals, but they also include reorganisations or retrenchments. External problems are often of an economic nature, like pending closure or loss of customers, but could also arise because of new legislation or drastic technological changes (Hofstede 2001: Exhibit 1.5). Organisational culture could be considered the by-product of the adaptation that follows upon these difficulties; viewed this way, organisational culture is a product of social ecology.

When considering organisations, three major components can be distinguished that 'work' together to generate the desired output. These aspects are structure, culture and processes and they are dynamically interrelated (e.g. see Hofstede 2001; Van Hoewijk 1988), which means that they all influence and are influenced by each other. Together they also provide the context in which behaviour, and hence also safety related behaviour, takes place.[8]

Organisational structure can be defined as 'the division of authority, responsibility, and duties among members of an organization' (Whittington and Pany 2004). The structure primarily outlines the formal organisation – i.e. how the work should be done and by whom. From the point of view of management an efficient structure facilitates both effective coordination and communication. With regard to the structure of organisations several scholars have proposed taxonomies of which Mintzberg's is perhaps the most well-known (Mintzberg 1979; 1980; 1983). These taxonomies offer solutions for structuring organisations in relation to their mission, main output(s) and environment. Apart from the 'organisational' structure all 'physical' structure can also be subsumed under this heading, e.g. the buildings, the hardware and the technology the organisation uses, as well as the various systems the organisation uses to carry out its processes in a uniform way and to control these.

---

8   Hofstede makes a distinction between strategy, structure, control and culture (Hofstede 2001: 408 ff.). It is not difficult to translate his 'controls' into the 'processes' of the present model. Moreover, I see his 'strategy' as the outcome of processes at the highest (strategic) level of the organisation, therefore this element in his model could be considered redundant.

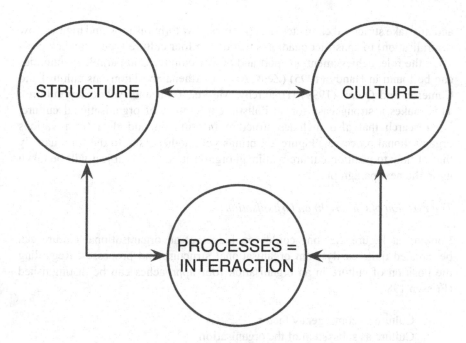

**Figure 2.2   The development of organisational culture based on a model
by Berger and Luckmann (1966)**

The culture is the basic assumptions, the underlying tacit convictions, described as a group's shared understanding above. For instance, 'We need a lot of supervisors because our people need to be watched constantly'. Such a conviction will be found reflected in the structure of the organisation and therefore also on the work floor.

The processes are the actual processes and interactions going on in the entire organisation. These processes are often formally described in the structure. Task execution at all levels might be according to what has been laid down in the structure, but this does not have to be the case. For instance, some supervisors do not watch constantly, or do not correct workers, although they see them make mistakes or violations. The reason for this might be structural – the wrong man in the right place – or cultural – the convictions of a group of people do not match up to the structure.

The tri-partition can be projected onto the various steps of the development process of culture described above, where 'processes' match with the first two steps of sensemaking and agreement, 'structure' with the step of formalisation and institutionalising and 'culture' with the remaining steps of collective agreement and internalising.

An important implication of Figure 2.2 is that an organisation's culture cannot be isolated from its structure or its processes. Harrison and Stokes (1992) actually take structural characteristics (high vs. low formalisation and high vs. low centralisation) to construct quadrants that define four culture types ('archetypes') – i.e. the role, achievement, support and power culture. Comparable taxonomies can be found in Handy (1995) (Zeus, Apollo, Athena and Dionysus culture) and Cameron and Quinn (1999) (Hierarchy, Market, Clan and Adhocracy culture). This makes a strong case for an holistic exploration of organisational culture, i.e. research that also includes structure information and data from various organisational processes. Figure 2.2 brings yet another issue to the fore, namely that of how to position culture within an organisation. This subject will be taken up in the next paragraph.

The Position of Culture in an Organisation

Looking at Figure 2.2, one could conclude that an organisational culture can be isolated quite easily from organisational structure and processes. Regarding the position of culture in an organisation four approaches can be distinguished (Frissen 1986):

1. Culture as contingency factor
2. Culture as subsystem of the organisation
3. Culture as an aspect system of the organisation
4. The organisation as a cultural phenomenon

A somewhat similar classification can be found with Smircich (1983), who states that an organisational culture can be viewed either as an independent or external variable; as an internal variable within an organisation; or as a root metaphor representing a collective view on life and experience.

Hofstede can be considered a proponent of the first approach, in that he considers an organisational culture to be primarily a product of national culture. Organisations within a country only differ in their 'practices', i.e. the outer layers of the culture onion, not so much in their values. When culture is considered a subsystem it is seen as functioning relatively independently next to other subsystems and can therefore be singled out easily for any further analysis. When culture is regarded as an aspect system, culture cannot be separated so easily, because it is ingrained in many subsystems of the organisation. Finally, when immersing oneself deeply in an organisation, one will probably get the impression that the organisation not so much has a culture but basically is a culture, i.e. is a cultural phenomenon. In this view culture is considered a 'root metaphor' (ibid.), 'a way of looking at life within a collectivity' (Martin 2002: 42), and culture stops being a research variable. In so doing, we have arrived at the fourth approach.

Frissen (1986) presents his approaches not as mutually exclusive but rather as successive stages of research into an organisation's culture. When starting a project, culture is often considered as something influencing an organisation (culture as a contingency factor). One then tries to isolate culture and study it in more detail (culture as a subsystem) and in its manifestations (culture as an aspect system). When the investigation is both deep and broad, culture will be encountered not so much as a part or aspect but as something the organisation invariably is – culture as a root metaphor. Once again, however, one should be wary of reification, which is sometimes hard to resist (Avruch 1998). In the next paragraph the concept of organisational culture will be examined further, looking at the issue of diversification of culture, i.e. culture and subcultures.

Cultures and Subcultures

Organisational cultures can be defined as having one unifying culture. Several scholars, like Schein, held this position for some time and this notion might have been inspired by research done on national or indigenous cultures. Organisational culture research conducted with standardised questionnaires often implies a common set of dimensions or scales on which such cultures primarily differ and hence also contains the notion of a single culture, although with local nuances. Moreover, the word 'organisational' already seems to imply a large, monolithic entity and certainly not something that is disintegrated or even fragmented.

However, organisations are quite open systems with leaders changing places rather often. Furthermore, many organisations are spread over more than one building or location. So most members of the organisation do not have a chance to interact and develop much together as a collective. Additionally, members bring along their own cultures – for instance, their national culture, their regional culture, their professional or occupational culture, their religious culture and their (socioeconomic) class culture. It is therefore quite possible that no specific organisational culture develops, especially when organisational setbacks are comfortably absent (Guldenmund et al. 2006), assuming that such upheaval initiates the (re-)formation of a culture. However, local subcultures might also develop, for instance based on the professional background of members or some challenging events a certain group had to face in the past. When members have a similar educational background, they do not even have to interact to share common cultural features. This seems to be the case in Schein's interpretation of subcultures (Schein 1996) and might also underlie Jones and James' findings in the US Navy (Jones and James 1979).

Nowadays, the notion of a unitary organisational culture has lost popularity in favour of a view promoting differentiation (Martin 2002; Richter and Koch 2004). In this view, (an organisational) culture is not considered unitary but consists of multiple subcultures. A quite radical view is based on social constructivism and proposes that culture is predominantly dynamic and much more defined situationally, i.e. a fragmentation view (Martin 2002). The heart of the matter lies perhaps between these latter two views, in that culture in the form of basic assumptions will be the ultimate result of continual interaction between group members, partly shown situationally and partly shown universally.

Additionally, whatever the point of view – i.e. integration, differentiation or fragmentation – within a group or population, culture can very well be 'socially distributed' (Avruch 1998: 18 ff.). That is, individuals belonging to a particular culture do not share their cultural content perfectly; this phenomenon has already been hinted at in the discussion of Figure 2.1 above. Additionally, culture can be 'psychologically distributed' within a group, meaning that cultural content can be deeply ingrained in one individual whereas in another it is only superficial (ibid.).[9]

One could question whether the characteristics of culture that have been discussed earlier apply in equal force to the concept of organisational culture. As already commented above, the word 'organisational' seems to suggest a large entity. The word 'culture' elicits the work and paradigms of anthropologists and sociologists and invites their ontology and epistemologies into the realm of organisation research. Indeed, initially the concept of culture as applied to organisations seemed rather attractive and provided explanations for certain phenomena that went unexplained previously. But organisational research also supplies managers with new ideas and ideologies and organisational culture also became something they wanted to manage, to control. Consequently, organisational culture became another instrument with which managers tried to pull things their way. When that happened some scholars of organisational culture pulled out, because they did not feel like contributing to yet another management tool (Salzer-Morling 2003).

A similar process can be recognised in the development of the safety culture concept. The first anthropologist has already stood up and summoned the research community to more sensitivity and (safety) managers to more modesty regarding the assessment and control of (safety) culture (Haukelid 2008). The current discussion will now be narrowed down to the safety culture concept, which will be held up against the theoretical light of (organisational) culture that has been kindled in the previous discussion. In the next few paragraphs the assessment of safety culture will be reviewed and analysed.

Organisational Culture and Safety

Ever since INSAG coined the term 'safety culture' to denote the far from optimal conditions and decision processes at the Chernobyl nuclear power plant (International Nuclear Safety Advisory Group 1986), it has become part of the standard explanatory safety vocabulary. Safety culture became a term with which people all around the globe explained everything they could not explain or understand otherwise. Whether the concept itself remained fuzzy, did not seem to matter much. However, this fuzziness is both its strength and its weakness. Indeed, (groups of) people sometimes seem to perform in dark mysterious ways (Kets de Vries 1999) and, when groping for an explanation, a concept such as

---

9   See also Schein's experiences with POWs, discussed above.

safety culture is highly attractive. A similar (initial) attraction can be identified in the development of the organisational culture concept (Salzer-Morling 2003); a discussion of the weakness of such a concept will be taken up below.

As with culture and organisational culture, safety culture has been defined by different authors differently, although many seem to refer to the same notion of shared basic assumptions, a shared understanding of reality (Antonsen 2009). How safety culture is studied will be discussed next, organised according to the three major approaches, the academic, the analytical and the pragmatic approach. For each of these approaches the dominant paradigm, the primary research methods and some example studies will be given.

Academic or Anthropological Approach

The primary research methodology of cultural anthropology is field research (ethnography), which is qualitative in nature. Its purpose is to describe and understand a culture rather than evaluate it and, hence, it is non-normative or value free. Moreover, the subject is never fitted onto some researcher's pre-existing notions. Because of these characteristics, it is not well-suited for comparative research. Applied to organisations, culture is considered as something an organisation *is*, rather than *has*. This approach is labelled 'academic' for it is employed almost exclusively by academics and it is hardly used outside the scientific realm (Hofstede 1991: 180), although the International Atomic Energy Agency (IAEA) is currently advocating a safety culture self-assessment (SCSA) for its member states, involving just such an approach. Schein has adopted this approach in what he calls 'clinical research' (Schein 1987). The term clinical already betrays the fact that some evaluation is taking place, but this is more in terms of a discrepancy between a given organisation's ambitions or intentions and what it actually accomplishes. In terms of safety this can become pertinent when a company claims to put safety as its number one priority, but nevertheless has many accidents.

The research method can be narrative research, a phenomenological study, a study using grounded theory, an ethnography or a case study (Creswell 2007), or various combinations thereof. Ideally, the research starts with a problem definition or an issue turned into a problem to focus the investigation; for instance, the discrepancy between safety priority and performance mentioned above. Research techniques include interviews, observations, document studies and whatever else the company brings forth that may hold clues for its underlying assumptions (see Guldenmund 2010 for an overview). What is important, however, is that information is collected with sufficient context, so that it can be interpreted accurately.

Whatever research method is chosen (case study, grounded theory, etc.) the results are (almost) never quantified because it is meaning and interpretation and not some numerical abstractions and calculations that drive the research. Moreover, numbers are never taken as data abstracted from an objective world, which would be in conflict with the research paradigm. The result is a 'thick description' (Geertz 1973), or a 'theory' of the culture of an organisation (cf. Glaser and Strauss 1967). When the description or theory turns out to be incomplete or 'wrong', the theory is adjusted to accommodate the contrasting empirical findings. Falsification can occur when another researcher with the same data comes to a different description or theory. In this approach safety culture is considered to be a nominal variable.

Current safety culture literature is still not well endowed with qualitative studies. This might be due to both publication policies, i.e. encouragement of quantitative rather than qualitative studies, and limitations regarding length of papers. Books describing such studies are equally absent. Moreover, methods are limited to either studies building on grounded theory (e.g. Berends 1995; Stave and Törner 2007; Walker 2008) or case studies (e.g. Brooks 2008; Farrington-Darby et al. 2005; Guldenmund 2008; Meijer 1999).

Analytical or Psychological Approach

This approach is the study of safety culture through (self-administered) questionnaires, which is the primary research instrument of (social and organisational) psychologists. This approach could be considered 'analytical' in that it considers safety culture an attribute of an organisation, i.e. something an organisation *has*, rather than *is* (cf. Hofstede 1991, but see also the discussion above) and isolates parts of it that are considered important or indicative to assess.

The field of safety culture is very much dominated by questionnaire studies; possibly because surveys are deceptively simple to use; probably also, because questionnaires are so popular with organisational psychologists. In various papers this approach has been disqualified as culture research and has been placed under the heading of safety climate (Collins and Gadd 2002; Glendon and Litherland 2001; Guldenmund 2000). Safety climate is considered to be a transient psychological variable, much less stable than safety culture.

Questionnaire studies generally follow this routine. First, potential concepts or facets of interest are identified that together make up the construct; this could be the result of a qualitative study. Based on these a questionnaire is composed using questions that cover the pertinent concepts best. This is at first an assumption which is tested in a subsequent survey where the questionnaires are put to an appropriate population. Subsequent data analyses should reveal whether the assumed concepts are actually present in the responses. The concepts are often conceived as dimensions spanning a multidimensional space; (sub-)cultures then become positions in that space. Additional analysis methods can model various relationships between the concepts that make up the culture construct and other numerical variables from outside the questionnaire. This way culture is caught in a web of concepts.

Paradigmatically, this appears to be a positivistic, (semi-)quantitative approach, because the questionnaire results are numerical as the questions are often answered on a numbered response scale (e.g. a Likert-scale or a semantic differential).[10] However, the analytical approach also has qualitative – that is, interpretive – elements to it. For instance, although the questionnaire should have a solid theoretical underpinning (as reflected in the chosen concepts), a subsequent analysis could go beyond these concepts and aim for new and (or) improved ones. Nevertheless, the final goal is to develop a robust set of general concepts (factors, dimensions, scales, facets) on which organisations can be assessed and, if necessary, compared. These latter characteristics make the analytical approach, in contrast with the previous academic approach, well-suited for comparative research. Such comparisons are, in principle, non-normative; that is, the mean scores do not have an evaluative sign to them, although the underlying individual responses might be based on such evaluations, preferences or perceptions (cf. Hofstede 2001: 15 ff.).

There are several important aspects to this approach, however, that are sometimes overlooked. For one thing, the numbers obtained from the rating scales are basically at the ordinal level of measurement. When such numbers are treated as though they are at a higher measurement level, there should at least be checks to see whether this assumption is justified. For another, although safety climate is not culture, it is still an emergent property of a group and therefore the within-group agreement, i.e. the coherence, should be tested (e.g. Zohar and Luria 2005). There are several indices available for this purpose, see Bliese (2006) for an overview.

The analytical approach can be considered a research methodology, which can be employed in either a case study or a (comparative) survey encompassing several organisations. Its research technique is a standardised questionnaire that is typically self-administered. It can be administered either group-wise, for instance at the start of a company training session, or sent to the worker's home addresses.

To summarise, viewed from the analytical perspective culture is a multidimensional construct and different cultures can be positioned at diverse positions in that space. These dimensions are either given beforehand or determined through analysis. An organisation's position in the culture space is calculated using questionnaire responses, often by using the mean as a descriptor of a dimension. There is abundant literature about research applying the analytic approach, aimed at the development of a questionnaire (e.g. Berends 1995; DeDobbeleer and Béland 1991; Díaz-Cabrera et al. 2007; Human Engineering Ltd. 2005; Kines et al. 2011), a case study (e.g. Guldenmund 2008; Havold 2005; Reiman and Oedewald 2004), a comparative study (e.g. Nielsen et al. 2008; Reiman et al. 2005; Zohar and Luria 2005), or modelling relationships (e.g. Cheyne et al. 1998; Johnson 2007; Neal and Griffin 2006).

---

10 There is a way of putting the questionnaire to qualitative use. The analysis then is not aimed at spanning a multidimensional space and projecting cases into it. The responses are used to generate themes, which are used in subsequent (qualitative) research (for example, see Guldenmund 2008).

Pragmatic or Experience-based Approach

There is yet another approach that can be distinguished in safety culture research. While the previous approaches could be considered descriptive, the pragmatic approach is normative. From an academic, interpretative point of view a culture can be neither 'good' nor 'bad'; such evaluations having been replaced by a relativist position. From the 'academic' perspective cultures are largely functional and have meaning in relation to their context and history. However, an organisational culture might be considered dysfunctional in relation to its future, for instance in relation to particular ambitions or goals. Such ambitions can be about many things, and therefore also about safety. For example, an organisation's ambition might be to have 'zero' accidents but serious accidents might still occur occasionally.

This normative approach has been labelled pragmatic because its content is not so much the result of empirical research on cultures but is rather based on experience and expert judgement. In practice, the pragmatic approach concentrates on both the structure and processes or interactions of an organisation, which, because of their dynamic interplay, will influence the culture in the long run (see Figure 2.2). Applied approaches concentrating on processes often focus on desired behaviour and the correction of deviations (e.g. DuPont's STOP™ or ProAct Safety's Lean Behavior-Based Safety^{SM}). It is thought that a change in behaviour will result in subsequent cultural adjustments. According to cognitive dissonance theory (Eagly and Chaiken 1993: 469 ff.), attitudes and thoughts about particular behaviours will change in the long run when the two are incongruent and the desired behaviour is rewarded.

Typically, what an organisation should do to obtain an advanced or mature status is prescribed in detail; that is, what processes should be implemented, supported by an accompanying structure. Geller's Total Safety Culture (Geller 1994) is a prime example of this approach, and the IAEA requirements and characteristics for nuclear power plants are of a similar nature (International Nuclear Safety Advisory Group 1991). Descriptive approaches towards culture such as the ones already discussed are of less relevance here, because it is not the organisation's current status but deviations from a predefined norm that are assessed and considered. However, knowledge of the current status might result in dissatisfaction with management which can be helpful in providing the organisation with a sense of urgency for change. Moreover, such knowledge also provides information on what structure and processes are suitable given the current status.

Lately, stages or levels of organisational maturity with regard to safety management have become fashionable (e.g. Energy Institute undated; Lardner 2004; Parker et al. 2006; Westrum 2004). Each level describes common local attitudes and behaviours in relation to safety, especially in relation to incident and accident prevention, reporting, investigation and solutions. An initial diagnosis of the current organisational status in relation to these attitudes and behaviours might be prepared. However, the main objective is to ascend the safety maturity hierarchy.

This might be accomplished by following the behavioural approach above, i.e. an emphasis on processes and behaviours in these, or with more structural adaptations. It is again assumed that culture will follow in the wake of these interventions. This approach assumes, rather implicitly, that safety culture is something an organisation has, or does not have; that is, mature 'generative' or 'cooperating' organisations have 'it', whereas immature 'pathological' or 'emerging' (Energy Institute undated; Lardner 2004; Westrum 2004) organisations do not.

The level of development of an organisation is assessed through behaviourally anchored rating scales, with either overt or covert ordinal scales. These assessments are always done in groups for two important reasons. Firstly, it is a group's shared opinions one is after, not the mean score of a group of employees. Secondly, it is not so much the rating but the ensuing discussion that follows because of this rating process that is considered the most important outcome. Nevertheless, scores are calculated and reported back to the organisation.

From the point of view of the interpretative academic approach the inferences that are made about an underlying culture solely based on descriptions of behaviour are committing a mortal sin. According to this approach it is impossible to infer such meanings based on observed behaviour. Geertz, quoting the philosopher Ryle, illustrates this nicely by comparing a wink with a twitch and with a parody of a wink: all three look much the same, but have quite different meanings (Geertz 1973: 6 ff.).

To summarise, regarding the matter of safety culture and its assessment, there are several aspects that require particular attention:

1. From an academic viewpoint, culture is a value-free concept (a nominal variable) whereas safety is not. The required purpose of safety culture assessments is not descriptions but evaluations, preferably with recommendations on how the underlying culture can be improved to support safety (more).
2. Safety is about behaviour, whereas culture is about the meaning of behaviour. The relationship of culture with behaviour is partly dependent on the strength with which the core assumptions are held. Hence, knowledge about the direction of assumptions is not sufficient; also their intensity is important for behaviour.
3. The assessment of culture is therefore complicated and certainly not straightforward. Behaviour has become the major focus with allusions to an underlying culture. In the end, the actual meaning of the observed behaviour becomes much less important than the behaviour itself.

Influencing Organisational Safety Culture

Most organisational safety culture assessments are not carried out for their own sake. Management is either interested in a diagnosis to compare it with a previous one or to benchmark with peer organisations, or the diagnosis is followed by a so-called gap analysis, where the present status of culture is compared with an ideal or optimal one, resulting in one or more recommendations to improve the current status. Enough has been said about safety culture assessment in the previous section. With regard to influencing culture, some final remarks will be made.

When discussing Figure 2.1, it has already been observed that the output of the development process is never an intended outcome, but rather the product of many different forces working on the group at various points in time. In that sense, the resulting culture is not so much an optimal 'textbook culture', but rather something that seems to work in this group under these conditions. When the group remains successful in its accomplishments, the accompanying set of basic assumptions is reinforced and strengthened.

This does not mean that any attempts can be made to influence the current safety culture, by influencing its development process. To choose particular influences, Figure 2.1 can be taken as a starting point and interventions can be chosen that might influence the several steps within the model. In Figure 2.3 generic interventions are suggested, that might influence these various steps. Carrying out multiple interventions at the same time does seem to be more effective in influencing this process, rather than doing a single one or a few in succession (Hale et al. 2010).

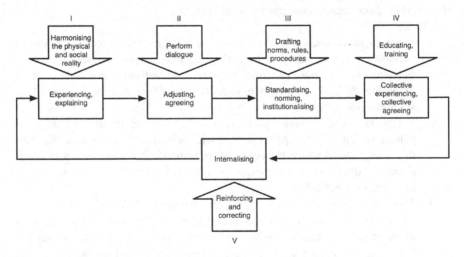

Figure 2.3   Generic influences on the development process of (organisational safety) culture

The generic influences in Figure 2.3 are as follows.

1. Harmonising the physical and social reality. The first step in the model describes the sensemaking processes of the individual members of the group, their understanding of reality. Possible influences that might affect this process are aimed at ensuring that parts of the physical and social reality are comparable across the organisation e.g. similar technology, instruments, personal protection equipment, workplace layout, campaigns, etc. with regard to the physical reality. With regard to the social reality possible influences are leadership styles, rituals (e.g. meetings, celebrations), (systems for) recognition, etc. The aim of these interventions would be to create similar meanings, a comparable 'reality' across the organisation.

2. Perform dialogue. This is a crucial step in the development of culture and efforts should be made to perform this dialogue across the organisation to ensure that consensus is reached to the extent that most people in the organisation have a comparable understanding of the reality they act on. Possible examples of interventions are STOP-GO cards or Last Minute Risk Assessments, rules for approaching and correcting people, reporting of unsafe situations, etc.[11] It should be noted that steps one and two of the culture development process are iterative and remain so until a particular degree of consensus is reached (or a consensus is enforced, but this usually means that the ensuing rules are not sufficiently supported by the majority and are therefore also not internalised).

3. Drafting norms, rules and procedures. At this stage the consensus on (parts of) reality that has been reached at stage two can be formalised and institutionalised. Because of the shared consensus, the ensuing rules are recognised and understood by the majority of the group. Importantly, some rules are not formalised to the extent that they not are written down, yet they function as such within a group.

4. Educating, training. After formalising the rules, they can be trained or otherwise disseminated amongst the members of the group. New members will often start at this step, although when things do not make sense to them, they might also speak up and processes at steps one and two will become pertinent too. When an organisation is operating for some time, new members will have less and less impact and they will either agree with the rules (or pretend they agree), or leave the company. This situation can change when the existence of the company, and therefore of the group, is threatened, or when the group is otherwise facing a significant challenge.

---

11   Stopping the production process should result in a dialogue about what is safe and what is not. The same goes for reports of unsafe situations.

5. Reinforcing, correcting. For the basic assumptions of a culture to be secure, i.e. to become genuine basic assumptions, they have to be reinforced for some time. After a while, they become self-explanatory and objective to the extent that members of the organisation cannot imagine understanding or approaching reality otherwise. This basic understanding is again influencing the sensemaking of step one.

Patient Safety Culture

Patient safety culture is a reasonably new development of the organisational safety culture construct. As with other types of culture, the addition 'patient safety' serves as a qualifier for the general concept of organisational culture. By using this qualifier the proposition is made that an organisational culture can be conducive, or unfavourable, to patient safety.

Within hospitals several distinct groups operate, with different functions, tasks and experiences and with different educational backgrounds. Through interaction between them and with patients, common systems of meaning will develop in teams, wards, departments and even hospitals. Because these interactions will be intense and often emotion-laden, several of these meanings around patient safety will be profound. Moreover, public attention to patient safety will also have shaped these meanings. Overall, the notion of patient safety culture appears to have face validity.

The exploration of patient safety culture through the exclusive use of questionnaires is contrary to the concept of safety culture. This approach has been labelled 'analytical' above. While questionnaires certainly have merit in the study of culture, the insight they provide is shallow compared to the results of the qualitative 'academic' approach. Moreover, what might be an important (cultural) meaning in one hospital might be different or absent in another. Questionnaires do not adapt themselves to such local differences or nuances, whereas an ethnographic field study does. Furthermore, surveys collect thoughts, attitudes and perceptions rather than the basic assumptions, which are thought to underlie these, and actual practice. It should be clear that the study of patient safety culture should comprise both approaches, so the validity of the concepts used in the questionnaire, can be tested across hospitals within and between countries.

Summary

Culture is a prerequisite for human beings to be able to live, to understand their surroundings, to work together. There are many definitions around and these often differ in their wording, but not so much in their essence (e.g. Antonsen 2009). Culture is the result of a process based on sensemaking and interaction and adjustment within a group, yet it is never an intended result but rather a *modus vivendi* developed by the group while operating together. Culture transcends the individuals that share the culture, as it is passed on and, relatively, enduring.

Conceptually, culture can be thought of as a group's shared understanding of reality, as a way of looking at and experiencing that reality and all the things that happen in it. Culture research is aimed at describing the lens through which a particular group experiences its reality. Culture can be studied at different levels, with the level of the nation considered to be the highest one. The construct can be modelled as something consisting of an invisible and relatively intangible core that is projected onto one or more outer layers, which are taken as the manifestations of the core. The core represents the basic assumptions of the group that help its members understand reality. Built onto these are various norms, rituals, institutions, symbols and behaviours which are particular expressions of the core.

Three ways of approaching culture are available, the academic, the analytical and the pragmatic approach. The academic approach makes use of qualitative techniques and results in a thick description of a culture which is value-free. The analytical approach is based on self-administered questionnaires and makes comparisons between (sub-)cultures possible. Finally, the pragmatic approach uses developmental hierarchies to describe cultures. Organisations are supposed to aim for the highest steps on these hierarchies.

The culture development process might be used to formulate general intervention strategies that could influence the different steps of this process. In general, using several interventions at the same time might be more effective than carrying them out in succession or doing a few.

Finally, the concept of patient safety culture appears to have face validity. However, its study should not limit itself to the use of climate questionnaires but should rather encompass qualitative studies that could provide the questionnaire with a more solid conceptual backing.

References

Antonsen, S. 2009. *Safety Culture: Theory, Method and Improvement.* Farnham: Ashgate.

Avruch, K. 1998. *Culture & Conflict Resolution.* Washington, DC: United States Institute of Peace Press.

Bem, D.J. 1970. *Beliefs, Attitudes and Human Affairs.* Belmont, CA: Brooks/Cole Publishing Company.

Berends, J.J. 1995. *Developing and Using a Widely Applicable Measurement Tool for Safety Culture.* (Unpublished interim report), Eindhoven University of Technology, Eindhoven.

Berger, P.L. and Luckmann, T. 1966. *The Social Construction of Reality: A Treatise in the Sociology of Knowledge.* Garden City, NY: Anchor Books.

Bliese, P.D. 2006. *Multilevel Modeling in R (2.2). A Brief Introduction to R, the Multilevel Package and the NLME Package.* Washington, DC: Walter Reed Army Institute of Research.

Bloch, M.E.F. 1998. *How We Think They Think: Anthropological Approaches to Cognition, Memory, and Literacy.* Oxford: Westview Press.

Brooks, B. 2008. The natural selection of organizational and safety culture within a small to medium sized enterprise (SME). *Journal of Safety Research*, 39(1), 73–85.

Cameron, K.S. and Quinn, R.E. 1999. *Diagnosing and Changing Organisational Culture: Based on the Competing Values Framework.* Reading, MA: Addison-Wesley.

Castellucci, V.F., Carew, T.J. and Kandel, E.R. 1978. Cellular analysis of long-term habituation of the gill-withdrawal reflex of Aplysia Californica. *Science*, 202, 1306–8.

Cheyne, A., Cox, S., Oliver, A. and Tomæs, J.M. 1998. Modelling safety climate in the prediction of levels of safety activity. *Work & Stress*, 12(3), 255–71.

Collins, A.M. and Gadd, S. 2002. *Safety Culture: A Review of the Literature.* Sheffield: Health and Safety Laboratory, Human Factors Group.

Creswell, J.W. 2007. *Qualitative Inquiry and Research Design. Choosing among Five Traditions* (2nd edition). Thousand Oaks, CA: Sage Publications Inc.

Deal, T.E. and Kennedy, A.A. 1982. *Corporate Cultures.* Reading, MA: Addison-Wesley.

DeDobbeleer, N. and Béland, F. 1991. A safety climate measure for construction sites. *Journal of Safety Research*, 22, 97–103.

Díaz-Cabrera, D., Hernández-Fernaud, E. and Isla-Díaz, R. 2007. An evaluation of a new instrument to measure organisational safety culture values and practices. *Accident Analysis & Prevention*, 39(6), 1202–11.

DuPont undated. *DuPont™ Stop™ Program.* Available at: http://www.training.dupont.com/dupont-stop/ (last accessed on 1 June 2014).

Eagly, A.H. and Chaiken, S. 1993. *The Psychology of Attitudes.* Fort Worth: Harcourt Brace Jovanovich.

Energy Institute undated. *Hearts and Minds Programme.* Available at: http://www.eimicrosites.org/heartsandminds/index.php (last accessed on 1 June 2014).

Farrington-Darby, T., Pickup, L. and Wilson, J.R. 2005. Safety culture in railway maintenance. *Safety Science*, 43, 39–60.

Frissen, P. 1986. Organisational culture: An overview of approaches (in Dutch). *M&O, Tijdschrift voor Organisatiekunde en Sociaal Beleid*, 6, 532–44.

Geertz, C. 1973. *The Interpretation of Cultures.* New York: Basic Books.

Geller, E.S. 1994. Ten principles for achieving a Total Safety Culture. *Professional Safety*, September, 18–24.

Glaser, B.G. and Strauss, A.L. 1967. *The Discovery of Grounded Theory: Strategies for Qualitative Research.* Chicago: Aldine.

Glendon, A.I. and Litherland, D.K. 2001. Safety climate factors, group differences and safety behaviour in road construction. *Safety Science*, 39, 157–88.

Guldenmund, F.W. 2000. The nature of safety culture: A review of theory and research. *Safety Science*, 34(1–3), 215–57.

Guldenmund, F.W. 2008. Safety culture in a service company. *Journal of Occupational Health and Safety – Australia and New Zealand*, 24(3), 221–35.

Guldenmund, F.W. 2010. *Understanding and Exploring Safety Culture*. Oisterwijk: BOXPress.

Guldenmund, F.W., Ellenbroek, M. and van den Hende, R. 2006. Organisational culture research in a true Dutch company (in Dutch). *Tijdschrift voor Toegepaste Arbowetenschap*, 19(2), 24–32.

Hale, A.R., Guldenmund, F.W., van Loenhout, P.L.C.H. and Oh, J.I.H. 2010. Evaluating safety management and culture interventions to improve safety: Effective intervention strategies. *Safety Science*, 48, 1026–35. doi: 10.1016/j. ssci.2009.05.006.

Handy, C. 1995. *Gods of Management: The Changing Work of Organizations*. New York: Oxford University Press.

Hardin, R. 2009. *How Do You Know? The Economics of Ordinary Knowledge*. Princeton, NJ: Princeton University Press.

Harrison, R. and Stokes, H. 1992. *Diagnosing Organizational Culture*. San Francisco, CA: Pfeiffer & Co.

Haukelid, K. 2008. Theories of (safety) culture revisited – An anthropological approach. *Safety Science*, 46(3), 413–26.

Havold, J.I. 2005. Safety-culture in a Norwegian shipping company. *Journal of Safety Research*, 36(5), 441–58.

Hofstede, G.R. 1991. *Cultures and Organisations: Software of the Mind*. London: McGraw-Hill.

Hofstede, G.R. 2001. *Culture's Consequences* (Second edition). London: Sage Publications.

Human Engineering Ltd. 2005. *Development and Validation of the HMRI Safety Culture Inspection Toolkit*. London: Health and Safety Executive.

International Nuclear Safety Advisory Group 1986. *Summary Report on the Post-Accident Review Meeting on the Chernobyl Accident*. Vienna: IAEA.

International Nuclear Safety Advisory Group 1991. *Safety Culture*. Vienna: International Atomic Energy Agency.

Johnson, S.E. 2007. The predictive validity of safety climate. *Journal of Safety Research*, 38(5), 511–21.

Jones, A.P. and James, L.R. 1979. Psychological climate: Dimensions and relationships of individual and aggregated work environment perceptions. *Organizational Behavior and Human Performance*, 23, 201–50.

Kandel, E.R. and Schwartz, J.H. (eds). 1985. *Principles of Neural Science* (Second edition). New York: Elsevier.

Keesing, R.M. 1981. *Cultural Antropology. A Contemporary Perspective* (Second edition). Fort Worth: Holt, Rinehart and Winston Inc.

Kets de Vries, M.F.R. 1999. *Struggling with the Demon – Essays on Individual and Organizational Irrationality* (in Dutch). Amsterdam: Uitgeverij Nieuwezijds.

Kines, P., Lappalainen, J., Mikkelsen, K.L., Olsen, E., Pousette, A., Tharaldsen, J., Tómasson, K. and Törner, M. 2011. Nordic Safety Climate Questionnaire (NOSACQ-50): A new tool for diagnosing occupational safety climate. *International Journal of Industrial Ergonomics*, 41(6), 634–46. doi: 10.1016/j.ergon.2011.08.004.

Kroeber, A. and Kluckhohn, C. 1952. *Culture: A Critical Review of Concepts and Definitions*. New York: Meridian Books.

Lardner, R. 2004. Mismatches between safety culture improvement and behaviour-based safety. Paper presented at the *NeTWork 2004 Workshop 'Safety Culture and Behavioural Change at the Workplace'*, Blankensee, Berlin, 9–11 September.

Martin, J. 2002. *Organizational Culture: Mapping the Terrain*. Thousand Oaks, CA and London: Sage Publications.

Meijer, S.D. 1999. *The Gordian Knot of Organisational and Safety Culture* (in Dutch). (PhD thesis), Eindhoven University of Technology, Eindhoven.

Mintzberg, H. 1979. *The Structuring of Organizations*. Englewood Cliffs: Prentice Hall.

Mintzberg, H. 1980.Structure in fives: a synthesis of the research on organisation design. *Management Science*, 26(3), 322–41.

Mintzberg, H. 1983. *Structures in Five: Designing Effective Organisations*. Englewood Cliffs, NJ: Simon & Schuster.

Neal, A. and Griffin, M.A. 2006. A study of the lagged relationships among safety climate, safety motivation, safety behavior, and accidents at the individual and group levels. *Journal of Applied Psychology*, 91, 946–53.

Nielsen, K.J., Rasmussen, K., Glasscock, D. and Spangenberg, S. 2008. Changes in safety climate and accidents at two identical manufacturing plants. *Safety Science*, 46(3), 440–49.

Parker, D., Lawrie, M. and Hudson, P.T.W. 2006. A framework for understanding the development of organisational safety culture. *Safety Science*, 44(6), 551–62.

Pederson, J.S. and Sorensen, J.S. 1989. *Organisational Cultures in Theory and Practice*. Aldershot, UK: Avebury & Gower.

ProAct™ Safety undated. *Lean Behaviour-based Safety*. Available at: http://www.proactsafety.com/leanbbs (last accessed on 1 June 2014).

Reiman, T. and Oedewald, P. 2004. Measuring maintenance culture and maintenance core task with CULTURE-questionnaire – A case study in the power industry. *Safety Science*, 42(9), 859–89.

Reiman, T., Oedewald, P. and Rollenhagen, C. 2005. Characteristics of organizational culture at the maintenance units of two Nordic nuclear power plants. *Reliability Engineering & System Safety*, 89(3), 331–45.

Richter, A. and Koch, C. 2004. Integration, differentiation and ambiguity in safety cultures. *Safety Science*, 42, 703–22.

Rousseau, D.M. 1990. Assessing organisational culture: The case for multiple methods (Chapter 5). In B. Schneider (ed.), *Organisational Climate and Culture*. Oxford: Jossey-Bass.

Salzer-Morling, M. 2003. Cultivating culture: Mats Alvesson, Organisationskultur och ledning, Liber förlag, Malmö, 2001; Joanne Martin, Organizational Culture: Mapping the Terrain, Sage, London, 2002; Martin Parker, Organizational Culture and Identity, Sage, Thousand Oaks, 2000. *Scandinavian Journal of Management*, 19(3), 385–392.

Sanders, G. and Neuijen, B. 1987. *Organisational Culture: Diagnosis and Influencing* (in Dutch). Assen: Van Gorcum.

Schein, E.H. 1987. *The Clinical Perspective in Fieldwork.* Sage University Paper series on Qualitative Research Methods, Beverly Hills: Sage Publications.

Schein, E.H. 1990. Organisational culture. *American Psychologist*, 45(2), 109–19.

Schein, E.H. 1992. *Organizational Culture and Leadership* (Second edition). San Francisco: Jossey-Bass.

Schein, E.H. 1996. Three cultures of management: The key to organisational learning. *Sloan Management Review*, Fall 1996, 9–20.

Schein, E.H. 1999. Empowerment, coercive persuasion and organizational learning: Do they connect? *The Learning Organization*, 6(4), 163–72.

Schein, E.H. 2004. *Organizational Culture and Leadership* (Third edition). San Francisco: Jossey-Bass.

Smircich, L. 1983. Concepts of culture and organizational analysis. *Administrative Science Quarterly*, 28, 339–58.

Spencer-Oatey, H. 2000. *Culturally Speaking: Managing Rapport through Talk across Cultures.* London: Continuum.

Spradley, J.P. and McCurdy, D.W. 1975. *Anthropology: The Cultural Perspective.* New York: John Wiley & Sons Inc.

Stave, C. and Törner, M. 2007. Exploring the organisational preconditions for occupational accidents in food industry: A qualitative approach. *Safety Science*, 45(3), 355–71. doi: 10.1016/j.ssci.2006.07.001.

Trompenaars, F. and Hampden-Turner, C. 1997. *Riding the Waves of Culture: Understanding Cultural Diversity in Business* (Second edition). London: Nicholas Brieley Publishing Ltd.

Van Hoewijk, R. 1988. The meaning of organisational culture: An overview of the literature (in Dutch). *M&O, Tijdschrift voor Organisatiekunde en Sociaal Beleid*, 1, 4–46.

Walker, A. 2008. A qualitative investigation of the safety culture of two organisations. *The Journal of Occupational Health and Safety – Australia and New Zealand*, 24(3), 201–12.

Westrum, R. 2004. A typology of organisational cultures. *Quality and Safety in Health Care*, 13(Suppl II), ii22–ii27. doi: 10.1136/qshc.2003.009522.

Whittington, R. and Pany, K. 2004. *Principles of Auditing and Other Assurance Services* (Fourteenth edition). Boston, MA: McGraw-Hill.

Zohar, D. and Luria, G. 2005. A multilevel model of safety climate: Cross-level relationships between organization and group-level climates. *Journal of Applied Psychology*, 90(4), 616–28.

Chapter 3

Patient Safety Culture and Organisational Behaviour: Integrating Error, Leadership and the Work Environment

Zenobia Talati and Mark Griffin

Overview of Patient Safety Culture

In 1999, the Institute of Medicine released a report estimating that between 44,000 and 98,000 deaths occur in American hospitals each year due to medical errors. Errors which do not result in death may result in injury or a prolonged stay with increased costs to the individual, to the hospital and the community. Most worryingly, many of these events are preventable. The main suggestion from the Institute of Medicine's report was to learn from errors and improve the system rather than blame individuals for errors; implement mandatory or voluntary error reporting systems and develop strong safety leadership. These activities are also the focus of this chapter.

The aim of this chapter is to review the nature of patient safety culture (PSC), how healthcare organisations can develop a stronger PSC and how a strong PSC can help reduce adverse events such as injuries or death. We discuss the role of error management in achieving this outcome. This involves being open to errors and attempting to learn from them. We will make links between PSC and other areas of research and other industries with strong positive safety cultures. Finally, we will discuss the role played by leaders in creating a PSC.

Definition of Organisational Climate/Culture

Organisational culture has been defined as

> a pattern of shared basic assumptions learned by a group as it solved its problems of external adaptation and internal integration, which has worked well enough to be considered valid and, therefore, to be taught to new members as the correct way to perceive, think, and feel in relation to those problems. (Schein 2010: 18)

Thus an organisation's culture represents the underlying beliefs, norms and values held by its employees. On the surface, these values and beliefs manifest themselves in people's actions and in organisational artefacts such as training and hiring practices and HR systems. These attitudes often inform, and are therefore closely linked to, employee behaviour (Clarke 2006). As the above definition implies, culture is transmitted to new employees. This is done through the process of socialisation. Gundry and Rousseau (1994) studied how critical incidents in the workplace can influence newcomers' perceptions of the organisational culture. Critical incidents were described to participants as events which made an impression on them and gave them some insight into what it would be like to work for that company. Newcomers learned behavioural norms from these critical incidents by observing how these incidents were received within the organisation. This demonstrates that the process of assimilation is a complex one which is affected by a wide range of factors outside of organisationally controlled events such as orientations and training sessions.

Safety climate/culture   The concept of organisational culture has been applied to safety to describe an organisation's safety culture. Safety culture describes the values, beliefs and attitudes which employees hold about their organisation in relation to safety which influence their commitment to safety (Guldenmund 2000). An organisation's culture can be a major source of motivation and information relevant to safety practices (Nahrgang et al. 2008). While some researchers have drawn a distinction between safety climate and safety culture, the terms are commonly used interchangeably (Denison 1996; Mearns and Flin 1999). This is especially the case in the Patient Safety Culture literature (Flin 2007). Thus, for the purposes of this chapter, we consider research on both culture and climate.

An organisation's safety culture can create norms that employees refer to when deciding how to behave in certain situations (Clarke 2006; Quick et al. 2008). In this way, safety culture can affect safety behaviours and lead to positive outcomes. Indeed, numerous studies conducted in high risk industries have shown that a positive safety culture leads to greater levels of safe behaviours (Cooper and Phillips 2004; Neal and Griffin 2006), increased motivation to engage actively in safety behaviours rather than just comply with them (Griffin and Neal 2000) and fewer occupational injuries (Barling et al. 2002). The culture continues to exert influence as it is passed on through successive generations of employees. Through the process of socialisation, new employees learn about and adopt the culture of the organisation.

Conceptions of PSC

Antecedents and Outcomes

Halligan and Zecevic (2011) reviewed 137 studies on safety culture within healthcare settings (including patient safety culture). Based on these studies, they described what previous researchers considered to be components of safety culture. These included safety leadership, teamwork, shared beliefs about safety, open communication, organisational learning and non-punitive reporting of error. This was further supported by Sammer et al. (2010) who also determined that leadership, teamwork, communication and learning were important aspects of safety culture. The model shown in Figure 3.1 describes the proposed linkages between safety climate and outcomes (Flin 2007). As can be seen, a combination of external organisational and internal personal factors influence whether an injury will occur.

Hudson (2003) put forward a model of cultural maturity, based on the model initially proposed by Westrum (2004), proposing that organisations could be categorised into different stages of development: pathological (where individuals are only concerned about their personal needs), reactive (where organisations deal with safety incidents after they have occurred), bureaucratic/ calculative (where individuals are concerned more about their department), proactive (when the workforce take safety into their hands) and generative (when individuals are concerned more about the mission of their organisation).

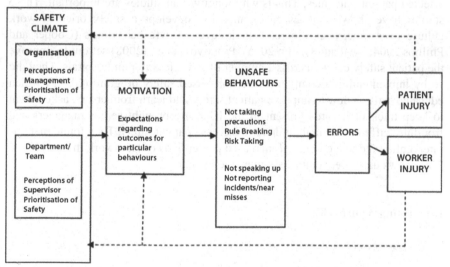

Figure 3.1   Link between safety climate and outcomes
Source: Flin 2007.

These cultures are thought to be put in place by management through the values and attitudes which they project through the organisation. They are thought to relate to an organisation's safety culture in terms of how readily individuals will communicate about and share safety information and the manner in which errors are dealt with.

Huang et al. (2007) surveyed hospital staff at four ICUs within the same hospital and found that PSC varied greatly between ICUs. Additionally, perceptions of PSC differed according to position in the department. That is, doctors tended to report a more positive safety culture, while nursing directors over and under estimated the safety culture of their teams. This study shows that culture should be measured at the unit, rather than the organisational, level. Difference in perceptions of PSC between doctors and nurses has also been reported by Thomas et al. (2003). There is evidence to suggest that it is more informative to measure PSC through nurses as they have the greatest amount of contact with patients. This line of reasoning will be expanded on later in the chapter.

In general, organisations with a safety oriented work culture tend to have employees who are more motivated to engage in safe behaviours (Clarke 2006). This also applies specifically to PSC. Studies have found a correlation between PSC and safety outcomes such as workplace injuries, medication errors and urinary tract infections (Felknor et al. 2000; Hofmann and Mark 2006). Furthermore, this relationship remains valid after controlling for severity of patient illness and other risk factors (Stone et al. 2007). These studies provide initial evidence of a link between PSC and safety outcomes. However, given the correlational nature of these studies it is hard to determine the extent to which safety climate directly affected patient outcomes. This is why longitudinal studies are important. These studies have shown that, as an organisation develops a safety oriented work culture over time, teams increase their level of safe behaviours (Cooper and Phillips 2004; Neal and Griffin 2006). Pronovost et al. (2008) set out to improve the patient safety culture across a number of ICUs in different hospitals. They did so by implementing a comprehensive unit-based safety program (CUSP) which educated staff on how to improve patient safety and learn from errors, asked them to keep track of hazards present in the ICU and brought senior managers and frontline staff together in their drive to increase patient safety. They found that this intervention led to a climate of increased perceptions of teamwork in most ICUs. This shows that safety culture is amenable to change.

Integrating Approaches

Error Management Culture

It is clear that there is a negative perception of error in hospitals. A fear of discussing errors is probably ingrained into healthcare providers' memories early in their careers when they are in medical school and there is a lot of pressure not to make errors. Leape (1994) argues that errors are assumed to be due to negligence or incompetence. As he describes it, 'Efforts at error prevention in medicine have characteristically followed what might be called the perfectibility model: if physicians and nurses could be properly trained and motivated, then they would make no mistakes. The methods used to achieve this goal are training and punishment' (Leape 1994: 1852). Part of many healthcare organisations' cultures is the underlying belief that individuals are largely responsible for their behaviours. Thus, when errors occur, people are often quick to blame the individual. However, given that healthcare providers often work in multidisciplinary teams and with complex equipment, it is hard to say that only one individual is responsible for an error (Carroll and Quijada 2004). Nonetheless, this emphasis on individual autonomy is passed on through the process of socialisation from senior members of staff to junior members (Saks and Ashforth 1997). As it continues to be passed down, it becomes part of the organisation's culture. This is further propagated by the increased likelihood that the organisation will attract and recruit employees who align with this culture (Schneider et al. 1995). This culture is then maintained throughout the physician's career with harsh punishments for those who do commit errors.

Hospital staff not only avoid making errors but also avoid discussing them. In their survey of medical house officers, Wu et al. (2003) found that just over half (54 percent) of the house officers discussed errors with their supervisors. A much larger number (88 percent) discussed errors with non-supervisory medical colleagues. Similar results were reported by Sexton et al. (2000) who surveyed over 1,000 hospital staff working on surgical and anaesthetic wards. Over half of the staff surveyed said they found it difficult to talk about errors in their workplace. The main reasons for this were fear of damaging their reputation, fear of being charged with malpractice, high expectations from the patient's family and society in general, fear of disciplinary action from a licensing board and fear of losing their job. Barach and Small (2000) reviewed the barriers to reporting incidents. They found that the main barriers reported by medical staff were fear of punishment – legally, from their peers, from their organisation and from medical boards. With so much perceived to be at risk when an error is made, it is unsurprising that staff are reluctant to talk about errors.

The inability to talk about errors for longer than is absolutely necessary, leads to what Tucker and Edmondson (2003) describe as first order problem solving. This is when someone implements a quick fix to resolve a problem caused by a lack of supplies, resources or information for example. However, this does nothing to address the underlying cause of the problem. At times, the quick fix to one problem may even cause other problems later down the track. Since the problem is generally not discussed again after it has occurred, no changes are made and it continues to occur. Second order problem solving, which involves attempting to address the underlying cause of the problem, was found to occur far less among nurses (Tucker and Edmondson 2003). Defaulting to first order problem solving was partly attributed to the nurses being under time pressures, using rules of thumb to fix problems and feeling that they were responsible for any problems that occurred during their shift. Some nurses in this study even reported feeling a sense of satisfaction when they were able to solve problems successfully using first order problem solving. This is because they would immediately receive feedback which acts as a positive reinforcer of their behaviour. Over time however, a frequent reliance on first order problem solving (as opposed to getting to the root cause of the problem the first time) is thought to lead to burnout.

Chang and Mark (2011) found that a more positive learning climate led to fewer medical errors. That is, when employees were more willing to disclose their errors, discuss them openly with others and think about what factors could have led to that error, they produced fewer errors as a result. This effect was moderated by the number of Registered Nurses (RNs) in the nursing team. When there was a poor learning climate, a larger number of RNs led to fewer medical errors and a lower number of RNs led to more medical errors. But when there was a good learning climate, the number of RNs did not affect the rate of errors. This suggests that the total number of RNs is not as important as the learning climate which they work in. This is especially important for hospitals experiencing nursing shortages.

Error management culture emphasises that errors are an important source of learning. However, this approach seems unsuitable for a high risk environment such as healthcare where errors can have life threatening consequences. Researchers are well aware of how counter intuitive this is. They would argue that while an organisation should do everything it can to prevent errors, they will never be completely avoidable. Thus, if errors are going to occur anyway, they should be dealt with in a manner that allows the organisation to improve its safety performance rather than just punishing individuals (which will have little impact on safety performance).

Katz-Navon et al. (2009) examined more closely the conflict that can occur in organisations with both a high error management culture and a strong safety culture. In their study they measured active learning climate which is similar to error management in that both place an emphasis on learning from errors and continuous improvement. They also measured the extent to which the organisation had a climate which prioritised safety. These two climates can conflict since 'the active learning process requires exploration, risk taking, and tolerance of mistakes, whereas a priority of safety emphasizes control and requires working by acceptable channels, planning, procedures, and rules' (Katz-Navon et al. 2009: 1202). They hypothesised that too much emphasis on one climate would result in poor safety outcomes since the organisation needs to achieve a balance between making safety a priority and learning from mistakes. They found that errors were lowest when the priority given to a safety climate was moderate and there was a high active learning climate.

One consequence of error which is often not immediately considered is the impact it can have on the person who made the error. When physicians commit errors, they are likely to experience emotional distress (Schwappach and Boluarte 2010). This can lead to burnout and decreased empathy for patients. Burnout has been linked to suboptimal patient care (Shanafelt et al. 2002). In their review, Schwappach and Boluarte (2010) found that the higher a physician was on depersonalisation (a subscale of burnout), the more likely it was that they would commit another error three months later. The authors describe this as a vicious cycle whereby distress from the error leads to poor patient care, which then leads to more errors. One way to break the cycle would be to provide support for the person who has committed the error. If this helps to reduce their emotional distress, they may be better able to provide patient care.

Near Misses

When there is less fear of blame and punishment, people become more open to discussing errors and near misses (Barach and Small 2000). Near misses are an important source of information as they are accidents that almost happened. Since they differ only slightly from accidents, they can teach us a lot about causes of error in a workplace. This is described by (Reason 1990) in his Swiss cheese model seen in Figure 3.2. There are a variety of factors that affect whether or not an accident occurs. The factors influencing a situation are likened to slices of Swiss cheese which contain a number of holes. These holes represent weaknesses which could lead to errors. Accidents occur when the holes in each 'slice' (or variable) align, while near misses occur when most, but not all, of the factors necessary for an accident are present. In this way, near misses can provide crucial information about unidentified risk factors.

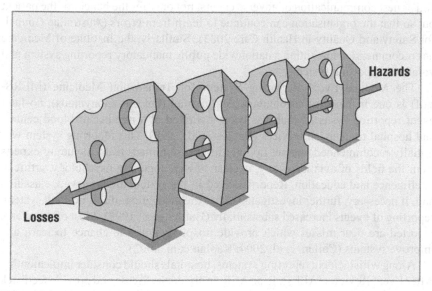

Figure 3.2   Swiss cheese model
Source: Reason 2000.

Incident Reporting Systems

Ninety percent of staff in Sexton et al.'s (2000) study felt that patient safety would be improved if their hospital were to implement a confidential system for staff to report errors. This would presumably encourage open communication of errors and organisational learning, which are two components of a positive patient safety culture (Halligan and Zecevic 2011). Open communication of errors and organisational learning are also the foundations of an error management culture (Van Dyck et al. 2005). An error management culture is one in which people recognise that errors are inevitable and can be a source of information. Thus there is a greater emphasis on working out what changes should be made to avoid repeating the error, rather than working out who is to blame for the error.

A lack of error reporting is one of the biggest barriers to hospitals creating an error management culture and improving their PSC. The current culture in many hospitals is to attribute errors to incompetence or bad luck (Dekker 2011). When people believe these to be the sources of error, they have little motivation to report or learn from errors. In the case of the incompetent medical practitioner, the solution is to provide more training or replace the practitioner. If an error is thought to be caused by bad luck, people may develop the mentality that there is little that can be done to anticipate it or prevent it from happening again. Many government bodies have advocated open disclosure as common practice in healthcare organisations (Joint Commission on Accreditation of Healthcare Organizations 2001; Australian Council for Safety and Quality in Health Care 2003). For example, in its report, the Australian Commission on Safety and Quality in Healthcare recommends open and honest communication of adverse events, not only for the benefit of the patient but so that the organisation can continue to learn from errors (Australian Council for Safety and Quality in Health Care 2003). Similarly, the Institute of Medicine has recommended instituting a nationwide public mandatory reporting system and voluntary reporting systems.

The Medical Event Reporting System for Transfusion Medicine (MERS-TM) is one instance of a voluntary, confidential (but not anonymous), no-fault event reporting system which has been instituted at a number of blood centres and hospital transfusion services (Battles et al. 1998). This reporting system was initially recommended and set up by a multidisciplinary team including experts from the fields of aviation safety, nuclear power, cognitive psychology, artificial intelligence and education. Reports lodged in the system are reviewed, classified and, if necessary, further investigated. After the instigation of this no-fault system, reporting of events increased substantially (Battles et al. 1998). Most of the events reported are near misses which provide hospitals with the chance to learn and improve systems (Callum et al. 2004; Kaplan et al. 2002).

Along with incident reporting systems, hospitals should consider implementing After Event Reviews (AERs). In an AER, individuals discuss and provide feedback on critical events with the aim of detecting and learning from errors. Not only does this provide a forum to learn from errors, it also helps create a mindset that errors are something to be discussed rather than ignored. Ellis et al. (2006) examined the impact of failure focused AERs on learning and performance in a subsequent task. They found that having participants think about and discuss failures after either a failed or a successful task increased performance on a subsequent task. The benefit of conducting an AER after a successful event is that people are required to think about how they performed in all aspects of a task, rather than just considering whether the final outcome was a positive or a negative one. This is important since a procedure with a positive outcome can still contain errors which could reoccur and lead to negative consequences the next time.

Similarly, a root cause analysis can be used to learn from errors. While initially used in other high risk industries, many researchers have advocated their use in healthcare (Carroll et al. 2002). A root cause analysis begins with a report which details all the critical events that occurred in the lead up to the incident. Next, the individuals undertaking the analysis should determine what factors were present in the incident that were not present at other times, when no incidents occurred. Finally, they are required to verify (using evidence rather than inference) which of the proposed contributing factors could be responsible for the incident. A meta-analysis on root cause analysis showed that it generally has positive effects on safety (Percarpio et al. 2008). For this kind of analysis to be successful, however, there needs to be an error management culture in place whereby employees can openly discuss potential contributing factors to the incident and search for evidence to verify this. This is illustrated by one study which found that healthcare providers would often not report an incident if they could classify it in such a way that they weren't required to report it (Tamuz and Thomas 2006). That is, a potentially dangerous event would go unreported if staff could classify away the incident. This was thought to be occurring because the reporting system focused more on creating accountability for errors rather than organisational learning.

Customer Service Climate

Boundary Employees

A patient safety climate could be conceptualised similarly to a climate for service. In a climate for service, customer satisfaction is the top priority. Similarly, in a patient safety culture, customer safety is the top priority. Given the similar focus on customer outcomes, research into patient safety could be enhanced through learning from the climate for service literature. A climate for service is defined by Schneider et al. (1998) as the extent to which an organisation puts in place policies, procedures and reward systems that encourage a high level of customer service. One important finding to emerge from this literature is the importance of frontline workers as an influence on customer perceptions of the organisation and as a source of information on customer perceptions. These frontline workers, known as boundary employees, are those employees who have close contact with customers (Parkington and Schneider 1979). They are a valuable source of information as they are more aware of customers' perceptions of service quality.

The ability to listen to and use customer feedback is important in maintaining a climate for service (Schneider et al. 1998). Boundary employees and customers tend to have similar perceptions of the quality of customer service provided by a company (Schneider et al. 1980). This is because there is a reciprocal relationship between customer perceptions and employee perceptions of climate for service (Schneider et al. 1998). For example, employee behaviour towards customers influences customer perceptions of the company. But at the same time, companies use customer feedback to modify their procedures. In this way, both employees and customers have an influence on each other.

When thinking of a hospital context, nurses often act as the boundary employees. In a survey of over 400 hospital staff Cook et al. (2004) found that over 90 percent of physicians, administrators and nurses believed that patient care was the responsibility of the nurses. As Tucker and Edmondson (2003) point out, nurses work closely with patients, are required to have good knowledge about patient needs and are often lower in the chain of command. This is similar to the situation of boundary employees in customer service organisations. Like boundary employees in other industries, nurses have a more accurate perception of the organisational climate, especially in relation to customer outcomes. This can be seen in the research showing that there is a stronger link between readmission rates and frontline staff's (nurses and physicians in this study) perceptions of safety climate than there is to higher management's perceptions of safety climate (Hansen et al. 2011).

Foundations for Customer Service

Another concept discussed in the climate for service literature is the foundation that is required for a company to be high in customer service climate. The foundations for customer service are the various aspects of the work environment which facilitate work output for employees. Examples of this include having sufficient access to resources, high quality training, supportive supervision and quality service within the organisation. The *foundation* for customer service thus leads to a strong *climate* for customer service. Schneider et al. (1998) found that employees who work for a company with good foundations of customer service are better able to focus their efforts on satisfying customer demands. When this is not the case, employees spend their time chasing up issues within the company and have less time to devote to customer service. Based on their results, the authors concluded that these foundations are necessary (but not sufficient) for a climate of service. Trying to encourage good service is unlikely to work if the foundations are not in place. Similarly, if nurses spend a large amount of their time trying to solve intra-organisational problems (e.g. through the first order problem solving mentioned above) they will have less time to dedicate to their patients.

Impact of Nurse Work Environment on Patient Safety

The impact of a nurse's work environment on patient safety outcomes was studied by Spence Laschinger and Leiter (2006). Nurse work environment was measured through Lake's (2002) five worklife factors; quality of nurse leaders, participation in organisational affairs, adequate staffing, support for a nursing (versus medical) model of patient care (in which there is an emphasis on the psychosocial and emotional aspects of patient care) and good nurse/physician relationships. Lake's (2002) five worklife factors are similar to those described by Schneider et al. (1998) (i.e. access to high quality resources, training, supervision and service within the organisation) and thus can be conceptualised as the foundations required for a patient safety culture. Spence Laschinger and Leiter (2006) found that all of the factors significantly correlated with patient safety outcomes. Adequate staffing and the nursing model of patient care were the strongest direct predictors of patient safety outcomes. Burnout was a partial mediator of this relationship. One limitation of this study however, was that patient safety outcomes were measured through self-report items asking nurses to recall how frequently over the last year they encountered falls, nosocomial infections, medication errors and patient complaints. This is an issue as some nurses may have under-reported incidents due to the culture of not discussing errors (as mentioned above). Even if they provided honest answers to the anonymous survey, it would still be very difficult accurately to recall these incidents several months later.

Armstrong and Laschinger (2006) conducted a similar study but used safety climate as the dependent variable. Although this is a self-report measure, nurses are more likely to provide accurate answers since the questions concern deeply ingrained attitudes rather than the occurrence of certain events. Furthermore, safety culture is a better reflection of how safe/unsafe a workplace is than accident and incident rates. This is because accidents and incidents can still occur as a result of random situational processes (rather than unsafe behaviour) in even the safest workplace. Conversely, an unsafe workplace may have a low accident rate but a high rate of near misses. Armstrong and Laschinger (2006) also measured nurse empowerment through constructs such as support and access to resources, which are closely linked to the nurse work environment factors. They found that the combination of empowerment and a positive work environment explained a large amount of the variance in perceived safety climate. These findings were later replicated by Armstrong et al. (2009). These studies provide support for the idea that aspects of the nurse work environment, such as access to information, support and resources and quality supervision, set the foundations for a positive safety climate.

Safety Culture in Aviation

There are many similarities between teams working in medical and aviation settings. Teams in both environments are at risk of producing serious errors which cost human lives, be it through human error (caused by fatigue, work overload or miscommunication) or through their interactions with complicated technology. However, the aviation industry has fully embraced the importance of a positive safety culture. This could be due to the fact that errors in the aviation industry are often on a larger scale and are generally more public (Helmreich 2000). The Institute of Medicine estimates that healthcare is more than a decade behind aviation and other high risk industries in its approach to safety (IOM 2000). It also states that medicine has failed to apply what has been learned from these industries. As those in the field of medicine attempt to catch up, there is a lot to be learned from this high risk industry.

One aspect of the aviation industry's culture which is not often seen in medicine is its treatment of error. The industry regulator has mandatory reporting requirements, as well as a confidential reporting system which follows up on incidents without identifying the person who reported the incident (Hudson 2007). Additionally, many companies recognise the importance of monitoring and learning from error so as to create an error management culture (as described previously). An example of this is the Line Operations Safety Audit (LOSA) procedure which is used to measure the threats and errors that can occur during flights (Helmreich et al. 2001). These errors, and the way in which they are dealt with, are systematically recorded by observers. This provides aviation companies with a wealth of data on common types of errors and on which errors occur more under various conditions. These threats and errors are then classified into different categories, and interventions are suggested which target the root cause of the error. These interventions are then incorporated into pilot training. For example, communication based errors require more training in teamwork (Helmreich et al. 2001). Figure 3.3 lists the 10 operating characteristics that are essential for a LOSA.

Box 3.1   Essential operating characteristics of a LOSA (ICAO 2002)

---

1. Jump-seat observations during normal flight operations
2. Joint management / pilot sponsorship
3. Voluntary crew participation
4. De-identified, confidential and safety-minded data
5. Targeted observation instrument
6. Trusted, trained and calibrated observers
7. Trusted data collection site
8. Data verification round tables
9. Data-derived targets for enhancement
10. Feedback of results to the line pilots

An observation system which records the errors that take place during normal medical operations could be of great benefit to the medical industry. In order to implement such a system, there needs to be a culture in place in which people feel they can discuss errors openly. Sexton et al. (2000) compared perceptions of error in hospital settings (1,000 hospital staff working surgical and anaesthetic wards) with that of the aviation industry (30,000 pilots). The authors found that there is a better error management culture in the aviation industry. The medical industry however, tended not to acknowledge errors. Without sufficient acknowledgement, errors cannot be handled properly and learned from.

Anaesthesia

Anaesthesia is one area of medicine which appears to be at the forefront of patient safety. This can be attributed to a number of factors. According to Gaba (2000) anaesthesiologists have adopted sophisticated patient monitoring technology, their own safety standards and guidelines, training and simulation programs and critical incident reporting techniques, all in an effort to increase patient safety. Training simulations provide new anaesthesiologists with a chance to make and learn from errors in a safe environment (Gaba et al. 2001). This early experimentation with errors probably instils an error management culture in new anaesthesiologists (Gaba 2000). Chopra et al. (1994) found that 'anaesthetists trained on a high fidelity anaesthesia simulator respond more quickly, deviate less from the accepted guidelines and perform better in handling crisis situations, such as malignant hyperthermia, during anaesthesia than those who are not trained on the simulator' (Chopra et al. 1994: 3). Similarly, critical incident reporting and analysis contributes to creating an error management culture by encouraging people to discuss errors openly and design new systems that incorporate suggestions for change (Cooper et al. 1978). Both training simulations and critical incident reporting are techniques adapted from the field of aviation.

Leading and Creating a Patient Safety Culture

Patient Safety Leadership

Leadership is another important determinant of patient safety culture and patient safety outcomes (Katz-Navon et al. 2009; Wong and Cummings 2007). This is because leaders are often the drivers of safety culture within the organisation. They set the example for employees through their behaviours (Maierhofer et al. 2000). They also shape subordinate behaviour through the provision of rewards or punishments (Hinze 2002). When leaders take time to discuss safety with employees, employees demonstrate less unsafe behaviour (Zohar and Luria 2003). Similarly, if leaders do not make safety a priority among their team, this will be a signal to their subordinates that safety is not important. This can be especially detrimental when the organisation is trying to encourage safe behaviour and the leader is providing conflicting messages. This is why the best way to promote PSC is to have both leaders and the organisation as a whole promoting safety culture. Katz-Navon et al. (2005) demonstrated this effect by showing that treatment errors were significantly reduced when there was congruence between the safety practices of managers with safety information flow and the perceived suitability of using safety procedures at work. Thus a leader's actions and spoken beliefs in relation to safety can influence whether employees behave in a safe manner.

Executive walk rounds may be one way to increase patient safety culture among nurses. During a walk round, senior members of staff visit different hospital wards and ask about the safety issues which occurred recently in the ward. Discussions around these safety issues are recorded and followed up until a resolution is reached (Frankel et al. 2003). Walk rounds have been found to lead to improvements in safety attitudes (Frankel et al. 2008; Thomas et al. 2005) greater teamwork and the greater perceived importance of safety (Campbell and Thompson 2007). However, there is no clear evidence that walk rounds lead to changes in safety behaviour. Thomas et al. (2005) hypothesised that there might be a spill over benefit whereby employees who did not participate in the walk rounds but worked in the wards where the rounds took place would show positive changes. Thus in their study, employees were encouraged to discuss their walk rounds with their co-workers. Contrary to expectations, Thomas et al. (2005) found that walk rounds only had an impact on the attitudes of those employees who participated in the walk rounds.

As the walk rounds are generally carried out by high level management employees, frontline staff may be reluctant to discuss errors with their superiors and this can make it hard to gather unbiased data. In contrast, the LOSA procedure used by the aviation industry avoids this issue by using third party observers. If an error management culture is present and there is trust between employees and their supervisors, staff should feel more open to discussing errors. It is important for employees to feel that they can trust their organisation to use the monitoring system to learn from errors rather than as a way to allocate blame (Klinect et al. 2003). Trust is an important feature of a generative safety culture (Hudson 2003). Trust can be gained through the use of a non-punitive anonymous error reporting system. For example using the MERS-TM, which was described above.

Psychological Safety and Voice

Poor communication is thought to be the one of the contributing factors to errors and near misses in healthcare settings. Residents in one study reported being hesitant to ask for help from superiors and not wanting to say something that would go against what their attending had said (Sutcliffe et al. 2004). Often this hesitancy stems from not wanting to look incompetent or feeling as though one's input won't be considered and thus considering it best to stay silent. This hesitancy to speak up led to the finding by Blatt et al. (2006) that residents only spoke up 14 percent of the time when they detected errors.

Team leader behaviour can determine whether team members feel comfortable speaking up about issues occurring at work. The more employees speak up, the greater the opportunity to implement new practices (Edmondson 2003). Leaders can encourage team members to speak up by being inclusive in their interactions with all team members (Nembhard and Edmondson 2006). This occurs when leaders break through the hierarchy and invite others to share their thoughts and opinions. Leader inclusivity is especially important in multidisciplinary teams such as those found in healthcare settings since team members have different areas of expertise and fit in to different levels of the hierarchy.

Nembhard and Edmondson (2006) found that, in these multidisciplinary healthcare teams, individuals with high status had higher psychological safety on average. However, the disparity in psychological safety between high and low status individuals decreased if these individuals had an inclusive physician leader. Furthermore, this psychological safety mediated the positive relationship between leader inclusiveness and greater engagement in quality improvement. Nembhard and Edmondson (2006) argued that creating an environment of psychological safety fosters a greater commitment to quality improvement since low status employees are more likely to participate in problem solving and suggest improvements. Psychological safety has also been shown to predict team learning (Edmondson 1999) and employee voice behaviours (Walumbwa and Schaubroeck 2009).

Voice is the behaviour shown by an employee who feels the need to speak out about something (even though it may challenge the status quo) in order to try and improve organisational processes (Detert and Burris 2007). Whether or not an employee chooses to speak out when they detect a potential error depends largely on how certain they are of what they want to say and on their relationship with the person with whom they need to voice their concerns (Blatt et al. 2006). Employees are more likely to speak up when they have supportive leaders (Detert and Burris 2007). For example, Edmondson (2003) studied how voice and trust affected interdisciplinary teams working together on minimally invasive cardiac surgery. This form of surgery involved coordination among the various healthcare professionals in the theatre and thus the ability to speak up when something was amiss that was important for success. Team leader coaching – that is, when leaders listen to employee concerns, encourage them to speak up and provide feedback – was strongly correlated with ease of speaking up.

Voice is an important requirement for safety culture as it indicates that employees feel comfortable bringing up and discussing errors rather than avoiding them. When expressed at the right time, it can have an immediate impact on preventing errors from occurring. It is also beneficial since it provides another avenue for learning and continual quality improvement as people discuss their observations of problems and suggestions for change. Greater learning can lead to positive outcomes for patients as it means that the healthcare practitioner is less likely to repeat previous errors. Psychological safety is important for achieving these positive outcomes as it provides team members with the security to discuss errors and work to fix them.

Magnet hospitals   Hospitals with a foundation for patient safety were described by Armstrong and Laschinger (2006) and Spence Laschinger and Leiter (2006) as magnet hospitals. These are hospitals which are able to attract and retain nurses (especially compared to other hospitals in times of nurse shortages) due to the participative and supportive nature of their work environments (Scott et al. 1999). Nurses working within magnet hospitals report greater autonomy and control and better relationships with physicians (Kramer and Hafner 1989). The aspects of magnet hospitals that most strongly related to patient safety culture are nursing foundation for care, manager ability and participation in hospital affairs (Armstrong et al. 2009). The culture in magnet hospitals leads to better patient safety outcomes among those receiving Medicare. Aiken et al. (1994) compared the mortality rate in 39 magnet hospitals with rates in 195 matched control hospitals (which were similar along dimensions such as the average daily census, occupancy rate and number of hospital beds). After controlling for differences in the predicted mortality of patients, they found that magnet hospitals had a lower mortality rate than their matched controls. This study effectively demonstrates that nurse work environment has a significant impact on the quality of care received by patients.

Conclusion

There is an alarmingly high rate of injuries and deaths in hospitals as a result of preventable errors. Accordingly, healthcare organisations are experiencing increasing pressure to tackle this issue on all fronts. One area of improvement which hospitals can focus on is their organisational culture. Organisational culture is a powerful driver of employee behaviour since it provides people with a set of norms to work by. As an organisation's culture becomes entrenched it affects employees' thoughts and behaviours. Throughout this chapter we have shown that a positive PSC can have a beneficial impact on patient outcomes.

The current culture in many hospitals is one in which people are highly negative towards errors. While errors lead to negative results, they also provide an important learning opportunity. Hospitals with an error management culture aim to learn from errors rather than repeat them. In this culture, employees are encouraged to report and discuss errors and near misses. Rather than avoiding the problem, this allows people to confront it. However, this kind of behaviour will only occur if the organisational culture and systems are in place to encourage it. Research in the area of climate-for-service looks at how an organisation's policies, procedures and reward systems can be the foundation for high quality customer service (or in our case, high quality patient care). In an error management culture, this foundation would include a voluntary or mandatory reporting system, after event reviews, supportive supervision, a positive nurse work environment, leader walk rounds etc.

The PSC in many hospitals today indicates that there is still a long way to go before this problem is properly addressed. Research from the field of aviation (another high risk industry relying on multidisciplinary teams) provides hope, however. There are several examples of procedures which have been successfully implemented in the aviation industry to create a positive safety culture. Recently some of these have been adapted to the field of patient safety. Anaesthesia is one area of medicine which seems to be a leader in PSC. With sophisticated training simulations and reporting systems, anaesthetists are taking steps in the direction of encouraging a positive error management culture. As the use of voluntary reporting systems and near miss analysis slowly becomes more widely accepted, we will hopefully see a shift in attitudes towards error and a greater desire to learn from it.

References

Aiken, L.H., Smith, H.L. and Lake, E.T. 1994. Lower medicare mortality among a set of hospitals known for good nursing care. *Medical Care*, 32(8), 771–87. doi: 10.2307/3766652.

Armstrong, K. and Laschinger, H. 2006. Structural empowerment, magnet hospital characteristics, and patient safety culture: Making the link. *Journal of Nursing Care Quality*, 21(2), 124–32.

Armstrong, K., Laschinger, H. and Wong, C. 2009. Workplace empowerment and magnet hospital characteristics as predictors of patient safety climate. *Journal of Nursing Care Quality*, 24(1), 55–62.

Australian Council for Safety and Quality in Health Care 2003. *Open Disclosure Standard: A National Standard for Open Communication in Public and Private Hospitals, Following an Adverse Event in Health Care*. Commonwealth of Australia: Publications no 3320.

Barach, P. and Small, S.D. 2000. Reporting and preventing medical mishaps: Lessons from non-medical near miss reporting systems. *British Medical Journal*, 320(7237), 759–63.

Barling, J., Loughlin, C. and Kelloway, E.K. 2002. Development and test of a model linking safety-specific transformational leadership and occupational safety. *Journal of Applied Psychology*, 87(3), 488–96.

Battles, J.B., Kaplan, H., Van der Schaaf, T. and Shea, C. 1998. The attributes of medical event-reporting systems. *Archives of Pathology & Laboratory Medicine*, 122(3), 132–8.

Blatt, R., Christianson, M.K., Sutcliffe, K.M. and Rosenthal, M.M. 2006. A sensemaking lens on reliability. *Journal of Organizational Behavior*, 27(7), 897–917.

Callum, J.L., Merkley, L.L., Coovadia, A.S., Lima, A.P. and Kaplan, H.S. 2004. Experience with the medical event reporting system for transfusion medicine (MERS-TM) at three hospitals. *Transfusion and Apheresis Science: Official Journal of the World Apheresis Association: Official Journal of the European Society for Haemapheresis*, 31(2), 133.

Campbell, D.A. and Thompson, M. 2007. Patient safety rounds: Description of an inexpensive but important strategy to improve the safety culture. *American Journal of Medical Quality*, 22(1), 26–33. doi: 10.1177/1062860606295619.

Carroll, J.S. and Quijada, M.A. 2004. Redirecting traditional professional values to support safety: Changing organisational culture in health care. *Quality and Safety in Health Care*, 13(suppl 2), 16–21.

Carroll, J.S., Rudolph, J.W. and Hatakenaka, S. 2002. Lessons learned from non-medical industries: Root cause analysis as culture change at a chemical plant. *Quality and Safety in Health Care*, 11(3), 266–9.

Chang, Y.K. and Mark, B. 2011. Effects of learning climate and registered nurse staffing on medication errors. *Nursing Research*, 60(1), 32–9.

Chopra, V., Gesink, B., De Jong, J., Bovill, J., Spierdijk, J. and Brand, R. 1994. Does training on an anaesthesia simulator lead to improvement in performance? *British Journal of Anaesthesia*, 73(3), 293–7.

Clarke, S. 2006. The relationship between safety climate and safety performance: A meta-analytic review. *Journal of Occupational Health Psychology*, 11(4), 315–27.

Committee on Quality of Health Care in America, Institute of Medicine 2000. *To Err is Human: Building a Safer Health System*. Washington, DC: The National Academies Press.

Cook, A.F., Hoas, H., Guttmannova, K. and Joyner, J.C. 2004. An error by any other name. *The American Journal of Nursing*, 104(6), 32–43.

Cooper, J.B., Newbower, R.S., Moore, J.W. and Trautman, E.D. 1978. A new anesthesia delivery system. *Anesthesiology*, 49(5), 310–18.

Cooper, M.D. and Phillips, R.A. 2004. Exploratory analysis of the safety climate and safety behavior relationship. *Journal of Safety Research*, 35(5), 497–512.

Dekker, S. 2011. *Patient Safety: A Human Factors Approach*. USA: Taylor and Francis Group.

Denison, D.R. 1996. What is the difference between organizational culture and organizational climate? A native's point of view on a decade of paradigm wars. *The Academy of Management Review*, 21(3), 619–54.

Detert, J.R. and Burris, E.R. 2007. Leadership behavior and employee voice: Is the door really open? *Academy of Management Journal*, 50(4), 869–84.

Edmondson, A. 1999. Psychological safety and learning behavior in work teams. *Administrative Science Quarterly*, 44(2), 350–83. doi: 10.2307/2666999.

Edmondson, A.C. 2003. Speaking up in the operating room: How team leaders promote learning in interdisciplinary action teams. *Journal of Management Studies*, 40(6), 1419–452. doi: 10.1111/1467-6486.00386.

Ellis, S., Mendel, R. and Nir, M. 2006. Learning from successful and failed experience: The moderating role of kind of after-event review. *Journal of Applied Psychology*, 91(3), 669.

Felknor, S.A., Aday, L.A., Burau, K.D., Delclos, G.L. and Kapadia, A.S. 2000. Safety climate and its association with injuries and safety practices in public hospitals in Costa Rica. *International Journal of Occupational and Environmental Health*, 6(1), 18–25.

Flin, R. 2007. Measuring safety culture in healthcare: A case for accurate diagnosis. *Safety Science*, 45(6), 653–67.

Frankel, A., Gandhi, T.K. and Bates, D.W. 2003. Improving patient safety across a large integrated health care delivery system. *International Journal for Quality in Health Care*, 15(suppl 1), i31–i40. doi: 10.1093/intqhc/mzg075.

Frankel, A., Grillo, S.P., Pittman, M., Thomas, E.J., Horowitz, L., Page, M. and Sexton, B. 2008. Revealing and resolving patient safety defects: The impact of leadership walkrounds on frontline caregiver assessments of patient safety. *Health Services Research*, 43(6), 2050–66. doi: 10.1111/j.1475-6773.2008.00878.x.

Gaba, D.M. 2000. Anaesthesiology as a model for patient safety in health care. *British Medical Journal*, 320(7237), 785–8.

Gaba, D.M., Howard, S.K., Fish, K.J., Smith, B.E. and Sowb, Y.A. 2001. Simulation-based training in anesthesia crisis resource management (ACRM): A decade of experience. *Simulation & Gaming*, 32(2), 175–93.

Griffin, M.A. and Neal, A. 2000. Perceptions of safety at work: A framework for linking safety climate to safety performance, knowledge, and motivation. *Journal of Occupational Health Psychology*, 5(3), 347–58.

Guldenmund, F.W. 2000. The nature of safety culture: A review of theory and research. *Safety Science*, 34, 215–57.

Gundry, L.K. and Rousseau, D.M. 1994. Critical incidents in communicating culture to newcomers: The meaning is the message. *Human Relations*, 47(9), 1063–88. doi: 10.1177/001872679404700903.

Halligan, M. and Zecevic, A. 2011. Safety culture in healthcare: A review of concepts, dimensions, measures and progress. *BMJ Quality & Safety*, 20(4), 338–43. doi: 10.1136/bmjqs.2010.040964.

Hansen, L.O., Williams, M.V. and Singer, S.J. 2011. Perceptions of hospital safety climate and incidence of readmission. *Health Services Research*, 46(2), 596–616. doi: 10.1111/j.1475-6773.2010.01204.x.

Helmreich, R.L. 2000. On error management: Lessons from aviation. *British Medical Journal*, 320(7237), 781–85. doi: 10.1136/bmj.320.7237.781.

Helmreich, R.L., Klinect, J.R. and Wilhelm, J.A. 2001. *System Safety and Threat and Error Management: The Line Operational Safety Audit (LOSA)*. Paper presented at the The Eleventh International Symposium on Aviation Psychology, Columbus, OH: The Ohio State University.

Hinze, J. 2002. Safety incentives: Do they reduce injuries? *Practice Periodical on Structural Design and Construction*, 7(2), 81–4.

Hofmann, D.A. and Mark, B. 2006. An investigation of the relationship between safety climate and medication errors as well as other nurse and patient outcomes. *Personnel Psychology*, 59(4), 847–69. doi: 10.1111/j.1744-6570.2006.00056.x.

Huang, D.T., Clermont, G., Sexton, J.B., Karlo, C.A., Miller, R.G., Weissfeld, L.A., Rowan, K.M. and Angus, D.C. 2007. Perceptions of safety culture vary across the intensive care units of a single institution. *Critical Care Medicine*, 35(1), 165–76.

Hudson, P. 2003. Applying the lessons of high risk industries to health care. *Quality and Safety in Health Care*, 12(suppl 1), i7–i12.

Hudson, P. 2007. Implementing a safety culture in a major multi-national. *Safety Science*, 45(6), 697–722.

ICAO 2002. *International Civil Aviation Organisation Line Operations Safety Audit* (Doc 9803 AN/761). Montreal: International Civil Aviation Organisation.

Institute of Medicine 2000. *To Err is Human: Building a Safer Health System*. Washington, DC: National Academies Press.

Joint Commission on the Accreditation of Healthcare Organisations 2001. *JCAHO Standard RI.1.2.2 1*.

Kaplan, H.S., Callum, J.L., Fastman, B.R. and Merkley, L.L. 2002. The medical event reporting system for transfusion medicine: Will it help get the right blood to the right patient? *Transfusion Medicine Reviews*, 16(2), 86–102.

Katz-Navon, T.A.L., Naveh, E. and Stern, Z. 2005. Safety climate in health care organizations: A multidimensional approach. *Academy of Management Journal*, 48(6), 1075–89.

Katz-Navon, T., Naveh, E. and Stern, Z. 2009. Active learning: When is more better? The case of resident physicians' medical errors. *Journal of Applied Psychology*, 94(5), 1200–209.

Klinect, J.R., Murray, P., Merritt, A. and Helmreich, R. 2003. *Line Operations Safety Audit (LOSA): Definition and Operating Characteristics*. Paper presented at the Proceedings of the 12th International Symposium on Aviation Psychology, Dayton, OH: The Ohio State University.

Kramer, M. and Hafner, L.P. 1989. Shared values: Impact on staff nurse job satisfaction and perceived productivity. *Nursing Research*, 38(3), 172–7.

Lake, E.T. 2002. Development of the practice environment scale of the nursing work index. *Research in Nursing and Health*, 25(3), 176–88.

Leape, L.L. 1994. Error in medicine. *JAMA: The Journal of the American Medical Association*, 272(23), 1851–7.

Maierhofer, N.I., Griffin, M.A. and Sheehan, M. 2000. Linking manager values and behavior with employee values and behavior: A study of values and safety in the hairdressing industry. *Journal of Occupational Health Psychology*, 5(4), 417–27.

Mearns, K. and Flin, R. 1999. Assessing the state of organizational safety – culture or climate? *Current Psychology*, 18(1), 5–17. doi: 10.1007/s12144-999-1013-3.

Nahrgang, J.D., Morgeson, F.P. and Hofmann, D.A. 2008. *Predicting Safety Performance: A Meta-analysis of Safety and Organizational Constructs.* Paper presented at the 23rd Annual Meeting of the Society for Industrial and Organizational Psychology, San Francisco, CA.

Neal, A. and Griffin, M.A. 2006. A study of the lagged relationships among safety climate, safety motivation, safety behavior, and accidents at the individual and group levels. *Journal of Applied Psychology*, 91(4), 946–53.

Nembhard, I.M. and Edmondson, A.C. 2006. Making it safe: The effects of leader inclusiveness and professional status on psychological safety and improvement efforts in health care teams. *Journal of Organizational Behavior*, 27(7), 941–66.

Parkington, J.J. and Schneider, B. 1979. Some correlates of experienced job stress: A boundary role study. *The Academy of Management Journal*, 22(2), 270–81.

Percarpio, K.B., Watts, B.V. and Weeks, W.B. 2008. The effectiveness of root cause analysis: What does the literature tell us? *Joint Commission Journal on Quality and Patient Safety*, 34(7), 391–8.

Pronovost, P.J., Berenholtz, S.M., Goeschel, C., Thom, I., Watson, S.R., Holzmueller, C.G., Lyon, J.S., Lubomski, L.H., Thompson, D.A., Needham, D., Hyzy, R., Welsh, R., Roth, G., Bander, J., Morlock, L. and Sexton, J.B. 2008. Improving patient safety in intensive care units in Michigan. *Journal of Critical Care*, 23(2), 207–21.

Quick, B.L., Stephenson, M.T., Witte, K., Vaught, C., Booth-Butterfield, S. and Patel, D. 2008. An examination of antecedents to coal miners' hearing protection behaviors: A test of the theory of planned behavior. *Journal of Safety Research*, 39(3), 329–38.

Reason, J. 1990. *Human Error.* New York, USA: Cambridge University Press.

Reason, J. 2000. Human error: Models and management. *British Medical Journal*, 320(7237), 768–70.

Saks, A.M. and Ashforth, B.E. 1997. Organizational socialization: Making sense of the past and present as a prologue for the future. *Journal of Vocational Behavior*, 51(2), 234–79. doi: 10.1006/jvbe.1997.1614.

Sammer, C.E., Lykens, K., Singh, K.P., Mains, D.A. and Lackan, N.A. 2010. What is patient safety culture? A review of the literature. *Journal of Nursing Scholarship*, 42(2), 156–65. doi: 10.1111/j.1547-5069.2009.01330.x.

Schein, E.H. 2010. *Organizational Culture and Leadership* (Fouth edition). San Fransisco: Jossey-Bass.

Schneider, B., Goldstein, H.W. and Smith, D.B. 1995. The ASA framework: An update. *Personnel Psychology*, 48, 747–74.

Schneider, B., Parkington, J.J. and Buxton, V.M. 1980. Employee and customer perceptions of service in banks. *Administrative Science Quarterly*, 25(2), 252–67.

Schneider, B., White, S.S. and Paul, M.C. 1998. Linking service climate and customer perceptions of service quality: Tests of a causal model. *Journal of Applied Psychology*, 83(2), 150–63.

Schwappach, D.L.B. and Boluarte, T.A. 2010. The emotional impact of medical error involvement on physicians: A call for leadership and organisational accountability. *Swiss Medical Weekly*, 139(1), 1–7.

Scott, J.G., Sochalski, J. and Aiken, L. 1999. Review of magnet hospital research: Findings and implications for professional nursing practice. *The Journal of Nursing Administration*, 29(1), 9–19.

Sexton, J.B., Thomas, E.J. and Helmreich, R.L. 2000. Error, stress, and teamwork in medicine and aviation: Cross sectional surveys. *BMJ: British Medical Journal*, 320(7237), 745–9.

Shanafelt, T.D., Bradley, K.A., Wipf, J.E. and Back, A.L. 2002. Burnout and self-reported patient care in an internal medicine residency program. *Annals of Internal Medicine*, 136(5), 358–67.

Spence Laschinger, H.K. and Leiter, M.P. 2006. The impact of nursing work environments on patient safety outcomes: the mediating role of burnout engagement. *Journal of Nursing Administration*, 36(5), 259–67.

Stone, P.W., Mooney-Kane, C., Larson, E.L., Horan, T., Glance, L.G., Zwanziger, J. and Dick, A.W. 2007. Nurse working conditions and patient safety outcomes. *Medical Care*, 45(6), 571–8.

Sutcliffe, K.M., Lewton, E. and Rosenthal, M.M. 2004. Communication failures: An insidious contributor to medical mishaps. *Academic Medicine*, 79(2), 186–94.

Tamuz, M. and Thomas, E.J. 2006. Classifying and interpreting threats to patient safety in hospitals: Insights from aviation. *Journal of Organizational Behavior*, 27(7), 919–40.

Thomas, E.J., Sexton, J.B. and Helmreich, R.L. 2003. Discrepant attitudes about teamwork among critical care nurses and physicians. *Critical Care Medicine*, 31(3), 956–9 910.1097/1001.CCM.0000056183.0000089175.0000056176.

Thomas, E., Sexton, J.B., Neilands, T.B., Frankel, A. and Helmreich, R.L. 2005. The effect of executive walk rounds on nurse safety climate attitudes: A randomized trial of clinical units. *BMC Health Services Research*, 5(1), 28–36.

Tucker, A.L. and Edmondson, A.C. 2003. Why hospitals don't learn from failures: Organizational and psychological dynamics that inhibit system change. *California Management Review*, 45(2), 55–72.

Van Dyck, C., Frese, M., Baer, M. and Sonnentag, S. 2005. Organizational error management culture and its impact on performance: A two-study replication. *Journal of Applied Psychology*, 90(6), 1228–40.

Walumbwa, F.O. and Schaubroeck, J. 2009. Leader personality traits and employee voice behavior: Mediating roles of ethical leadership and work group psychological safety. *Journal of Applied Psychology*, 94(5), 1275–86.

Westrum, R. 2004. A typology of organisational cultures. *Quality & Safety in Health Care*, 13(2), 22–7.

Wong, C.A. and Cummings, G.G. 2007. The relationship between nursing leadership and patient outcomes: A systematic review. *Journal of Nursing Management*, 15(5), 508–21. doi: 10.1111/j.1365-2834.2007.00723.x.

Wu, A.W., Folkman, S., McPhee, S.J. and Lo, B. 2003. Do house officers learn from their mistakes? *Quality and Safety in Health Care*, 12(3), 221–6.

Zohar, D. and Luria, G. 2003. The use of supervisory practices as leverage to improve safety behavior: A cross-level intervention model. *Journal of Safety Research*, 34(5), 567–77.

Chapter 4

Safety Culture Dimensions, Patient Safety Outcomes and Their Correlations

Kenji Itoh, Henning Boje Andersen and Kim Lyngby Mikkelsen

Introduction

Risk Management and Safety Culture in Healthcare

Safety and risk management research in healthcare has adopted as its dominant trend the 'systems oriented' approach, modelled largely on previous research in safety critical industries such as aviation and nuclear power. The systems view entails that the focus is not primarily on the mechanisms of individual human error but on the factors that shape human performance (Rasmussen 1986; Reason 1993, 1997). In an organisational context, such factors are, of course, those that are within the control of the organisation. For instance, it has been suggested that quality and safety are affected not only by operators' professional and technical competence and skills, but also by their attitudes to and perceptions of their job roles, their organisation and management (Helmreich and Merritt 1998). Such employee attitudes and views are important elements which shape 'safety culture' – and its related notion, 'safety climate'. Indeed, survey studies have shown that staff attitudes are important indices of safety performance not only in human-machine system domains such as railway operations and construction (e.g. Itoh and Andersen 1999; Itoh et al. 2004; Silva et al. 2004) but also in healthcare (e.g. Colla et al. 2005; Itoh and Andersen 2010).

The term 'safety culture' was launched in the late 1980s in the aftermath of the Chernobyl nuclear power accident in 1986 – at that time to signal what was wrong in the organisational culture that allowed and even encouraged unsafe practices. Since then, a large number of studies have developed models, measures and tools and instruments for safety culture, particularly in nuclear power and aviation. Subsequently, a number of safety culture studies have also appeared in healthcare, developing and adapting methods and techniques that often have their pedigree in industry (aviation, nuclear power etc.) The adaptation and further development have taken place in the recognition that healthcare is a safety critical domain. In recent years – and for several prior to the seminal US report on the impact of error in healthcare (Kohn et al. 2000) – a number of studies measuring and diagnosing safety culture in specific organisations and work units have been published (e.g. Gershon et al. 2000; Helmreich and Merritt 1998; Singer et al. 2003). For instance, 97 percent of acute and primary care organisations in the UK are reported to view safety culture as a central task for clinical management and one third of them have been using safety culture instruments for management purposes (Mannion et al. 2009). There have also been large scale surveys conducted for measuring healthcare safety culture in many countries, e.g. the USA (Singer et al. 2009; Sorra et al. 2009) and Japan (Itoh and Andersen 2008). Thus, safety culture assessment has been become widely used not only for research, but also for practical application where it is seen as an essential management tool in the modern healthcare organisation.

In a traditional risk management approach, attempts to attain safety have focused on eliminating risks, preventing unexpected events from taking place and protecting against unwanted outcomes when they happen (Hollnagel 2009). From this perspective, safety culture assessment can be typically applied to analysing risks which potentially exist in an organisation, particularly as systemic latent factors, which must then be eliminated. More broadly including safety culture assessment, risk analysis methods can be largely classified into two types: reactive and proactive approaches. In a *reactive* approach, on the one hand, cases or events that have occurred are retrospectively analysed. Accident investigation and analysis of incident cases are typical examples of a reactive approach applied to risk management. On the other hand, applying a *proactive* approach, we can *look ahead* to something related to an organisation's risks or unwanted states and take actions to remove or mitigate such risks *before* an actual adverse event happens. This approach allows the organisation to identify 'weak points', for instance, in the attitudes, norms and practices of the target groups or the entire organisation; in turn, knowledge of 'weak points' may be used to guide the planning and implementation of intervention programmes and directed at enabling them to develop improved patient safety practices and safety management mechanisms. Safety culture assessment is one of the promising methods for application of the proactive approach to risk management.

Of course, this approach is based on the assumption that safety culture correlates with safety outcomes. Therefore, it is of critical importance to confirm the 'culture-outcome' link, which is one of the requirements for a safety culture scale, understood as *criterion validity* – a more comprehensive summary of the required properties can be found in other literature (e.g. Itoh et al. 2012). For this applied purpose of safety culture, this chapter specifically looks at dimensions of safety culture, how to measure safety outcomes, and the safety culture-outcome link through an examination of case studies, primarily drawn from Japanese hospitals. Before stating these issues in detail, we will, in the rest of this section, briefly argue notions of safety culture (and safety climate).

Concept of Safety Culture and Safety Climate

Numerous definitions have been proposed of safety culture (e.g. Flin et al. 2000; Pidgeon and O'Leary 1994; Zohar 1980). The most widely accepted definition is the one proposed by the Advisory Committee on the Safety of Nuclear Installations (ACSNI 1993: 23): safety culture is 'the product of individual and group values, attitudes, perceptions, competencies and patterns of behaviour that determine the commitment to, and the style and proficiency of, an organisation's health and safety management'. Thus, safety culture is coupled not only with management's commitment to safety, its communication and leadership style and the overt rules for error reporting but also to employees' motivation, perception of errors and attitudes towards management and factors that have an impact on safety, e.g. stress, workload, fatigue, risk taking and violations of rules/procedures.

Safety culture can be conceptualised as a three-layer structure, following Schein's (1992) work on organisational culture: (1) 'basic assumptions', i.e. a core of largely tacit underlying assumptions taken for granted by the entire organisation; (2) 'espoused values', i.e. values and norms that are embraced and adopted; and (3) 'artefacts' that include tangible and overt items and acts such as procedures, inspections and checklists. Safety culture is overlapped with the inner layer of Schein's model as part of the overall culture of the organisation that affects members' attitudes and perceptions related to hazards and risk control (Cooper 2000). Thus, culture concerns shared symbolic and normative structures that (a) are largely *tacit* (implicit, unconscious), (b) are *stable over time* and (c) can be assigned a meaning only by reference to surrounding symbolic practices of the cultural community.

In contrast, safety climate is viewed as the middle layer of Schein's model, and accordingly is governed by safety culture and contextual and possibly local issues. Therefore, safety climate has been characterised as reflecting the surface manifestation of safety culture, and it refers to employees' *context-dependent* attitudes and perceptions about safety related issues (Flin et al. 2000; Glendon and Stanton 2000). In particular, we may sometimes use the term safety climate when emphasising the need for referring to *local, changeable* and *explicit* attitudes and perceptions.

As such, culture and climate are theoretically distinguished in terms of how stable, tacit and interpretable these shared values, attitudes, etc. are. However, since the term safety culture is more commonly used and sounds more natural than safety climate, we shall, having no need to distinguish between 'underlying' and 'overt' attitudes and perceptions, mostly use the term safety culture in its inclusive sense. In the rest of this chapter, following this convention, we shall, unless precision is required, refer to 'safety culture and climate'.

Dimensions of Safety Culture

Variety of Safety Culture Dimensions

A number of safety culture dimensions (scales, factors, components or aspects) have been proposed, ranging diversely from psycho-social aspects (satisfaction, motivation, morale, team atmosphere, etc.) to behavioural and attitudinal factors regarding job, management, organisation, incident reporting and other safety related issues. Many of these dimensions have been elicited by applying multivariate analysis techniques such as factor analysis and principal component analysis to questionnaire data. Some dimensions proposed by different researchers are quite similar – though the terms may differ – while others differ very much. The number of dimensions also differs among studies, ranging from 2 to 16, according to Guldenmund (2000).

In healthcare, a number of studies have also been seeking to develop dimensions of safety culture (e.g. Gershon et al. 2000; Smits et al. 2008; Sorra and Nieva 2004; Turnberg and Daniell 2008). According to the review by Halligan and Zeceivic (2011), the following four tools have been most frequently used: (1) Agency for Healthcare Research and Quality (AHRQ)'s Hospital Survey on Patient Safety Culture (HSOPS; Sorra and Nieva 2004); (2) Safety Attitudes Questionnaire (SAQ; Sexton et al. 2006); (3) Patient Safety Culture in Healthcare Organisations Survey (PSCHO; Singer et al. 2007); and (4) Modified Stanford Instrument (MSI; Ginsburg et al. 2009). Safety culture dimensions used in these four instruments are listed in Table 4.1. Dimension labels varied considerably across these safety culture instruments.

Table 4.1   Safety culture dimensions in instruments most widely used

	HSOPS (Sorra and Dyer 2010)	SAQ (Sexton et al. 2006)	PSCHO (Singer et al. 2007)	MSI (Ginsburg et al. 2009)
1	Communication openness	Teamwork climate	Senior managers' engagement	Organisational leadership for safety
2	Feedback & communication about error	Job satisfaction	Organisational resources for safety	Unit leadership for safety
3	Frequency of events reported	Perceptions of management	Overall emphasis on safety	Perceived state of safety
4	Handoffs & transitions	Safety climate	Unit safety norms	Shame & repercussions of reporting
5	Management support for patient safety	Working conditions	Unit recognition & support for safety efforts	Safety learning behaviour
6	Non-punitive response to error	Stress recognition	Fear of shame	–
7	Organisational learning – Continuous improvement	–	Provision of safe care	–
8	Overall perceptions of patient safety	–	Learning	–
9	Staffing	–	Fear of blame	
10	Supervisor/manager expectations & actions promoting safety	–	–	
11	Teamwork across units	–	–	
12	Teamwork within units	–	–	

The variation in safety culture dimensions has several sources: first, researchers have employed different questionnaires (so they have different sets of question items to respondents); second, surveys have been made of different fields or professional groups with different hazard levels, work conditions, recruitment criteria and training requirements, as well as different macro-level factors such as regulation regimes, health systems and national cultures; and third, when aggregating a group (factor) of question items, the choice of a label is essentially a subjective interpretation. However, although the labels of dimensions often vary, it is clear that similar dimensions (according to the meaning of the labels) can be seen across the different sets of dimensions proposed by different research groups.

Generalising from a number of dimensions suggested by former studies, safety culture dimensions can be classified into two parts: (1) general, 'core' elements that are commonly applicable to any healthcare context, e.g. field, organisational type and country, and (2) 'nation-dependent' elements that are specific to a national culture or the country's healthcare system.

Core Dimensions of Safety Culture

There have been attempts to explore 'core' dimensions of safety culture by generalising from diverse dimensions that have appeared in the literature. According to reviews of nine instruments by Colla et al. (2005), the following five dimensions appeared in most of the studies they reviewed: leadership, policies and procedures, staffing, communication and reporting. More recently, Halligan and Zecevic (2011), who surveyed 113 articles, reported that the following six dimensions have frequently been cited: leadership commitment to safety; open communication founded on trust; organisational learning; non-punitive approach to adverse event reporting and analysis; teamwork; and shared belief in the importance of safety. Similarly, Flin et al. (2006), surveying 12 instruments, identified *management commitment to safety* as the most frequently measured dimension of safety culture in healthcare which, along with *safety systems* and *work pressure* – which are related to job demands and workload, form the three 'core' themes of healthcare safety culture. Integrating the results of former studies as well as essential concepts, we would suggest the following seven labels as 'core' elements of safety culture in healthcare: (1) management commitment to safety; (2) safety systems; (3) work pressure; (4) communication; (5) teamwork; (6) leadership and (7) non-punitive (or blame-free) approach.

Nation-dependent Elements

As described in greater detail in the following case study, a Japanese study (Itoh and Andersen 2008) elicited two nation-dependent dimensions specific to this country's context, namely 'respect for seniority' and 'member-conflicting' attitudes. The former dimension seems to be affected by the national culture, particularly a conception of Confucianism, and Japan's traditional employment system, i.e. seniority system (e.g. Brunello and Ariga 1997; Yuki and Yamaguchi 1996). Regarding the discipline of

Confucianism, Japanese citizens have been taught from early childhood that seniors must be respected, admired, looked after kindly, etc. In the seniority system as an employment convention, an employee's position or job rank, as well as salary, will have increased according to age and experience in the current organisation.

According to well-known, pioneering work on national culture by Hofstede (1991; 2001), the Japanese culture is characterised by a strong trend of masculinity as well as that of uncertainty avoidance. Masculinity as an organisational culture factor is tied to competition among colleagues as well as equity and performance (Hofstede 1991). Therefore, it may be natural to speculate that another dimension, which could be labelled 'member-conflicting attitudes', can be elicited as dependent on the specific national culture of Japan.

In general, nation-dependent dimensions like those mentioned above may not be only specific to one country, e.g. the Japanese context, but also apply to other countries which have the same or similar background behind a particular element of safety culture. For instance, a cultural component related to 'seniority dependency' may possibly become a nation-dependent dimension in other Asian countries that are also strongly affected by Confucianism, such as China, Korea and Taiwan. Indeed, a similar safety culture factor, which was labelled 'seniority system', was suggested by a Chinese survey (Gu and Itoh 2011) which applied the same questionnaire (Chinese translation) that was used in the Japanese survey. The Chinese study also elicited other nation-dependent factors, 'authoritarianism' and 'pursuit of ideal work setting', but those did not appear in the Japanese study because of different cultural elements.

Case Study of Safety Culture Assessment: Japanese Example

In this subsection, we illustrate a safety culture construct which was made up of both 'core' and 'nation-dependent' dimensions in the above-mentioned Japanese survey (Itoh and Andersen 2008). In this study, a safety culture questionnaire was developed, adapted from Robert Helmreich's ORMAQ (Operating Room Management Attitudes Questionnaire; Helmreich and Merritt 1998). We collected more than 20,000 (an 84 percent response rate) staff responses (doctors, nurses, pharmacists and clinical engineers) from 84 hospitals all over Japan. Principal component analysis was applied to all staff responses, yielding 12 safety culture factors (see Table 4.2 for their factor labels).

Besides the two nation-dependent factors mentioned previously, the factors elicited from the Japanese sample covered or conceptually overlapped with most 'core' elements of safety culture. For instance, the Japanese construct included a dimension on *work pressure* – which was labelled 'attitudes to work pressure'. Most of other dimensions had the same labels as or were closely related to the core elements of safety culture: *leadership, communication, blame-free, teamwork, safety awareness, workload* and *stress, competence* and *job satisfaction*. In addition, essential elements of national culture (Hofstede 1991; 2001) were included: *power distance*, and *collectivism-individualism*.

When assessing safety culture by the use of scales such as the above, we need to determine for each factor whether an increase or a decrease of its score indicates a 'better' safety culture. Most of these factors have a more or less self-evident orientation. Thus, factors such as communication, job satisfaction, safety awareness and blame-free atmosphere are undoubtedly oriented in a positive direction so that it is preferable to attain higher scores, whereas others, such as power distance and member-conflicting attitudes, are oriented in a negative direction, i.e. lower scores indicate a 'better' safety culture. However, three of the factors – attitudes to work pressure, competence awareness and respect for seniority – may be seen as two-sided: too much is 'bad' and so is too little. For instance, a strong awareness of work pressure may, on the one hand, indicate positive attitudes so that hospital staff are likely to work hard and do their best when they are in a demanding work situation; on the other hand, *too* strong attitudes to work pressure may be based on unrealistic views of the effects of stress and workload, views that may in turn incur a higher risk (e.g. failure to initiate workload sharing during a critically difficult situation).

Partial results in the Japanese survey are shown in Table 4.2 in terms of the percentage of positive respondents for each factor of safety culture as well as significance levels across four professional groups – although every factor was significantly different. As an overall trend, almost all Japanese healthcare professionals have a strong awareness of communication as well as strong respect for seniority and senior members. There is a small power distance as well as a blame-free atmosphere within a hospital.

Table 4.2   Differences in safety culture across professional groups

Safety culture factors	Doctor	Nurse	Pharma.	Engr.	Total	p
Communication	93%	94%	96%	96%	94%	***
Job satisfaction	56%	33%	39%	44%	35%	***
Small power distance	83%	84%	87%	79%	83%	***
Leadership	46%	40%	33%	35%	40%	***
Safety awareness	57%	73%	59%	58%	71%	***
Collectivism-individualism	49%	50%	49%	46%	49%	***
Member-conflicting attitudes (weak conflicting attitudes)	4%	6%	3%	8%	6%	***
Team/stress management	55%	55%	50%	56%	55%	**
Blame-free atmosphere	94%	90%	95%	92%	90%	***
Attitudes to work pressure[†]	31%	19%	18%	27%	21%	***
Competence awareness[†]	55%	44%	43%	45%	44%	***
Respect for seniority[†]	91%	73%	81%	82%	75%	***

Source: Adapted from Itoh and Andersen 2008.
Notes: Pharama.: Pharmacist, Engr.: Clinical engineer. Figures: percentage of positive respondents to the factor label. [†]: Two sided orientation (both too little and too much are 'bad'). *: $p < 0.05$, **: $p < 0.01$, ***: $p < 0.001$.

Regarding occupational differences, doctors indicate a much higher level of job satisfaction, stronger awareness of own competence, stronger respect for seniority and more positive perceptions of leadership than the other groups. In contrast, nurses' job satisfaction and respect for seniority are the lowest of the four professional groups. However, their safety awareness is far stronger than the others, and they are less liable to take member-conflicting attitudes. Pharmacists' and clinical engineers' perceptions of safety culture seem alike and are at intermediate levels between doctors and nurses for most dimensions.

Table 4.3   Safety culture perceptions by doctors' specialities

Safety culture factors	Physician	Surgeon	Anaesthe.	Total[‡]	p
Communication	94%	94%	93%	93%	–
Job satisfaction	55%	59%	55%	55%	–
Small power distance	85%	83%	70%	83%	–
Leadership	43%	53%	26%	46%	***
Safety awareness	53%	61%	60%	57%	–
Collectivism-individualism	49%	48%	34%	49%	–
Member-conflicting attitudes (weak conflicting attitudes)	5%	4%	5%	4%	–
Team/stress management	56%	56%	51%	55%	–
Blame-free atmosphere	95%	94%	91%	94%	–
Attitudes to work pressure[†]	28%	38%	33%	31%	*
Competence awareness[†]	54%	56%	54%	55%	–
Respect for seniority[†]	93%	91%	91%	91%	–

Source: Adapted from Itoh and Andersen 2008.
Notes: *: $p < 0.05$, **: $p < 0.01$, ***: $p < 0.001$. Figures: percentage of positive respondents to the factor label. [†]: Two sided orientation (both too little and too much are 'bad'). [‡]: The number of responses in 'Total' includes that of other specialties. Anaesthe.: Anaesthesiologist.

Cultural Differences by Specialties and Work Units

Staff perceptions of safety culture by their specialties (doctor) and work units (nurse) are shown in Tables 4.3 and 4.4 in terms of percentages of positive responses to each safety culture factor. Regarding doctor perceptions, it can be seen from Table 4.3 that Japanese hospitals share a homogeneous professional safety culture across different specialties with almost no significant difference for each dimension. Stating a small specialty difference, anaesthesiologists' perceptions are slightly more negative than the other groups', for example in leadership. A plausible reason for their negative perceptions may be their strict views of safety requirements: Professionally and by tradition they focus strongly on patient safety, more so than other specialties; therefore, their criteria for an acceptable safety culture may well be more demanding than those of other specialty doctors.

Table 4.4   Safety culture perceptions by nurses' work units

Safety culture factors	Internal	Surge.	ICU	OR	Outpt.	Psychi.	Paedia.	Mixed	Total[‡]	p
Communication	93%	93%	94%	94%	96%	93%	93%	93%	94%	***
Job satisfaction	29%	30%	27%	29%	41%	36%	32%	32%	33%	***
Small power distance	84%	84%	85%	84%	81%	81%	83%	86%	84%	***
Leadership	42%	40%	36%	33%	38%	40%	46%	41%	40%	***
Safety awareness	72%	73%	71%	70%	75%	73%	77%	75%	73%	***
Collectivism-individualism	49%	50%	50%	47%	52%	42%	50%	51%	50%	***
Member-conflicting attitudes (weak conflicting attitudes)	7%	6%	6%	9%	6%	8%	7%	6%	6%	***
Team/stress management	55%	54%	49%	51%	59%	57%	55%	56%	55%	***
Blame-free atmosphere	90%	90%	92%	92%	88%	86%	90%	92%	90%	***
Attitudes to work pressure[†]	17%	15%	19%	20%	28%	21%	20%	18%	19%	***
Competence awareness[†]	42%	43%	44%	42%	47%	36%	41%	44%	44%	***
Respect for seniority[†]	73%	73%	76%	76%	74%	62%	74%	75%	73%	***

Source: Adapted from Itoh and Andersen 2008.
Notes: *: $p < 0.05$, **: $p < 0.01$, ***: $p < 0.001$. Figures: percentage of positive respondents to the factor label. Internal: Internal medicine ward, Surge.: Surgery ward, ICU: Intensive care unit, OR: Operating room, Outp.: Outpatient clinic, Psychi.: Psychiatry ward, Paedia.: Paediatrics ward, Mixed: Mixed ward. [†]: Two sided orientation (both too little and too much are 'bad'). [‡]: The number of responses in 'Total' includes other work units.

In contrast, classifying nurses' work units into eight groups (see Table 4.4), reveals highly significant differences for all the dimensions. Among the eight groups, on the one hand, nurses in the outpatient clinic and the psychiatric ward are more strongly aware of team/stress management and have a higher level of job satisfaction; on the other, they perceive a larger power distance and, to a lesser degree, a blame-free atmosphere. At the other extreme, nurses in OR and ICU (these two units have been acknowledged as high hazard settings) are likely to have stronger perceptions of a blame-free atmosphere and smaller power distance, but their job satisfaction and awareness of team and stress management are lower than the other work-unit groups. Nurses in these units interact with patients less frequently than in other inpatient wards – most of the time, patients in OR and ICU are anaesthetised, drowsy or sleeping. The local culture of inpatient wards, categorised as internal medicine or surgery types, generally lies in between the above-mentioned two extreme groups.

Organisational Differences in Safety Culture

Organisational differences in safety culture factors are summarised in Table 4.5 in terms of mean, maximum and minimum values and its range (max–min) of percentage of positive respondents over 19 and 79 hospitals surveyed for the doctor and nurse samples, respectively. As can be seen from this table, there are large variations in safety culture among the Japanese – we also surmise that large organisational variations exist in other countries. The hospital differences in safety culture are much greater than those by specialties/work units mentioned in the last subsection. In particular, large hospital variations exist in job satisfaction (e.g. 41 percent of hospital difference and 14 percent of work-unit difference in the nurse sample), leadership (58 percent vs. 13 percent), safety awareness (34 percent vs. 7 percent) and team/stress management (40 percent vs. 10 percent). In contrast, hospital differences are relatively small – but much larger than work-unit differences – in communication (14 percent vs. 3 percent), blame-free atmosphere (20 percent vs. 6 percent) and power distance (27 percent vs. 5 percent), all of which were rather positively perceived by healthcare employees in Japan, as mentioned previously.

Such large organisational variations in safety culture might stem from differences in style and procedures of management in general and risk management in particular, including error reporting systems, safety training and attention to safety procedures.

Table 4.5   Safety culture variations across hospitals

Safety culture factors	Doctor				Nurse			
	Mean	Max.	Min.	Range	Mean	Max.	Min.	Range
Communication	93%	97%	82%	16%	94%	99%	84%	14%
Job satisfaction	56%	81%	35%	46%	34%	55%	14%	41%
Small power distance	82%	93%	75%	18%	84%	94%	66%	27%
Leadership	48%	81%	29%	52%	40%	70%	12%	58%
Safety awareness	56%	68%	37%	31%	74%	92%	58%	34%
Collectivism-individualism	51%	62%	35%	27%	50%	67%	35%	32%
Member-conflicting attitudes (weak conflicting attitudes)	4%	14%	0%	14%	6%	20%	3%	17%
Team/stress management	55%	71%	41%	31%	56%	72%	33%	40%
Blame-free atmosphere	93%	100%	86%	14%	90%	98%	78%	20%
Attitudes to work pressure[†]	31%	50%	14%	36%	21%	42%	10%	32%
Competence awareness[†]	55%	69%	41%	28%	44%	68%	30%	38%
Respect for seniority[†]	92%	100%	84%	16%	73%	85%	56%	29%

Source: Adapted from Itoh and Andersen 2008.
Notes: Figures: percentage of positive respondents to the factor label. [†]: Two-sided orientation (both too little and too much are 'bad').

Measuring Safety Performance: Attitudes, Behaviours, Actions and Outcomes

Methods for Measuring Safety Performance

There are a number of methods that have been used for measuring safety performance in a specific organisation and its subsets or work units, such as departments and wards. Common methods such as observation, interviews and questionnaire surveys have not only been applied to achieve this, but special methods have also been developed. One of the special methods for this purpose is the *Global Trigger Tool*, which was developed by the Institute for Healthcare Improvement (Griffin and Resar 2009; IHI 2006). Some methods can make use of data that have been collected originally for other purposes and exploit the data to measure the safety performance of an organisation or its work units. A typical example of such data is the incident or near-miss report, which is collected primarily for the purpose of organisational learning. In addition, some indicator sets have been proposed to assess safety-related states in a hospital. The most widely used safety indicator system is the AHRQ's Patient Safety Indicators (PSI), which include specific types of complications and adverse events, and their occurrence rates will be calculated from reported cases as indicators (Mardon et al. 2010).

Each of the safety performance measurement methods can be characterised as a combination of the following three facets: (a) data source; (b) identification procedure and (c) data category. Typical data sources for safety performance in healthcare are patient case records, incident reports, field data (for example, obtained by observation), and self-reported data (typically elicited by the use of a questionnaire or interview).

There are several data categories related to safety performance, for instance, safety outcomes such as accidents and adverse events; incidents (and near-misses); errors; and safety-related actions, behaviours and attitudes. There are several levels of thoroughness of identification procedures for safety performance, ranging from complete review of a particular medical document to just counting something. For instance, detecting adverse events by *chart review* is characterised by the complete review of adverse events from a vast amount of patients' case records, while the Global Trigger Tool could be a method for quantifying adverse outcomes through limited, heuristic review (by the use of relevant clues) from the same data source – but typically limited, sampled case records – a method which will be illustrated later in this section. In the other extreme of the thoroughness axis, the reporting rate of incidents can be achieved just by counting incident cases (or errors) from incident reports.

From this classification, it can be easily guessed that data accuracy and efforts and time for assessment will be diverse, depending on application methods. In general, there is an efficiency-thoroughness trade-off for accuracy-time consumption, reliability-productivity, etc. On the one hand, if you adopt a time-consuming, thorough method with a limited time constraint for assessment of safety performance, you can analyse only a small amount of data in a short interval, say only one month's case records, although the estimated short-term safety performance should be rather reliable. But are you sure that that particular month is a representative period for the entire year? Or is it an exceptionally safe month having fewer adverse events than usual? On the other hand, if you apply a quick, efficient method, which may produce a poor estimation of safety, to long-term data, e.g. for three years, you could estimate a safety level for your hospital on the basis of an entire year, and you could further know the transition of safety levels by comparing each year's level during the three years. But are you confident of the estimated results – they may not be reliable enough. Therefore, we need to know limitations and expected reliability as well as the required time for each method of assessing safety performance. Based on such knowledge and understanding of the context to be applied, we will determine what data source, data category and measurement procedure we should use, taking into account the efficiency-thoroughness trade-off.

Reliability of Safety Performance Measures

One of the most intuitive and strongest measures of safety performance involves the rate of accidents or adverse events in an organisation. This type of data can be evaluated repeatedly and regularly for an entire organisation and its work units, such as clinical departments and wards, but it may also allow a comparison between work units or between organisations in the same domain. At the same time, there are many reasons why we should be wary of using these kinds of data – even within one and the same domain and the same types of specialties or tasks. First, such data may be essentially dependent on external factors and may not reflect internal processes – for instance, university hospital clinics may be more likely to admit patients who are more ill and may therefore experience a higher rate of adverse events (Baker et al. 2004). Second, accident data may be of dubious accuracy due to under-reporting by some organisations and over-reporting by others (Glendon and McKenna 1995).

Reporting of near-misses and incidents may possibly be a useful measure of safety performance – this method is very quick and efficient in obtaining related indices – though its overriding goal is not to derive reliable statistics about rates of different types of incidents. Incident reporting has typically been used for healthcare staff to report information on events which have led to unintended harm or potential harm to patients (Giles et al. 2006). Its primary objective is to help organisations learn continuously from such negative experiences (Barach and Small 2000). Thus, Helmreich (2000) stressed that one should not seek to derive rates from incident reporting but rather focus on the valuable lessons it contains. But the wide variation observed across, for instance, hospitals and individual departments may have much more to do with local incentives and local 'reporting culture' than with actual patient risks (Cullen et al. 1995). Therefore, when making comparative studies between organisations – and even between work units within a single organisation – we have serious difficulties in interpreting the incident rate. This may either be interpreted as a measure of *risk* – the greater the rate of reported incidents of a given type, the greater is the likelihood that a patient injury may take place; *or* it may be taken as an index of the inverse of risk, that is, as an index of *safety* – that is, the more that staff are demonstrating willingness to report, the greater is their sensitivity to errors and learning potential, and so the greater is the safety in their department (Edmondson 1996, 2004).

Another important issue concerning incident indices is their reporting types. One is an actual reporting rate which will be calculated on the basis of the number of *incident reports* submitted by staff under study. The other type is a *self-reported* rate of incidents, which will be elicited as staff responses to (a subjective measure of) safety performance. Self-reported measures of incidents have often been said to have low reliability for the following reason: healthcare providers are likely (and/or want) to forget consciously or unconsciously their errors or mistakes and to report a smaller number of incidents than they have actually made. Therefore, when testing the criterion validity of safety culture scales, it has been suggested that safety performance data should be obtained independently by using different methods from the one used for measuring a safety culture (Flin et al. 2006). In this sense, it is preferable to use the actual records of reported incidents for this purpose than the self-reported indices.

To overcome this weak point of self-reported data, we propose eliciting safety-related attitudes and indications of actions by asking respondents to report their likely actions with regard to fictitious adverse event cases (vignettes). This method for safety performance measurement will be mentioned later in this section, illustrated with some case studies.

Measures from Incident Reporting

We have raised the question of whether one may interpret the rate of incident reporting of a given unit as an indicator of *safety* or *risk*. In many cases, making an error is actually different from documenting an error, which is based on capturing the error, discussing it, etc. – which in turn leads to learning from such an experience (Edmondson 1996, 2004). If one can assume that all employees apply identical criteria for the submission of incident reports (and that there is no case of holding back on reporting an event which must be reported – a trend of 'holding back' may be considered as an implicit part of submission criteria), where a work unit generates more reports this may be assumed to indicate that staff produced more errors in this hospital. That is, the reporting rate will of course predict the likelihood of errors or incidents. Thus, it is reasonable to consider that more accidents must take place where errors are made more frequently (if tasks and patient profiles are comparable). Therefore, in this hypothetical situation, the incident reporting rate (e.g. the number of reporting cases per staff per year) would be an index of 'risk'.

In this context, a crucial question is whether one can accept the above assumption about identical or at least similar reporting criteria. For instance, if there is no reason to uphold the assumption, the rate of incident reporting may reflect the staff's safety awareness or sensitivity to errors or incidents. Thus, it is natural to speculate that staff who have a higher sensitivity to errors – and the importance of learning from errors – will be more likely to submit a report of an incident that might be undetected or unreported by staff with a lower sensitivity or perhaps a lower appreciation of the importance of learning from negative experience. In this case, the rate of incident reporting can be interpreted as a measure of 'safety'. That is, relatively high rates of reporting may be indicative of units that attach importance to errors and learning – so we should expect these units actually to be safer (Edmondson 1996).

Looking at the Japanese healthcare context, a majority of reports (more than 90 percent in most hospitals) are submitted by nurses. Most cases reported by nurses are near misses and no-effect events and very few are submitted for cases causing an adverse outcome such as prolonged hospitalisation. In addition, there is widespread, common agreement in Japanese hospitals that nurses – there is no such agreement about doctors – faithfully report all events that cause a patient at least a temporary, minor effect such as slight fever, headache or discomfort (Itoh et al. 2005). Hospitals in Japan commonly use a 6-level severity classification of adverse events, ranging from 0 (near miss) to 5 (death) like the National Coordinating Council for Medication Error Reporting and Prevention Index for Categorising Errors in the United States (NCC MERP 2001). In this classification, cases at Levels 0 and 1 are near misses that have no adverse effect for the patient. The above-mentioned minor effect is categorised as Level 2, which nearly corresponds to Category E of the NCC MERP Index. If the above agreement about 'minor event' reporting is accepted, the reporting frequency of Level 2 or higher (Level 2+) cases may be considered as a plausible candidate indicator of *risk* for work places where all cases can be assumed to be reported, e.g. by nurses in Japan (Itoh et al. 2009).

In addition, it was suggested that the reporting rate of near-miss cases – in which an error did not affect the patient (Level 0; corresponding to Category B of the NCC MERP Index) – may be a candidate indicator of *safety*, as far as its application is limited to incident cases reported by nurses (Itoh et al. 2009). However, the reporting rate of near misses varies greatly across hospitals and this rate is generally rather low even from nurses. Indeed, according to our investigation of seven general hospitals in Japan (in 2008), the percentage of Level 0 cases reported by nurses ranged between 0.7 percent and 36 percent (mean: 13 percent) and that of Level 1 cases ranged between 28 percent and 75 percent (mean: 58 percent). Therefore, assuming that many reports of events at Level 1 are also derived from higher awareness of organisational learning, we would expand the original suggestion to use the rate of Level 0 and 1 cases as a hypothetical measure of safety.

Examples of applications of these incident reporting indices will be mentioned in the next section through illustration of their correlations with safety culture scores.

Staff Behaviour Related to Error Reporting

As we have argued in the previous subsection, staff attitudes and indications of behaviours related to patient safety, e.g. reporting one's own errors and interaction with patients who have been victims of such errors, can be estimated by using the approach of asking respondents to indicate their likely reactions in scenarios involving errors and patient harm or risk. Staff's self-reported safety-related reactions after an adverse event can be elicited as their responses to a vignette describing a fictitious incident case. In this method, respondents are first asked to read a fictitious adverse event case, assuming themselves to have been involved in the event. Then, they will rate the likelihood that they would engage in various actions on a Likert-type scale, e.g. using five or seven points, ranging from 'definitely yes' to 'definitely no'. An example (severe outcome) of a fictitious event, which was adopted from the cases used in the Danish survey (Andersen et al. 2002) is as follows:

> A 42-year old woman (married, one child, school teacher) is hospitalised in order to receive chemotherapy. The drug has to be given as a continuous infusion intravenously. There is no pre-mixed infusion available in the department and you have to prepare it yourself. While you are preparing the infusion, you are distracted. By mistake you prepare an infusion with a concentration 10 times greater than the prescribed level.

> You do not discover the error until you administer the same drug to another patient later that day. By this time the 42-year old patient has already received all of the high concentration infusion. You are aware that in the long term the drug may impair cardiac functioning. You realise that there is a significant risk that the patient's level of functioning will be diminished and that she probably won't be able to maintain her present work.

Self-reported staff reactions that would potentially be taken after the event may be stated as follows: (1) keep the event to oneself, (2) report the event to one's leader or the doctor in charge, (3) report the event to the local reporting system, (4) write in the patient's case-record about the event, (5) inform the patient about the event and future risk, (6) explain to the patient that the event was caused by one's mistake and (7) apologise about the event to the patient.

Hereafter, we will illustrate application of the above-mentioned method through a case study conducted in Japan. In this study, two other fictitious cases – i.e. mild and near-miss cases, which were also adopted from Andersen et al. (2002) – were also used in addition to the above-mentioned severe outcome example. Hospital staff's self-reported reactions were elicited from about 1,000 doctor and 18,000 nurse responses in 84 hospitals. The results are summarised in Table 4.6 in terms of the percentage of respondents who strongly or slightly agreed with each reaction item. There were significant differences between doctors and nurses and between outcome severities for most reactions.

Table 4.6 **Doctors' and nurses' self-reported reactions related to patient safety after adverse events**

Staff reactions		Outcome severity		
		Near miss	Minor case	Severe case
(Do not) keep the event to oneself	Doctor:	83%	87%	94%
	Nurse:	79%	94%	92%
Report to leader or doctor in change	Doctor:	78%	91%	96%
	Nurse:	80%	93%	92%
Report to the local reporting system	Doctor:	66%	73%	93%
	Nurse:	71%	91%	90%
Write in the patient's case record	Doctor:	–	76%	87%
	Nurse:	–	73%	71%
Inform patient about the event	Doctor:	–	88%	92%
	Nurse:	–	71%	71%
Admit the event caused by own error	Doctor:	–	52%	86%
	Nurse:	–	44%	57%
Apologise to the patient about the event	Doctor:	–	62%	89%
	Nurse:	–	66%	76%

Notes: Figures: percentage of positive respondents who agreed strongly or slightly to perform a given action.

As an overall trend, staff attitudes and action indications become more positive when the target event has a more serious outcome. However, even for the fictitious near-miss event, where no harm at all reaches the patient, there is a staff willingness to react positively, with about 80 percent of doctors and nurses indicated reporting their actions to their leader or the doctor in charge. This suggests that both doctors and nurses have positive attitudes to error reporting and favourable self-reported interaction with the patient. For instance, about 80 percent of hospital staff said they would not keep the event to themselves and would report the event to their leader or the doctor in charge even in the near-miss case. In general, nurses have more positive attitudes to error reporting, while doctors self-reports show they are more likely to interact positively with the patient, e.g. informing the patient about the event and future risk and apologising to the patient about the event. This may be influenced by job requirements that such interactions with the patient are considered primarily a doctor's task in Japan.

One of the advantages of a self-reporting approach is its applicability to assessment from another party's perspective. For instance, patient views of doctors' likelihood to perform safety-related actions can be elicited as follows: patients are asked to what extent they would expect a doctor to perform each reaction item, assuming they become a victim of the same fictitious adverse event. Then, comparisons between patient and staff responses allow us to identify similarities or differences between staff's self-reported attitudes and patient expectations of staff actions.

Table 4.7   Gap between doctors' indications and patients' expectations about doctor reactions after adverse events

	Inform event and risk		Admit own error		Express apology	
	Mild	Severe	Mild	Severe	Mild	Severe
Doctors	88%	84%	44%	69%	44%	81%
Patients	44%	44%	32%	36%	34%	41%

Notes: Figures: percentage of doctors who indicated that they will definitely or probably perform a given action, or percentage of patients who expected that doctors would definitely or probably perform a given action.

An example of the results of such an application are shown in Table 4.7. For this exemplar study, 920 responses were collected from inpatients and outpatients in a big university hospital located in Tokyo (Itoh et al. 2006). Patient responses were compared with those collected from 33 doctors working in this hospital. This table shows comparative results for the mild and severe outcome cases in terms of percentages of doctors who reported that they would – and patients who indicate that they think the doctors would – definitely or probably perform each of the three actions described. There were significant differences for all reaction items in both severity cases. A large gap can be seen between doctors' indications of their reactions and patients' expectations. Patients are more sceptical of staff reactions after an adverse event than doctors' responses suggest they need to be.

Contribution of Safety Culture to Patient Safety

Case Study in Japanese Hospitals

In this section, results obtained in a case study (Itoh and Andersen 2008, 2010) will be reviewed to illustrate the contribution of safety culture to safety outcome. In a case study, a questionnaire-based survey concerning staff reactions after the adverse event introduced in the last section, was conducted in addition to the safety culture survey. At the same time, incident reports for three years (2004–06) submitted by nurses were obtained from one of the hospitals (Hospital M) that participated in the safety culture survey. Hospital M was a private, acute-type general hospital, located in Tokyo. This hospital covered almost all clinical specialties and, at the time of the survey in 2006, it had about 500 inpatient beds, 160 full-time doctors and 360 full-time nurses. Nurses belonged to any one of 18 clinical work units: 14 inpatient wards, an outpatient clinic, operating room, kidney centre, and medical examination unit.

In addition to Hospital M, we obtained incident reporting statistics submitted by nurses from 10 hospitals that were also surveyed for safety culture (Itoh 2011).

Contribution to Reduction of Minor Outcome Events

First, we will present the results that were applied to the data collected from Hospital M. We calculated a mean annual reporting rate over the three years 2004–06 as well as a mean score for each safety culture factor (cf. Tables 4.2–4.5) for each of 18 work units. A rank-based correlation analysis (Spearman's rho) was applied to these cross-unit data and the analysis of results is shown in Table 4.8.

Significant correlations were detected with the reporting rate of Level 2+ cases for many dimensions of safety culture. The correlations are, with one exception, negative and show an unfavourable or 'less safe' orientation for the following dimensions: (weaker) awareness of communication, (weaker) team and stress management, (weaker) competence awareness, (lower) job satisfaction, (weaker) blame-free atmosphere, and (stronger) power distance. All these results confirm that a positive level of safety culture component contributes to an increased level of patient safety, applying a hypothetical 'risk measure' to the reporting rate of Level 2+ cases mentioned previously.

Table 4.8   Correlations of safety culture factors with incident reporting indices

Safety culture factors	Levels 0 & 1	Levels 2+
Communication	-0.457	-0.744**
Job satisfaction	-0.228	-0.518*
Power distance	0.368	0.552*
Leadership	0.131	0.089
Safety awareness	0.037	-0.280
Collectivism-individualism	-0.482*	-0.316
Member-conflicting attitudes	0.148	0.239
Team/stress management	-0.366	-0.582*
Blame-free atmosphere	-0.456	-0.576*
Attitudes to work pressure[†]	-0.486*	-0.632**
Competence awareness[†]	-0.513*	-0.678**
Respect for seniority[†]	-0.090	-0.153

Source: Hospital M nurses; 2004–06.
Notes: *: $p < 0.05$, **: $p < 0.01$. Figures: Spearman's rho. [†]: Two sided orientation (both too little and too much are 'bad').

As for the reporting rate of Level 0 and 1 cases, significant correlations were also observed with several safety culture factors. However, contrary to the 'safety measure' hypothesis of this index, there was the same trend shown in the reporting rate of Level 2+ cases: a positive score for each factor contributes to lower frequency of reporting no-effect cases. One possible reason for rejection of the hypothesis may be very few reports of Level 0 cases submitted even by nurses in this hospital: only 2 percent of cases at Level 0 and 88 percent at Level 1 were reported. Therefore, the rate of Level 0 and 1 cases might almost capture characteristics of cases at Level 1, which might be different from those of Level 0 cases, but rather closer to cases at Level 2. These results may also indicate that there are not many cases where nurses hold back reporting even a 'no-effect' event in *this* hospital.

Another test of criterion validity of the safety culture factors will be stated here through application to the other, 10-hospital sample. Since safety performance may not only depend on internal factors such as safety culture, but also reflect external processes such as job requirements and conditions (Baker et al. 2004), we selected data only from 'inpatient wards', where their external factors were more homogeneous than the entire nurse sample, which included non-ward units such as the operating room and outpatient clinic. Results of correlation analysis applying to the 10-hospital sample are shown in Table 4.9. Only a single factor, member-conflicting attitudes was significantly correlated with the reporting rate of Level 2+ cases. Accepting the 'risk measure' assumption of this reporting index, it is suggested that weaker member-conflicting attitudes contribute to higher level safety in inpatient wards of the hospital.

Table 4.9   Correlations of safety culture factors with incident reporting indices (in 2006), applying to the 'inpatient ward' sample

Safety culture factors	Levels 0 & 1	Levels 2+
Communication	-0.117	-0.550
Job satisfaction	0.400	-0.433
Power distance	0.117	0.583
Leadership	0.700*	-0.100
Safety awareness	0.817**	0.033
Collectivism-individualism	0.683*	-0.083
Member-conflicting attitudes	0.617	0.700*
Team/stress management	0.683*	-0.133
Blame-free atmosphere	0.133	-0.583
Attitudes to work pressure[†]	0.033	-0.433
Competence awareness[†]	0.750*	0.100
Respect for seniority[†]	0.533	0.183

Notes: * $p<0.05$, ** $p<0.01$. Figures: Spearman's rho. [†]: Two sided orientation (both too little and too much are 'bad').

It is notable that there were significantly positive correlations between the rate of Level 0 and 1 and a number of safety culture factors: leadership; safety awareness; competence awareness; collectivism-individualism and team/stress management.

High scores in these factors, each of which contributes to a higher reporting rate, seem to be desirable for risk management goals, e.g. stronger leadership and higher safety awareness. Unlike the analysis results applying to the Hospital M data mentioned above, the results elicited from the 10-hospital sample support the 'safety measure' hypothesis for the rate of Level 0 and 1 cases.

Contribution to Positive Staff Attitudes

In this subsection, we apply another type of safety performance data, i.e., self-reported staff attitudes to error reporting and interaction with the patient, to the test of criterion validity of the safety culture factors. For this purpose, we used the nurse sample of the Japanese data including more than 17,000 questionnaire responses collected from 82 hospitals (Itoh and Andersen, 2010). An example of results of correlation analysis is shown in Table 4.10 in terms of Spearman's rho, using the mild outcome case in the three vignettes offered – results for the near-miss and severe cases were quite similar to this case.

Table 4.10 Correlations of safety culture factors with self-reported staff reactions after an adverse event (mild case; nurses in 82 hospitals)

Staff reactions Dimension	Keep the event to myself	Report to leader or dr. in change	Report to the local system	Inform pt. about event	Apology to pt. about event
Communication	-0.171	0.073	0.061	0.045	0.009
Job satisfaction	-0.049	-0.069	-0.043	0.138	-0.031
Power distance	0.537**	-0.553**	-0.593**	-0.490**	-0.392**
Leadership	-0.135	0.105	0.192	0.302**	0.186
Safety awareness	-0.172	-0.030	0.066	0.193	0.103
Collectivism-Individualism	0.070	0.003	0.022	0.046	0.014
Member-conflicting attitudes	0.079	0.133	0.087	0.071	0.037
Team/stress management	0.001	-0.176	-0.112	0.095	0.104
Blame-free atmosphere	-0.548**	0.583**	0.647**	0.503**	0.379**
Attitudes to work pressure[†]	-0.021	-0.179	-0.145	0.023	-0.089
Competence awareness[†]	-0.006	0.008	0.049	0.134	0.076
Respect for seniority[†]	-0.272*	0.334**	0.343**	0.263*	0.214

Notes: *: $p < 0.05$, **: $p < 0.01$ [†]: Two sided orientation (both too little and too much are 'bad'). Figures: Spearman's rho.

One of the most influential factors is power distance, which was significantly correlated with self-reported likelihood for all reaction items. In addition, there was also a positive correlation between a blame-free atmosphere and the self-reported likelihood of taking *any* patient safety action. These results indicate that the smaller a power distance and the stronger a blame-free atmosphere is within a hospital, the larger is the number of staff who would take positive actions toward error reporting and interaction with the patient after an adverse event. Besides these factors, significant correlations were identified with one or some reaction items for some other safety culture factors.

From these results, it can be seen that a positive safety culture contributes to favourable staff attitudes related to patient safety. In particular, small power distance and blame-free atmosphere appeared as the most effective elements of safety culture for positive staff attitudes and behaviours to error reporting and interaction with the patient after the adverse event.

Safety Culture and Outcomes

In the previous sections we have described examples derived from several studies correlating a safety culture measure with various measures of incident occurrence. In this section we describe and discuss in a more general setting key problems and issues related to correlating safety culture measures with patient safety outcomes. The reason why this is an important methodological issue is easy to state: while we may develop and perform internal validation of any number of constructs relating to 'organisational' culture, measures of safety culture (or safety climate) must have some empirically based connection with safety. If not, there is no basis for arguing that the measure is really an index of safety – it may be an index of other aspects of team work or the ways in which organisations operate and configure themselves. Therefore, issues pertaining to the question of criterion validity – the ability of the measure to predict an external outcome – raise essential methodological questions that must be addressed.

Safety Outcomes

Measurement of safety is often the Achilles heel in both safety research and healthcare quality research. The predicament is our inability to define safety in positive terms rather than in negative terms – for instance as the absence of accidents or danger. Most often safety is defined as the absence of adverse events and errors that may lead to such events. Hence, we may define safety in terms of some measure of 'unsafety' or the inverse of safety, for instance measured as the number of adverse events per measurement unit (e.g. per bed-day). But such sharp, inverse safety measures are poor proxy measures of safety although, when measured with high reliability, they are usually thought of as the gold standard. In contrast, if we define safety (loosely) as a continuous process of safe actions in a safe environment (cf. 'a dynamic non-event', Weick and Sutcliffe 2001) tell us that hard outcome measures are poor proxy measures, not only because they are uni-dimensional but also because they are not process measures – i.e. the processes that ensure safety and prevent 'unsafe events' from happening. In the widely known framework of Donabedian with its trio – structure, process and outcome – risk is measured in the outcome category whereas safety is a feature of structure and process. Therefore, defining safety as the inverse of risk makes it inverse to risk not only in a formal sense but also inverse in a qualitative sense. Hard outcomes (adverse events such as injuries, mortality, infections etc.) are therefore also poor safety indicators in the sense that they do not give any direction on how to improve safety.

As argued above, there is a need for positively defined direct process measures of patient safety. Within occupational health the TR safety observation method offers a method that has been validated in several trials (Laitinen et al. 1999; Laitinen and Päivärinta 2010). The method requires that a number of processes essential for safety are observed during a workplace walk-through. The observers will make at least 100 observations within, at most one hour, in the target work place, recording the number of observations made, respectively, of incorrectly and correctly performed processes. The proportion of correct observation is then used as the safety measure. The reliability of the measure has been found to be very high and, moreover, the inverse measure, i.e. the proportion of incorrect observations, is highly correlated to the numbers of lost time injuries (op.cit). Another example from occupational safety is the recording of safety communication as a proxy measure of safety (Zohar and Luria 2003). The advantage of this measure is that it reflects the very core of what safety is founded in: a preoccupation with safety in the minds of the people whose actions are needed to maintain safety and avert accidents. Further, this measure of safety offers a useful opportunity for improvements through leader-based interventions: good leadership is crucial to safety, and the primary responsibility for safety rests on the shoulders of leaders. Moreover, this measure has been found to be significantly (inversely) related to the number of injuries in workplaces and has thus been shown to have external validity (Zohar and Luria 2003; Kines et al. 2010).

Safety culture (and/or climate) measures may themselves be regarded as proxy measures of safety in the sense outlined above. However, few studies have found these measures to be strongly related to hard risk outcomes such as injuries and accidents (Guldenmund 2000; The Health Foundation 2011). There are several reasons for this apparent lack of correlation. Safety climate or culture is, by definition, shared within a social unit (a work group), but such units are usually ill defined and small. Safety climate or culture is multi-faceted, and each facet is a construct, as described in the previous sections of this chapter, based on a few items from a questionnaire. Although the reliability and the intra-class correlation for the constructs can be acceptable, repeated measurements are typically infeasible, and when the questionnaire has been applied repeatedly, its responsiveness (the ability of the construct to reliable measure changes over time (de Vet et al. 2011)) is usually not reported but can be expected to be low. At the same time, since the relevant work groups (social units) are typically small, the number of risk outcomes (errors, incidents, accidents) specific for the particular social units will nearly always be small and therefore subject to random variation. On the other hand, hard risk outcomes are typically not reported separately for individual social units that each shares a safety climate but are likely to fluctuate over time and are uni-dimensional. Therefore, the influence of other structural, organisational and performance shaping factors such as work load and technology interfaces, which may be common across work units and which provide near simultaneous exposures, will tend to outweigh those of shared climates within the work groups. In addition, the causal direction of safety climate is most naturally thought of as a precursor (leading indicator) to risk outcomes, whereas there seem to be no studies that have shown that the causation also works in the opposite direction, i.e. that the experience of accidents will lead to a poor safety climate (The Health Foundation 2011; Clarke 2006).

Risk Outcomes

An ideal risk outcome measure must reliably and accurately measure the level and change in harm. Moreover, it must first lead to a method through which the key processes that are involved in causing harm may be identified, and second deliver approximately real time measures of the outcomes for which it is an indicator. However, no such risk outcome measure exists in healthcare; therefore, less ideal measures have been developed and applied.

The Institute for Healthcare's Global Trigger Tool (GTT) delivers one of the most trusted and widely used patient risk outcomes (Naessens et al. 2010). Even so, less than satisfactory performance of the tool has also been reported (Schildmeijer et al. 2012; Lipczak et al. 2011: Mattsson et al. 2013). The GTT is an outcome measure of harm caused by healthcare, but it is also a learning tool. It can offer an organisation insight into how its systems function to provide quality of care to patients. In the patient safety organisation the insight from the GTT can be one of several indicators needed for feedback in the process of continuous improvement. Used in the hands of an experienced review team in a stable setting, the GTT can offer a reasonable estimate of the level of harm as well as temporal trends. Using the GTT is time consuming but, when applied in tandem with an effective learning and safety improvement organisation it can nevertheless be an efficient use of time. In the near future, as manual chart review of closed patient charts is replaced by computerised review of electronic patient records while the patient is still hospitalised, the automatised GTT will become a more appropriate 'ideal' measure and may possibly also be refined into a tool with acceptable reliability.

Traditionally, risk outcomes have focused on mortality and complications, but the spectrum of outcome metrics has been expanded to include indicators such as re-admissions, functional status and quality of life (Lazar et al. 2013). Dissimilarities of patient populations complicate the interpretation of outcome metrics across geographic regions and facilities. Statistical approaches are used to make risk adjustments but this is limited by incomplete data to encompass all factors that might be of importance. Lack of rigorous data sampling methodology with strict inclusion and exclusion criteria etc. is another problem that hampers comparison across facilities. Outcome data is usually obtained from administrative data sets developed for billing or regulatory purposes and is created by clinical coders. But coding instructions and traditions may vary and benchmarking and economic benefits have been seen to motivate manipulation or the presentation of data in ways that are more favourable for an organisation (Carter and Jarman 2013; Taylor 2013). The coders must (ideally) review the clinical documentation in detail to discriminate between diagnoses of co-morbidity present on admission and diagnoses of complications acquired in the hospital. Readmission rates are equally complicated to obtain: what are the index admission or the first admission in a chain, the regional differences in psychosocial facilities and home care services etc. after discharge and are they related or unrelated to the primary admission, and so on?

As pointed out by Lazar et al. (2013) outcome measures need to be harmonised. They must be reliable, valid and consistent. Most outcome measures reflect a single dimension but ultimately they must be developed to reflect the continuum of care. While we might ideally want and even require measures of safety climate and culture to correlate with outcome measures, we have tried to describe in this section the many factors that will tend to cover or weaken such correlations.

Conclusion: Perspectives for Effective Healthcare Management

In this chapter, we have reviewed the concept of safety culture, in particular its factor structure or dimensions, and have illustrated actual measurement of safety culture through some case studies in the Japanese healthcare context. It was suggested from the results of these studies that much larger safety culture differences exist across different hospitals than across different departments, wards or specialties within any single organisation. A determinant of these differences may be tied to a number of structural and process factors of healthcare work and related activities, e.g. management and safety rules, routine procedures and protocols, leadership styles, workload, work conditions and risk management aids such as reporting system – which must be investigated to attain a higher level of patient safety.

Because one of the largest roles of safety culture assessment is its application to *proactive* risk management, we particularly highlighted the criterion validity of the safety culture scale in this chapter. As a required method for making such a test, we introduced some possible techniques for measuring safety performance, each of which uses different data sources in healthcare, e.g. medical records (the Global Trigger Tool), near-miss and incident reports and self-reported expressions of own attitudes. We have demonstrated a link between safety culture and safety performance based on a correlation with two types of performance measures from different sources through recent case studies.

Based on the results of these studies, it was suggested that the reporting rate of incidents causing at least minor effect can be used as a *risk* measure in Japan. However, this index may be applicable only to assessment of nurses' work units, i.e. incident reports submitted only by nurses, because there is no confirmation about 'no holding back reporting' cases at the minor effect level or higher levels from other professional groups, even in Japan. Therefore, there is a need to devise a safety performance measure which is universally applicable (if external conditions such as tasks and patient profiles are comparable) not only to nurses but also to other professional groups such as doctors.

Regarding the other reporting index, i.e. near-misses and no-effect cases, we could not obtain evidence as a *safety* indicator, although this index seemed more likely to show this trend than a risk measure in our case studies. Rather, it could be more plausible to conclude that this relationship depends on the hospital and moreover on the current status of hospital risk management and the maturity of its incident reporting system at the moment. It may be expected, however, that the reporting index of near-misses or no-effect cases could become an indicator of safety if a more mature reporting culture is fostered in Japanese hospitals – although such near-miss reporting may be rare internationally.

Finally, we would like to stress the need for a proactive application of safety culture assessment to hospital management. However, patient safety is not only affected by cultural factors but also contributed to by other types of factors such as systems and work factors. These factors may often influence structural parts of healthcare activities, work processes and conditions, some of which may in turn become the latent causal factors of adverse events. Such latent factors, of course, may not be completely identified by safety culture assessment and moreover, may not be perfectly addressed by a proactive approach. Therefore, we need to adopt a *hybrid* approach of retrospective analysis of adverse events and prospective assessment of organisational culture, while at the same time balancing pairs of conflicted elements, e.g. thoroughness and efficiency (of analyses, interventions, actions, etc.), safety and productivity, cost and benefit, and other trade-offs in a real work setting that has limited time, budgets and human resources.

Acknowledgements

This work was in part supported by Grant-in-Aid for Scientific Research A (No. 23241048), Japan Society for the Promotion of Science.

References

ACSNI 1993. *Advisory Committee on the Safety of Nuclear Installations: Human Factors Study Group Third Report: Organising for Safety.* Sheffield: HSE Books.

Andersen, H.B., Madsen, M.D., Hermann, N., Schiøler, T. and Østergaard, D. 2002. Reporting adverse events in hospitals: A survey of the views of doctors and nurses on reporting practices and models of reporting. *Proceedings of the Workshop on the Investigation and Reporting of Incidents and Accidents, IRIA,* 127–36. Glasgow, UK, July 2002.

Baker, G.R., Norton, P.G., Flintoft, V., Blais, R., Brown, A., Cox, J., Etchells, E., Ghali, W.A., Herbert, P., Majumdar, S.R., O'Beirne, M., Palacious-Derflingher, L., Reid, R.J. and Tamblyn, R. 2004. The Canadian adverse events study: The incidence of adverse events among hospital patients in Canada. *Canadian Medical Association Journal,* 170(11), 1678–86.

Barach, P. and Small, S. 2000. Reporting and preventing medical mishaps: Lessons from non-medical near miss reporting systems. *British Medical Journal,* 320, 759–63.

Brunello, G. and Ariga, K. 1997. Earnings and seniority in Japan: A re-appraisal of the existing evidence and a comparison with the UK. *Labour Economics,* 4, 47–69.

Carter, P. and Jarman, B. 2013. Who knew what, and when, at Mid Staffs? *BMJ* 6;346:f726. doi: 10.1136/bmj.f726.

Clarke, S. 2006. The relationship between safety climate and safety performance: A meta-analytic review. *Journal of Occupational Health Psychology,* 11(4), 315–27.

Colla, J.B., Bracken, A.C., Kinney, L.M. and Weeks, W.B. 2005. Measuring patient safety climate: A review of surveys. *Quality & Safety in Health Care,* 14, 364–6.

Cooper, M.D. 2000. Towards a model of safety culture. *Safety Science,* 36, 111–36.

Cullen, D.J., Bates, D.W., Small, S.D., Cooper, J.B., Nemeskal, A.R. and Leape, L.L. 1995. The incident reporting system does not detect adverse drug events: A problem for quality improvement. *Journal on Quality Improvement,* 21(10), 541–8.

Davis, R., Briggs, M., Arora, S., Moss, R. and Schwappach, D. 2013. Predictors of health care professionals' attitudes towards involvement in safety-relevant behaviours. *Journal of Evaluation in Clinical Practice,* 20(1), 12–19. doi: 10.1111/jep.12073.

de Vet, H.C.W., Terwee, C.B., Mokkink, L.B. and Knol, D.L. 2011. *Measurement in Medicine. A Practical Guide.* Cambridge: Cambridge University Press.

EASHW. 2004. Achieving better safety and health in construction. European Agency for Safety and Health at Work. Available at: https://osha.europa.eu/en/publications/reports/314, 65–72 (last accessed on 14 April 2014).

Edmondson, A.C. 1996. Learning from mistakes is easier said than done: Group and organizational influences on the detection and correction of human error. *Journal of Applied Behavioral Science*, 32(1), 5–28.

Edmondson, A.C. 2004. Learning from failure in health care: Frequent opportunities, pervasive barriers. *Quality & Safety in Health Care*, 13(Suppl. 2), ii3–ii9.

Flin, R., Burns, C., Mearns, K., Yule, S. and Robertson, E.M. 2006. Measuring safety climate in health care. *Quality & Safety in Health Care*, 15, 109–15.

Flin, R., Mearns, K., O'Connor, P. and Bryden, R. 2000. Measuring safety climate: Identifying the common features. *Safety Science*, 34(1–3), 177–92.

Gershon, R.R.M., Karkashian, C.D., Grosch, J.W., Murphy, L.R., Escamilla-Cejudo, A., Flanagan, P.A., Bernacki, E., Kasting, C. and Martin, L. 2000. Hospital safety climate and its relationship with safe work practices and workplace exposure incidents. *American Journal of Infection Control*, 28(3), 211–21.

Giles, S., Fletcher, M., Baker, M. and Thomson, R. 2006. Incident reporting and analysis, in *Patient Safety: Research into Practice*, edited by K. Walshe and R. Boaden. Maidenhead: Open University Press, 108–17.

Ginsburg, L., Gilin, D., Tregunno, D., Norton, P.G., Flemons, W. and Fleming, M. 2009. Advancing measurement of patient safety culture. *Health Services Research*, 44(1), 205–24.

Glendon, A.I. and McKenna, E.F. 1995. *Human Safety and Risk Management*. London: Chapman and Hall.

Glendon, A.I. and Stanton, N.A. 2000. Perspectives on safety culture. *Safety Science*, 34, 193–214.

Griffin, F.A. and Resar, R.K. 2009. *IHI Global Trigger Tool for Measuring Adverse Events* (Second Edition). IHI Innovation Series white paper. Cambridge, MA: Institute for Healthcare Improvement.

Gu, X. and Itoh, K. 2011. A pilot study on healthcare safety climate and error reporting behavior in China. *Journal of Patient Safety*, 7(4), 204–12.

Guldenmund, F.W. 2000. The nature of safety culture: A review of theory and research. *Safety Science*, 34, 215–57.

Halligan, M. and Zecevic, A. 2011. Safety culture in healthcare: A review of concepts, dimensions, measures and progress. *BMJ Quality and Safety*, 10, 338–43.

Helmreich, R.L. 2000. On error management: Lessons from aviation. *British Medical Journal*, 320, 781–5.

Helmreich, R.L. and Merritt, A.C. 1998. *Culture at Work in Aviation and Medicine: National, Organizational and Professional Influences*. Aldershot: Ashgate.

Hingorani, M., Wong, T. and Vafidis, G. 1999. Patients' and doctors' attitudes to amount of information given after unintended injury during treatment: Cross sectional, questionnaire survey. *BMJ*, 318(7184), 640–41.

Hofstede, G. 1991. *Cultures and Organizations: Software of the Mind*. London: McGraw-Hill.

Hofstede, G. 2001. *Culture's Consequences: Comparing Values, Behaviors, Institutions and Organizations across Nations* (Second edition). Thousand Oaks, CA: Sage Publications.

Hollnagel, E. 2009. *The ETTO Principle: Efficiency-Thoroughness Trade-Off.* Farnham: Ashgate.

IHI 2006. *IHI Global Trigger Tool Guide* (Version 7). Cambridge, MA: Institute for Healthcare Improvement.

Itoh, K. 2011. Does incident reporting rate indicate risk or safety in healthcare? Implications from correlations with safety climate scores. *Proceedings of the International Conference on Healthcare Systems Ergonomics and Patient Safety, HEPS 2011.* Oviedo, Spain, June 2011, 201–4.

Itoh, K. and Andersen, H.B. 1999. Motivation and morale of night train drivers correlated with accident rates. *Proceedings of the International Conference on Computer-Aided Ergonomics and Safety.* Barcelona, Spain, May 1999 (CD-ROM).

Itoh, K. and Andersen, H.B. 2010. Dimensions of healthcare safety climate and their correlation with safety outcomes in Japanese hospitals. In B.J.M. Ale, I.A. Papazoglou and E. Zio (eds), *Risk, Reliability and Safety: Back to the Future.* London: CRC Press, 1655–63.

Itoh, K., Abe, T. and Andersen, H.B. 2005. A questionnaire-based survey on healthcare safety culture from six thousand Japanese hospital staff: Organisational, professional and department/ward differences. In R. Tartaglia, S. Bagnara, T. Bellandi and S. Albolino (eds), *Healthcare Systems Ergonomics and Patient Safety: Human Factor, a Bridge between Care and Cure.* London: Taylor & Francis, 201–7.

Itoh, K. and Andersen, H.B. 2008. A national survey on healthcare safety culture in Japan: Analysis of 20,000 staff responses from 84 hospitals. *Proceedings of the International Conference on Healthcare Systems Ergonomics and Patient Safety, HEPS 2008.* Strasbourg, France, June 2008 (CD-ROM).

Itoh, K., Andersen, H.B. and Madsen, M.D. 2012. Safety culture in health care. In P. Carayon (ed.), *Handbook of Human Factors and Ergonomics in Health Care and Patient Safety* (Second edition). Boca Raton, FL: CRC Press, 133–62.

Itoh, K., Andersen, H.B., Madsen, M.D., Østergaard, D. and Ikeno, M. 2006. Patient views of adverse events: Comparisons of self-reported healthcare staff attitudes with disclosure of accident information. *Applied Ergonomics*, 37, 513–23.

Itoh, K., Andersen, H.B. and Seki, M. 2004. Track maintenance train operators' attitudes to job, organisation and management and their correlation with accident/incident rate. *Cognition, Technology and Work*, 6(2), 63–78.

Itoh, K., Omata, N. and Andersen, H.B. 2009. A human error taxonomy for analysing healthcare incident reports: Assessing reporting culture and its effects on safety performance. *Journal of Risk Research*, 12(3–4), 485–511.

Kines, P., Andersen, L.P.S., Spangenberg, S., Mikkelsen, K.L., Dyreborg, J. and Zohar, D. 2010. Improving construction site safety through leader-based verbal safety communication. *Journal of Safety Research*, 41(5), 399–406.

Kohn, L.T., Corrigan, J.M., Donaldson, M.S. (Institute of Medicine) 2000. *To Err is Human: Building a Safer Health System*. Washington, DC: National Academy Press.

Laitinen, H. and Keijo Päivärinta, K. 2010. A new-generation safety contest in the construction industry – A long-term evaluation of a real-life intervention, *Safety Science*, 48(5), 68086, http://dx.doi.org/10.1016/j.ssci.2010.01.018.

Laitinen, H., Marjamäki, M. and Päivärinta, K. 1999. The validity of the TR safety observation method on building construction. *Accident Analysis and Prevention*, 31(5), 463–72.

Lazar, E.J., Fleischut, P., Regan, B.K. 2013. Quality measurement in healthcare. *Annual Review of Medicine*, 64, 485–96.

Lipczak, H., Neckelmann, K., Steding-Jessen, M., Jakobsen, E., Knudsen, J.L. 2011. Uncertain added value of Global Trigger Tool for monitoring of patient safety in cancer care. *Danish Medical Bulletin*, 58(11), A4337.

Mannion, R., Konteh, F.H. and Davies, H.T.O. 2009. Assessing organisational culture for quality and safety improvement: A national survey of tools and tool use. *Quality and Safety in Health Care*, 18, 153–6.

Mardon, R.E., Khanna, K., Sorra, J., Dyer, N. and Famolaro, T. 2010. Exploring relationships between hospital patient safety culture and adverse events. *Journal of Patient Safety*, 6(4), 226–32.

Mattsson, T.O., Knudsen, J.L., Lauritsen, J., Brixen, K. and Herrstedt, J. 2013. Assessment of the global trigger tool to measure, monitor and evaluate patient safety in cancer patients: Reliability concerns are raised. *BMJ Quality and Safety*, 22, 571–9. doi:10.1136/bmjqs-2012-001219.

Naessens, J.M., O'Byrne, T.J., Johnson, M.G., Vansuch, M.B., McGlone, C.M. and Huddleston, J.M. 2010. Measuring hospital adverse events: Assessing inter-rater reliability and trigger performance of the Global Trigger Tool. *International Journal for Quality in Health Care*, 22(4), 266–74. doi: 10.1093/intqhc/mzq026.

NCC MERP 2001. National Coordinating Council for Medication Error Reporting and Prevention Index for Categorizing Errors. Available at: http://www.necmerp.org/medErrorCatIndex.html (last accessed on 21 November 2012).

Pidgeon, N.F. and O'Leary, M. 1994. Organizational safety culture: Implications for aviation practice. In N.A. Johnston, N. McDonald and R. Fuller (eds), *Aviation Psychology in Practice*. Aldershot: Avebury Technical Press, 21–43.

Rasmussen, J. 1986. *Information Processing and Human-Machine Interaction: An Approach to Cognitive Engineering*. New York: Elsevier/North Holland.

Reason, J. 1993. Managing the management risk: New approaches to organizational safety. In B. Wilpert and T. Qvale (eds), *Reliability and Safety in Hazardous Work Systems*. Hove: Lawrence Erlbaum Associates.

Reason, J. 1997. *Managing the Risk of Organizational Accidents*. Aldershot: Ashgate.

Schein, E.H. 1992. *Organizational Culture and Leadership* (Second edition). San Francisco: Jossey-Bass.

Schildmeijer, K., Nilsson, L., Arestedt, K. and Perk, J. 2012. Assessment of adverse events in medical care: Lack of consistency between experienced teams using the global trigger tool. *BMJ Quality and Safety*, 21(4), 307–14. doi: 10.1136/bmjqs-2011-000279. Epub 2012 Feb 23.

Sexton, J.B., Helmreich, R.L., Nielands, T.B., Rowan, K., Vella, K., Boyden, J., Roberts, P.R. and Thomas, E.J. 2006. The Safety Attitudes Questionnaire: Psychometric properties, benchmarking data, and emerging research. *BMC Health Services Research*, 6, 44.

Silva, S., Lima, M.L. and Baptista, M.C. 2004. OSCI: An organisational and safety climate inventory. *Safety Science*, 42(3), 205–20.

Singer, S., Meterko, M., Baker, L., Gaba, D., Falwell, A. and Rosen, A. 2007. Workforce perceptions of hospital safety culture: Development and validation of the patient safety climate in healthcare organizations survey. *Health Services Research*, 42(5), 1999–2021.

Singer, S.J., Gaba, D.M., Falwell, A., Lin, S., Hayes, J. and Baker, L. 2009. Patient safety climate in 92 US hospitals: Differences by work area and discipline. *Medical Care*, 47(1), 23–31.

Singer, S.J., Gaba, D.M., Geppert, J.J., Sinaiko, A.D., Howard, S.K. and Park, K.C. 2003. The culture of safety in California hospitals. *Quality and Safety in Health Care*, 12(2), 112–18.

Smits, M., Christiaans-Dingelhoff, I., Wagner, C., van der Wal, G. and Groenewegen, P.P. 2008. The psychometric properties of 'Hospital Survey on Patient Safety Culture' in Dutch hospitals. *BMC Health Services Research*, 8, 230.

Sorra, J., Famolaro, T., Dyer, N., Nelson, D. and Khanna, K. 2009. *Hospital Survey on Patient Safety Culture: 2009 Comparative Database Report*. Agency for Healthcare Research and Quality, AHRQ Publication No. 09-0030, Rockville, MD.

Sorra, J.S. and Dyer, N. 2010. Multilevel psychometric properties of the AHRQ hospital survey on patient safety culture. *BMC Health Services Research*, 10, 199.

Sorra, J.S. and Nieva, V.F. 2004. *Hospital Survey on Patient Safety Culture*. Agency for Healthcare Research and Quality, Rockville, MD.

Taylor, P. 2013. Rigging the death rate. *The London Review of Books*. Available at: http://www.lrb.co.uk/v35/n07/paul-taylor/rigging-the-death-rate (last accessed on 30 April 2014).

The Health Foundation. 2011. *Research Scan: Does Improving Safety Culture affect Patient Outcomes?* November 2011. Available at: http://www.health. org.uk/public/cms/75/76/313/3078/Does%20improving%20safety%20 culture%20affect%20outcomes.pdf?realName=fsu8Va.pdf (last accessed on 30 April 2014).

Turnberg, W. and Daniell, W. 2008. Evaluation of a healthcare safety climate measurement tool. *Journal of Safety Research*, 39, 563–8.

Weick, K.E. and Sutcliffe, K.M. 2001. *Managing the Unexpected: Assuring High Performance in an Age of Complexity*. San Francisco: Jossey Bass.

Wilks, T. 2004. The use of vignettes in qualitative research into social work values. *Qualitative Social Work*, 3, 78.

Yuki, M. and Yamaguchi, S. 1996. Long-term equity within a group: An application of the seniority norm in Japan. In H. Grad, A. Blanco and J. Georgas (eds), *Key Issues in Cross-Cultural Psychology: Selected Papers from the 12th International Congress of the International Association for Cross-Cultural Psychology*. Lisse: Swets and Zeitlinger, 288–97.

Zohar, D. 1980. Safety climate in industrial organizations: Theoretical and applied implications. *Journal of Applied Psychology*, 65, 96–101.

Zohar, D. and Luria, G. 2003. *The Use of Supervisory Practices as Leverage to Improve Safety Behavior: A Cross-level Intervention Model*. NOIRS-2003: Intervention Evaluation Contest March, 2003. Available at: http://www.cdc.gov/niosh/noirs/2003/contest2003.html (last accessed on 30 April 2014).

Chapter 5

Looking Behind Patient Safety Culture: Organisational Dynamics, Job Characteristics and the Work Domain

Denham L. Phipps and Darren M. Ashcroft

Introduction

In order to make sense of a patient safety culture (PSC) assessment, it is helpful to understand how safety culture relates to the work setting within which it occurs. Behind PSC lie features of the healthcare organisation that affect patient safety. In this chapter, we will turn the spotlight onto such features, beginning with a consideration of how PSC can emerge from beliefs, goals and activities at all levels of the organisation. We will then review empirical data from our research with community pharmacists, which focus on the relationship between safety climate and job characteristics. Finally, we will briefly consider how formative analysis of the 'work domain' can provide additional insights into PSC.

What Do We Mean by a 'Safety Culture'? A View from Open System Theory

One way to view healthcare is as an open sociotechnical system (e.g. Buckle et al. 2003; Waterson 2009). In simple terms, a sociotechnical system is a collection of technical and social elements that come together in a dynamic manner to create work processes and products. Such a system is 'open' when it interacts with the wider environment; indeed, many organisations are shaped by (and possibly influence) broader social, political, economic and technological developments (Geysen and Verbruggen 2003; Rasmussen 1997). Hence, safety is seen as a product of an assumed system and safety culture as the capacity of the system to 'create' safety (cf. Carrillo 2011). Figure 5.1 illustrates this metaphor in the form of a feedback control loop. Feedback control loops are used to describe how variables within a dynamic system influence each other (e.g. Kontogiannis 2012; Marais, Saleh and Leveson 2006). In each loop, the arrows indicate which variables are influencing and which are being influenced. A plus or minus sign is used to indicate whether the influenced variable is being enhanced or diminished. A detailed but readable introduction to the use of feedback control loops can be found in Senge (1990).

Here, a workforce's engagement in safe practice is driven by the extent to which it perceives a discrepancy between the required and the current safety level. This discrepancy reduces as the level of engagement increases but increases as the required level of safety increases. Hence, at the heart of safety culture is a reciprocal relationship between the effort organisational members put into safety practice and their contentment with what is achieved through these efforts. This relationship can be thought of as being analogous to a thermostat, whose setting represents the required level of safety. While the analogy appears benign at first glance, it leads us to an important point: in order to improve safety, or to maintain a high standard of safety in the face of environmental changes, an organisation should be able to evaluate and, if necessary, alter its thermostat setting. In their theory of organisational learning, Argyris and Schön (1996) distinguish between 'single-loop' learning, in which an organisation develops itself only within the confines of its existing beliefs about the operating environment, and 'double-loop' learning, in which the organisation further ensures that its beliefs continue to be aligned with the environment. Single-loop learning can become maladaptive to the extent that an organisation becomes invested in achieving its assumed objectives regardless of how appropriate those objectives currently are. In 1987, a British underground train station was damaged by a fire which claimed 31 lives. The subsequent investigation identified as a contributory factor the operating organisation's denial that fire hazards (of which it was already aware) posed a serious risk; this led to the organisation making little effort to address the hazards (Fennell 1988). In the context of patient safety, Dixon-Woods et al. (in press) argue that one challenge for healthcare improvement programmes is to convince organisations that a problem actually exists and needs addressing.

Figure 5.1   A basic system model of safety culture

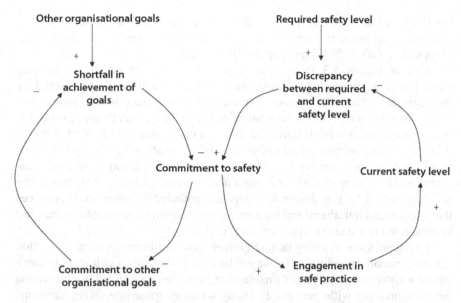

Figure 5.2   A conflict between commitment to safety and commitment to other organisational goals

The reader may also note the assumption in Figure 5.1 that an increase in the discrepancy between required and current levels of safety necessarily leads to an increase in safe practice. However, an intermediary variable can modify this relationship: the level of interest in, or commitment to, safety. Should the resulting feedback loop exist in isolation of any other organisational factors, then it would not be particularly contentious as far as safety is concerned. In practice, though, commitment to safety relates to other variables that are part of organisational life, and which could serve to reduce it. Figure 5.2 shows how such variables contribute to the original model; here, the need to achieve 'other' objectives (that is, ones unrelated to safety) conflicts with that of maintaining a required safety level. As we will see later in the chapter, a relationship of this type appears in British community pharmacies where safety and profitability are often perceived to be, at least in part, mutually exclusive.

Like commitment to safety, another element of the original model – engagement in safe practice – can interact with other variables that are present in working life. As shown in Figure 5.3, it could hinder the fulfilment of alternative goals that are unrelated to safety and which in turn provide an incentive to behave in ways that undermine safe practice. Such a relationship underlies certain types of procedural violation. For example, anaesthetists may reuse the contents of a drug ampoule, contrary to infection control guidelines, because they feel that discarding the surplus is needlessly wasteful (Phipps and Parker in press). (We hasten to add at this point that there is more to 'violations' in healthcare than the simplified account that we have given here. Some of the subtleties will be discussed later in the chapter, but a more comprehensive analysis is provided by Phipps et al. (2008, 2010) and Alper et al. (2012)).

Also in Figure 5.3 is what is known in control theory as a 'limiting condition': constraints on behaviour. We have included this variable to stand for the aggregate effect of workplace characteristics that induce staff to behave in a particular manner or hinder them from behaving in the manner that they should or would like to do. A well-documented example of a constraint on safe behaviour is related to the intrathecal administration of medication; the equipment used to do this was, until recently, typically designed in such a way as to enable the intrathecal administration of intravenous drugs in error (Lawton et al. 2009). The problem caused by this design flaw was compounded by operational practices that encouraged intrathecal and intravenous medication to be stored together and administered at the same time (Toft 2001).

A systems view of safety culture carries some implications for the way that it is understood and addressed. One implication is that safety culture is to some extent a matter of perception. Consider that, in Figures 5.1 to 5.3, each variable could be prefixed with 'perceived'. Doing so raises some interesting questions about an organisation's safety culture. For example: by whose perception is the 'required' level of safety determined and what informs that perception? From which point of view is safety seen to conflict with other organisational objectives and on what basis is the required balance between them determined? (Furthermore, are there alternative points of view in which there is no conflict between safety and other organisational objectives?) Similarly, what matters to organisational members in a discrepancy between engagement in safe practice and fulfilment of non-safety related goals? From whose perspective do constraints on their behaviour become apparent? McDonald, Waring and Harrison (2006), for example, found that doctors and managers within a hospital trust held differing views about the value of rules and procedures. Managers saw compliance with them as being important, while clinicians felt that the work was so complex that they needed to be able to exercise their professional judgement, which precluded the imposition of rules. Meanwhile, the railway workers in Clarke's (1999) study felt that adhering to rules would compromise their productivity to an extent that their managers would not approve of. Incidentally, such studies indicate one caveat in the analysis of violations described earlier: they assumed that safe practice is achieved by following rules and guidelines and that deviating from them constituted unsafe practice. Those carrying out tasks, however, may or may not believe this actually to be the case (see, for example, Smith and Alderson 2012).

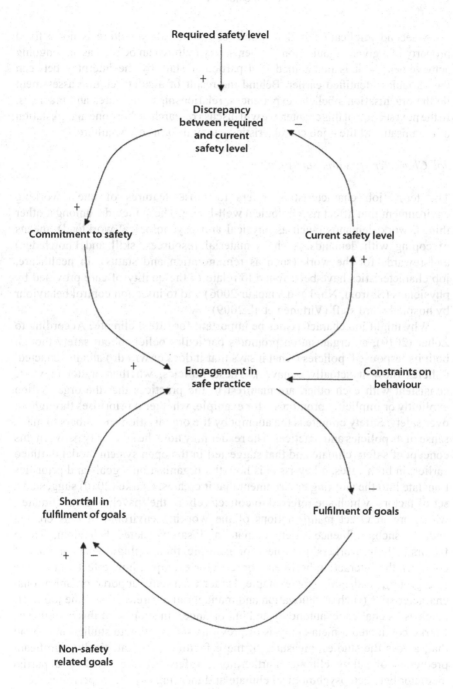

Figure 5.3   Factors affecting engagement in safe practice

A second implication is that 'good' or 'poor' safety culture is not a fixed property of a given organisation. Rather, safety culture can be seen as an 'ongoing achievement' – it is maintained in a particular state by the interplay between the variables identified earlier. Behind the result of a safety culture assessment lie the organisation's beliefs, experiences, relationships, structures and practices. In the next section of this chapter, we will discuss research on how one manifestation of organisational life – job characteristics – may inform safety culture.

Job Characteristics and Safety Culture

The term 'job characteristics' refers to those features of one's working environment that affect psychological well-being. These include. amongst other things, work demands (such as physical and psychological workload), means of coping with demands (such as material resources, skill and knowledge) and rewards for the work (such as remuneration and status). In healthcare, job characteristics have been found to relate to the quality of care provided by physicians (Shirom, Nirel and Vinokur 2006) and to infection control behaviour by hospital ward staff (Virtanen et al. 2009).

Why might job characteristics be important for safety climate? According to Zohar (2010), an organisation promotes particular beliefs about safety though both its 'espoused' policies (what it says that it does or will do) and its 'enacted' policies (how it actually behaves). These policies, whether or not they are consistent with each other, are manifest in the practices that the organisation explicitly or implicitly promotes – for example, whether it prioritises throughput over safety. Safety climate is the attempt by the organisation's members to make sense of its policies and practices. The reader may note the similarity between this concept of safety climate and that suggested in the open system model outlined earlier; in both cases, a key issue is how the organisation's goals and priorities translate into the working environment that it creates. Clarke (2010) suggested a set of factors, which she referred to collectively as the 'psychological climate', which are in effect manifestations of the working environment thus created and, as such, influence safety climate and safety-related behaviour. These factors include: workers' job roles (for example, their ambiguity and amount of conflict); the interaction between workers (for example, how effective they are as a group); leadership (for example, leader trust and support); organisational characteristics (such as innovation and managerial awareness) and the job itself (such as its challenge, autonomy and importance). In support of this hypothesis, Clarke conducted a meta-analysis of previous safety climate studies and found that, across the studies, measures of these factors were statistically significant predictors of safety climate. Furthermore, safety climate acted as a partial mediator between psychological climate and measures of safety performance.

Clarke's findings raise the possibility that we can further understand how safety climate arises, and hence how it can be improved, by understanding how to assess and modify job characteristics. Two theoretical starting points for such efforts are:

- *The demand-control-support model* (DCS; van der Doef and Maes 1999). This proposes that job strain – the personal experience of job stress – results from a combination of demands placed on the person by the job, the resources available to the person to manage these demands and the support provided by colleagues and supervisors;
- *The effort-reward imbalance model* (ER; Siegrist 1996). This proposes that job strain arises from a combination of the effort that a person is required to make in the course of his/her job and the rewards (psychological or material) that are obtained from it.

Of these two models, the DCS is the one that has been investigated the most with regard to organisational safety. For example, Hansez and Chmiel (2010) found that control and demand had differential effects on the commission of procedural violations, which was related in part to participants' beliefs in the level of managerial commitment to safety. Turner, Chmiel and Walls (2005) found that a high level of demand and low level of control (which, incidentally, typically predicts a high level of strain) was related to a low level of safety citizenship – the belief that safety improvement is part of one's job role. Conversely, Turner et al. (2012) found that control and social support are positively related to safety behaviour amongst hospital staff. In addition, social support moderates the negative effect of job demand on the safety ratings provided by rail workers (Turner et al. 2010). The ER model though, has, despite its theoretical plausibility, been the subject of comparatively fewer studies with regard to safety climate. Virtanen et al. (2009) used a measure of ER and found that the degree of mismatch between effort and reward (so-called 'effort-reward imbalance') predicted the rate of healthcare-acquired infections in a hospital ward. It would seem, then, that either one or both of these two models accounts for antecedents to patient safety and, in particular, influence patient safety climate.

Patient Safety and Community Pharmacies

The remainder of this chapter focuses on empirical work undertaken amongst community (retail) pharmacists who have been estimated to account for approximately 70 percent of registered pharmacists in Great Britain (Seston and Hassell 2008). Community pharmacies are privately owned businesses that are contracted to provide NHS services, including the dispensing of medication prescriptions. Like other settings, community pharmacies impose demands on the people who work in them: they have to maintain viability as a business as well as providing healthcare services. Therefore, community pharmacists' work can be subject to the job characteristics described in the previous section. For example, some concerns have been expressed about the workload under which pharmacy staff operate and the detrimental effect that this can have on safety (Phipps et al. 2009; Peterson et al. 1999).

In 2009 we conducted a cross-sectional safety climate survey of community pharmacists who were registered with the then professional regulator, the Royal Pharmaceutical Society of Great Britain (for further details about the study and its findings, see Phipps and Ashcroft (2011, 2012) and Phipps, Malley and Ashcroft (2012)). The survey used a sampling frame of 2,000 such pharmacists, selected at random from the register; this included respondents who were working in a variety of roles, ranging from owner/proprietors to locum (self-employed relief) pharmacists and in various sizes of organisation (from single-site independent pharmacies to large pharmacy chains). The survey included two self-report questionnaires on job characteristics. One was the Effort-Reward Imbalance Indicator (ERI; Siegrist et al. 2009), which measured:

- the respondent's perception of the effort required in his/her work ('effort')
- the material and psychological rewards gained ('reward'), which was made up of:
  - job security (expressed in the questionnaire as 'insecurity')
  - the esteem associated with one's role ('esteem')
  - promotion prospects ('promotion')
- the level of personal commitment given to work ('overcommitment')
- the ratio between effort and reward ('effort-reward ratio'), with scores greater than one indicating that effort exceeds reward (Lehr et al. 2010).

The other questionnaire was the Job Content Questionnaire (JCQ; Karasek et al. 1998). This measured the respondent's perceptions about:

- two aspects of work demand ('psychological demand' and 'physical demand')
- two aspects of job control ('skill decision' and 'decision authority')
- the support provided by coworkers ('support').

In addition to the ERI and JCQ, participants were asked to complete the Pharmacy Safety Climate Questionnaire (PSCQ; Phipps, de Bie et al. 2012). This measures respondents' perceptions of safety climate along four dimensions: 'organisational learning' (the pharmacy's willingness to engage in safety improvement); 'blame culture' (the propensity for pharmacy staff to be unfairly blamed for incidents); 'working conditions' (the extent to which the work environment supports safe working); and 'safety focus' (the extent to which safety is prioritised in day-to-day work). Participants were also invited to provide their own reflections about job characteristics and safety in their pharmacies, using a free-form text box; this gave us qualitative data to use alongside the quantitative data. Completed surveys were received from 860 participants (42.9 percent of the sampling frame) and the quantitative data from these were entered into SPSS version 15 for analysis. The demographic details of the sample and descriptive statistics from the questionnaire scores are reported in Phipps and Ashcroft (2011).

The ERI and JCQ scores were subjected to an exploratory factor analysis (EFA) to detect the presence of any common factors to be found amongst them. The factor analytical routine was maximum likelihood factor extraction with promax rotation and the eigenvalues greater than one criterion was used to determine the number of factors to be extracted. Bartlett's Test of Sphericity was statistically significant [$\chi^2(45)$ = 2647.79, p < .01] and the Kaiser-Meyer-Olkin Model of Sampling Adequacy was .83, indicating that an EFA could be conducted. Two factors were produced from the EFA, accounting for 56 percent of the variance in the dataset. These factors and their loadings are shown in Table 5.1. Factor 1 contains those scales that reflect the ability to cope with job demands: esteem; promotion; skill discretion; decision authority; coworker support. Factor 2 contains those scales that reflect job demands themselves: insecurity; effort; overcommitment; psychological demand; physical demand. In addition, insecurity and promotion have negative loadings on their opposing factors.

Table 5.1   Factor loadings for the job characteristics measures

Measure	Factor 1	Factor 2	Communality
Esteem	.58	–	.47
Insecurity	-.38	.32	.37
Promotion	.51	-.31	.51
Skill discretion	.83	–	.55
Decision authority	.78	–	.57
Coworker support	.41	–	.21
Effort	–	.83	.57
Overcommitment	–	.60	.38
Psychological demand	–	.85	.72
Physical demand	–	.40	.28
Eigenvalue at extraction	–	*4.04*	*1.56*
Percentage of variance accounted for	–	*34.97*	*11.28*

Note: Correlation between the factors = -.48.

In order to determine the relationship between these two factors and PSCQ scores, a canonical correlation analysis was carried out. Scores on the two factors were estimated for each respondent using the regression method. Two variate pairs were found in the analysis. The first had a correlation of .70 (accounting for 85.95 percent of the variance) and the second .37 (accounting for 14.05 percent). With both variate pairs included, the F ratio was statistically significant [$F(8,1416)$ = 88.62, p < .01, Wilks' λ = .44]; it was also significant when the first correlation was removed [$F(3,709) = 36.69$, $p < .01$, Wilks' $\lambda = .02$]. Therefore, both variate pairs accounted for a significant relationship between the factor scores and the PSCQ scales. Table 5.2 provides details about the pairs. The percentage of variance values indicates the extent to which each variate within the pair accounts for the 'set' of scales from which it is drawn (either the job characteristics or the safety climate scores). The percentage of redundancy values, meanwhile, indicates the extent to which each variate accounts for the other 'set' of scores. Across both pairs, the total variance accounted for in the job characteristics scales was 24.1 percent with 100 percent redundancy and the total variance accounted for in the safety climate scales was 77.46 percent with 31.35 percent redundancy. The variance accounted for by each variate pair indicates that the relation between the first pair is more substantial than that between the second and therefore the first pair was given greater emphasis in interpretation. Using a cutoff value of .30 for the correlation values in Table 5.2, all variables were correlated with the first pair.

Table 5.2   **Correlations, standardised coefficients and variance accounted for in the canonical variate pairs**

Measure	First pair		Second pair	
	Correlation	Coefficient	Correlation	Coefficient
Job characteristics				
Factor 1 (Coping)	.60	.95	.80	.54
Factor 2 (Demand)	-.50	-.87	.87	.65
Percent of variance	–	*14.73*	–	*9.37*
Percent of redundancy	–	*30.25*	–	*69.75*
Safety climate				
Organisational learning	.77	.31	.61	.92
Blame culture	-.70	-.15	-.20	-.10
Working conditions	.94	.69	-.32	-.99
Safety focus	.64	.01	.46	.23
Percent of variance	–	*59.40*	–	*18.06*
Percent of redundancy	–	*28.92*	–	*2.43*

The results indicate that a high level of coping and a low level of demand were associated with more favourable safety climate scores (high organisational learning, working conditions and safety focus and low blame culture). This is consistent with the view that a setting in which work demands are easily managed by staff is conducive to a strong safety climate. Interestingly, though, the second variate indicated that high levels of coping *and* demand were associated with less favourable scores on working conditions but more favourable scores on organisational learning and safety focus. This suggests the presence of a 'learning effect' with regard to safety climate: a challenging work environment can also foster a strong safety climate as long as staff have the personal and material resources to meet the challenge (see also Phipps, Malley et al. 2012).

The most obvious interpretation of the two job characteristics factors and their effects is that they are consistent with the job demands and resources approach suggested by Hansez and Chmiel (2010). Hence, the influence of both JCQ and ERI measures on safety can be reinterpreted in terms of a demand-resources framework (Phipps and Ashcroft 2011). The current findings suggest that both main effects and interactions could occur in the prediction of safety outcomes from these factors.

Combining the quantitative data with the qualitative data provided further insights into the relationship between job characteristics and patient safety (Phipps and Ashcroft 2012). Initially, we divided the respondents into groups using cluster analysis. We used a series of hierarchical and non-hierarchical clustering procedures to determine the number of clusters in the sample, given the JCQ, ERI and PSCQ scores of each participant. With each respondent assigned to a cluster, we then examined the qualitative data provided by respondents in each cluster to identify similarities and differences between them.

We identified four clusters from the questionnaire responses. One (which we labelled 'disenfranchising pharmacies') was characterised by a low level of control on the JCQ (little control over how they do their work), a mean effort-reward ratio of 1.36 on the ERI (effort exceeding reward), a low level of overcommitment on the JCQ (able to distance themselves from their work), a low level of organisational learning on the PSCQ (the pharmacy has little commitment to improving safety) and a low level of safety focus on the PSCQ (the pharmacy does not see safety as a priority). Examples of comments made by these respondents included:

> I no longer work for [X]. Having told the non-pharmacist manager that it was dangerous working in the pharmacy at the staff levels we had, I was told 'yes, dangerous for our bonus'.

> Locuming is the way to go [...] Just go in, do my job, come out and don't have to worry about running the shop.

The next cluster (which we labelled 'perilous pharmacies') had high demand (high psychological and physical demands), low control (little control over how they do their work) and low support (little support from colleagues) on the JCQ. It had a mean effort-reward ratio of 2.10 (effort is greater than reward) and high overcommitment (unable to distance themselves from their work) on the ERI. On the PSCQ, it had low organisational learning (the pharmacy had little commitment to improving safety), high blame culture (a tendency to single out individuals for blame), a low score for work characteristics (work characteristics are not conducive to safe working) and low safety focus (safety is not seen as a priority). Comments from these respondents included:

The workload is simply beyond a joke [...] [My employer] is now happy to pay compensation for mistakes rather than employ extra pharmacists because the former is cheaper.

Being a pharmacist in a busy branch is probably the worst job in the world and having to work at such high speed is scary. You go home worrying whether you have checked everything accurately.

The third cluster (which we labelled 'safety-focused pharmacies') had low demand (low psychological and physical demands), high control (much control over how they do their work), and high support (much support from colleagues) on the JCQ. The ERI had a mean effort-reward ratio of 1.15 (effort is greater than reward). The PSCQ scores showed high organisational learning (much commitment to improving safety), low blame culture (little tendency to single out individuals for blame), high work characteristics (work characteristics are conducive to safe working) and high safety focus (safety is seen as a priority). Comments from this cluster included:

We [...] try to provide the best service for our patients, this service is not motivated by profit margins!

I have excellent support from the company, a generous staffing profile and multiple pharmacist cover of good quality.

Finally were the respondents in the fourth cluster – the 'challenging pharmacies'. These were characterised by high demand (high psychological and physical demands), a mean effort-reward ratio of 1.63 (effort is greater than reward) and high over-commitment (unable to distance themselves from their work). Comments from these respondents included the following:

I have good staff but they and [I] are put under continuous pressure mainly from the company but also from the public.

Work demands and pressures are steadily increasing. Demands to perform [consultations], when one has insufficient time, and having to [do] CPD in one's own time, after a hard week's work leaves one feeling demoralised and demotivated.

The findings of our survey demonstrate the conceptual model outlined in the previous section: they show that safety climate can be understood in the context of job characteristics, which in turn are the product of organisational dynamics. For example, pharmacies that are considered to prioritise safety are the ones that ensure that pharmacy staff have sufficient resources to cope with their demands. Hence, a key issue to consider with regard to safety culture improvement is the nature of job roles, tasks, technology and organisational objectives (see also Sujan

2012; Yeung and Dixon-Woods 2010). With that in mind, we will conclude the chapter by briefly considering a method for eliciting and explaining such features.

Understanding the Pharmacy 'Work Domain'

Work domain analysis (WDA) comes from a suite of analytical methods, known collectively as cognitive work analysis (Vicente 1999). The basic premise of the approach is that work activity should be understood in terms of its 'constraints' – that is, the limitations and demands that the work environment places on behaviour due to its physical, social and psychological characteristics. Work domain analysis is used to identify the structural and functional relationships present within a work setting. These are presented in a multi-level map of the work setting (Figure 5.4) that links its highest level features (purposes and priorities) to its lowest level features (processes, tasks and physical objects).

In healthcare, work domain analysis was initially applied to the identification of user requirements for patient monitoring systems (e.g. Hajdukiewicz, Vicente, Doyle, Milgram and Burns 2001; Miller 2004). However, it has since been applied to the analysis of critical incidents in a given setting (Lim 2008; Lim, Anderson and Buckle 2008a, 2008b; see also Xu 2007). Its particular benefit in such an application is to show how a particular incident fits into the context of the work system within which it occurs; hence, it complements other methods, such as the London Protocol which deals more comprehensively with potential error- and violation-provoking factors but gives a more limited view of the work system under examination (Lim et al. 2008a, 2008b).

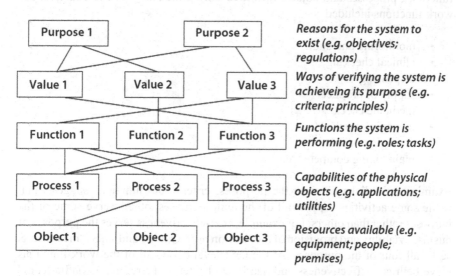

Figure 5.4   Different levels of the work setting in a WDA
Source: Adapted from Naikar et al. 2005.

In order to examine the systemic influences on patient safety in community pharmacies, we conducted a WDA of this setting (Phipps and Ashcroft 2010; Phipps, Tam and Ashcroft, in press. This used the method described in Vicente (1999) and Naikar, Hopcroft and Moylan (2005). Our data sources were documentary evidence from pharmacy training and design guidelines and observation of staff activity at various community pharmacies (including independent single-site pharmacies and branches of small and large pharmacy chains). From the WDA, we identified four main purposes of community pharmacy:

- business viability
- promoting health and clinical services
- provision of medication
- supporting the use of medication.

In order to achieve each of these purposes, pharmacy staff are required to work to a number of criteria:

- quality and safety
- cost-effectiveness
- legal standards
- ethical standards.

Meeting these criteria in turn depends on the effective use of a number of work functions, processes and objects identified during the analysis. For example, the work functions included:

- monitoring and audit
- clinical checking
- dispensing
- supply of products
- medicine check
- managing stock
- managing people
- maintaining competence.

Examining the links between the purposes, criteria and functions suggests that while some activities support all of the main purposes, others serve some of the purposes rather than others. For example, cost-effectiveness serves the purposes of business viability and provision of medication only. Quality and safety, meanwhile, serve all four of the purposes. At the next level down, all of the work functions serve both cost-effectiveness and quality and safety – therefore, a conflict could arise if the functions are being carried out in a manner that favours one of these criteria over the other.

The insights obtained from the WDA provide an alternative perspective on the notion of safety culture suggested in this chapter. As mentioned previously, some authors have defined safety culture as the product of staff perceptions about the organisation's priorities and how these affect their day-to-day work (Clarke 2010; Zohar 2010). In this view, safety climate reads across to the relationship between the priorities and values in the abstraction hierarchy and how these are respectively served by the lower-order elements. To take this point further, if a strong safety culture depends on the organisation possessing 'collective mindfulness' about the effect of its policies and practices on safety, as suggested by authors such as Sutcliffe (2011) and Hopkins (2002), then WDA would be beneficial in identifying the features and relationships of which the organisation should be mindful.

The philosophy behind work domain analysis is that within the constraints that are intrinsic to the work setting, practice may develop in any manner, depending on practitioner knowledge and local contingencies. Therefore, the instantiation of a work system is seen as dynamic and evolving (Vicente 1999). A similar view – expressed in the concept of emergence – is to be found in systems thinking approaches to organisational behaviour (Checkland 1981; White 1995). In practical terms, it highlights the need to consider how well the organisation's functions are served by design factors such as equipment, staffing, training and procedures – the very things that could comprise limiting conditions in the sense of the system model outlined earlier (in Figure 5.3; see also Moray 1994).

Summary

What lies behind the result of a patient safety culture assessment? In this chapter we have examined the notion of a 'safety culture' from a systems thinking perspective. In doing so, we depict it as the interpretation of the dynamic relationship between goals, priorities and activities within an organisation. Therefore it is not so much a fixed, essential quality as an 'ongoing achievement' between the organisational members (whether 'management' or 'frontline') in the context of the broader environment.

The goals, priorities and activities are manifest in job characteristics – those features of organisational life and work that directly affect its members. Therefore, one line of enquiry within safety culture research is how it is affected by job characteristics. We presented evidence to suggest that such a relationship exists within British community pharmacies. In basic terms, safety climate varies according to the levels of psychological and physical demands and the personal and material resources available to meet these demands.

Finally, we presented a method for analysing the 'work domain' within which healthcare activity takes place. This method can be used to identify the links between an organisation's goals, priorities, activities and physical features. In so doing, it provides a way of eliciting the factors that shape the formation and development of a safety culture and identifies parts of the organisation that can be altered in order to stimulate culture change.

References

Alper, S.J., Holden, R.J., Scanlon, M.C., Patel, N., Kaushal, R., Skibinski, K., Brown, R.L. and Karsh, B.-T. 2012. Self-reported violations during medication administration in two paediatric hospitals. *BMJ Quality and Safety*, 21, 408–15.

Argyris, C. and Schön, D.A. 1996. *Organizational Learning II: Theory, Method and Practice*. Wokingham: Addison-Wesley.

Buckle, P., Clarkson, P.J., Coleman, R., Lane, R., Stubbs, D., Ward, J., Jarrett, J. and Bound, J. 2003. *Design for Patient Safety: A System-wide Design-led Approach to Tackling Patient Safety in the NHS*. London: Department of Health/Design Council.

Carrillo, R.A. 2011. Complexity and safety. *Journal of Safety Research*, 42, 293–300.

Checkland, P.M. 1981. *Systems Thinking, Systems Practice*. Chichester: Wiley.

Clarke, S. 1999. Perceptions of organizational safety: Implications for the development of safety culture. *Journal of Organizational Behavior*, 20(2), 185–98.

Clarke, S. 2010. An integrative model of safety climate: Linking psychological climate and work attitudes to individual safety outcomes using meta-analysis. *Journal of Occupational and Organizational Psychology*, 83, 553–78.

Dixon-Woods, M., McNicol, S. and Martin, G. 2012. Ten challenges in improving quality in healthcare: Lessons from the Health Foundation's programme evaluations and relevant literature. *BMJ Quality and Safety*, 21, 876–84.

Fennell, D. 1988. *Investigation into the King's Cross Underground Fire*. London: HMSO.

Geysen, W. and Verbruggen, J. 2003. The acceptance of systemic thinking in various fields of technology and consequences on the respective safety philosophies. *Human Factors and Ergonomics in Manufacturing*, 13(3), 231–42.

Hajdukiewicz, J.R., Vicente, K.J., Doyle, D.J., Milgram, P. and Burns, C.M. 2001. Modelling a medical environment: An ontology for integrated medical informatics design. *International Journal of Medical Informatics*, 62, 79–99.

Hansez, I. and Chmiel, N. 2010. Safety behavior: Job demands, job resources and perceived management commitment to safety. *Journal of Occupational Health Psychology*, 15(3), 267–78.

Hopkins, A. 2002. *Safety Culture, Mindfulness and Safe Behaviour: Converging Ideas?* Working paper, National Research Centre for OHS Regulation, Australian National University.

Karasek, R., Brisson, C., Kawakami, N., Houtman, I., Bongers, P. and Amick, B. 1998. The job content questionnaire: An instrument for internationally comparative assessments of psychosocial job characteristics. *Journal of Occupational Health Psychology*, 3(4), 322–55.

Kontogiannis, T. 2012. Modeling patterns of breakdown (or archetypes) of human and organizational processes in accidents using system dynamics. *Safety Science*, 50(4), 931–44.

Lawton, R., Gardner, P., Green, B., Davey, C., Chamberlain, P., Phillips, P. and Hughes, G. 2009. An engineered solution to the maladministration of spinal injections. *Quality and Safety in Health Care*, 18, 492–5.

Lehr, D., Koch, S. and Hillert, A. 2010. Where is (im)balance? Necessity and construction of evaluated cut-off points for effort-reward imbalance and overcommitment. *Journal of Occupational and Organizational Psychology*, 83, 251–61.

Lim, R.H.M. 2008. *A Systems Approach to Medication Safety in Care Homes: Understanding the Medication System, Investigating Medication Errors and Identifying the Requirements of a Safe Medication System.* PhD thesis, University of Surrey.

Lim, R.H.M., Anderson, J. and Buckle, P. 2008a. Analysing care home medication errors using work domain analysis. In S. Hignett, B. Norris, K. Catchpole, A. Hutchinson and S. Tapley (eds), *IPS2008: Proceedings of the Improving Patient Safety Conference.* Loughborough: Ergonomics Society, 147–51.

Lim, R.H.M., Anderson, J. and Buckle, P. 2008b. *Analysing Care Home Medication Errors: A Comparison of the London Protocol and Work Domain Analysis.* Presented at the 52nd Human Factors and Ergonomics Society Annual Meeting, New York, 22–26 September.

Marais, K., Saleh, J.H. and Leveson, N.G. 2006. Archetypes for organizational safety. *Safety Science*, 44, 565–82.

McDonald, R., Waring, J. and Harrison, S. 2006. Rules, safety and the narrativisation of identity: A hospital operating theatre case study. *Sociology of Health and Illness*, 28(2), 178–202.

Moray, N. 1994. Error reduction as a systems problem. In M.S. Bogner (ed.), *Human Error in Medicine.* Hillsdale, NJ: Lawrence Erlbaum Associates, 67–91.

Naikar, N., Hopcroft, R. and Moylan, A. 2005. *Work Domain Analysis: Theoretical Concepts and Methodology.* Report DSTO-TR-1665, Defence Science and Technology Organisation.

Peterson, G.M., Wu, M.S.H. and Bergin, J.K. 1999. Pharmacists' attitudes towards dispensing errors: Their causes and prevention. *Journal of Clinical Pharmacy and Therapeutics*, 24, 57–71.

Phipps, D.L. and Ashcroft, D.M. 2010. What are the features of the community pharmacy work system? *International Journal of Pharmacy Practice*, 18 (Supplement 1), 42–3.

Phipps, D.L. and Ashcroft, D.M. 2011. Psychosocial influences on safety climate: evidence from community pharmacies. *BMJ Quality and Safety*, 20, 1062–8.

Phipps, D.L. and Ashcroft, D.M. 2012. An investigation of occupational subgroups with respect to patient safety culture. *Safety Science*, 50(5), 1290–98.

Phipps, D.L., de Bie, J., Herborg, H., Guerriero, M., Eickhoff, C., Fernandez-Llimos, F., Bouvy, M.L., Rossing, C., Mueller, U. and Ashcroft, D.M. 2012. Evaluation of the pharmacy safety climate questionnaire in European community pharmacies. *International Journal for Quality in Health Care*, 24(1), 16–22.

Phipps, D.L., Malley, C. and Ashcroft, D.M. 2012. Job characteristics and safety climate: The role of effort-reward and demand-control-support models. *Journal of Occupational Health Psychology*, 17(3), 279–89.

Phipps, D.L., Noyce, P.R., Parker, D. and Ashcroft, D.M. 2009. Medication safety in community pharmacy: A qualitative study of the sociotechnical context. *BMC Health Services Research*, 9, 158.

Phipps, D.L. and Parker, D. (in press). A naturalistic decision-making perspective on anaesthetists' rule-related behaviour. *Cognition, Technology and Work*. doi: 10.1007/s10111-014-0282-2.

Phipps, D.L., Parker, D., Meakin, G.H. and Beatty, P.C.W. 2010. Determinants of intention to deviate from clinical guidelines. *Ergonomics*, 53(3), 393–403.

Phipps, D.L., Parker, D., Pals, E.J.M., Meakin, G.H., Nsoedo, C. and Beatty, P.C.W. 2008. Identifying violation provoking factors in a healthcare setting. *Ergonomics*, 51(11), 1625–42.

Phipps, D.L., Tam, W.V. and Ashcroft, D.M. (in press). Integrating data from the UK National Reporting and Learning System with work domain analysis to understand patient safety incidents in community pharmacy. *Journal of Patient Safety*. doi: 10.1097/PTS.0000000000000090.

Rasmussen, J. 1997. Risk management in a dynamic society: A modelling problem. *Safety Science*, 27(2/3), 183–213.

Senge, P.M. 1990. *The Fifth Discipline: The Art and Science of the Learning Organization*. London: Random House.

Seston, L. and Hassell, K. 2008. *Pharmacy Workforce Census 2008: Main Findings*. London: Royal Pharmaceutical Society of Great Britain.

Shirom, A., Nirel, N. and Vinokur, A.D. 2006. Overload, autonomy and burnout as predictors of physicians' quality of care. *Journal of Occupational Health Psychology*, 11(4), 328–42.

Siegrist, J. 1996. Adverse health effects of high-effort/low-reward conditions. *Journal of Occupational Health Psychology*, 1(1), 27–41.

Siegrist, J., Wege, N., Pühlhofer, F. and Wahrendorf, M. 2009. A short generic measure of work stress in the era of globalization: Effort-reward imbalance. *International Archives of Occupational and Environmental Health*, 82, 1005–13.

Smith, A. and Alderson, P. 2012. Guidelines in anaesthesia: Support or constraint? *British Journal of Anaesthesia*, 109(1), 1–4.

Sujan, M.A. 2012. A novel tool for organisational learning and its impact on safety culture in a hospital dispensary. *Reliability Engineering and System Safety*, 101, 21–34.

Sutcliffe, K.M. 2011. High reliability organizations. *Best Practice and Research Clinical Anaesthesiology*, 25, 133–44.

Toft, B. 2001. *External Inquiry into the Adverse Incident that Occurred at Queen's Medical Centre, Nottingham, 4th January 2001*. London: Department of Health.

Turner, N., Chmiel, N., Hershcovis, M.S. and Walls, M. 2010. Life on the line: Job demands, perceived co-worker support for safety, and hazardous work events. *Journal of Occupational Health Psychology*, 15(4), 482–93.

Turner, N., Chmiel, N. and Walls, M. 2005. Railing for safety: Job demands, job control, and safety citizenship role definition. *Journal of Occupational Health Psychology*, 10(4), 504–12.

Turner, N., Stride, C.B., Carter, A.J., McCaughey, D. and Carroll, A.E. 2012. Job demands-control-support model and employee safety performance. *Accident Analysis and Prevention*, 45, 811–17.

van der Doef, M. and Maes, S. 1999. The job demand-control(-support) model and psychological well-being: A review of 20 years of empirical research. *Work and Stress*, 13(2), 87–114.

Vicente, K.J. 1999. *Cognitive Work Analysis: Toward Safe, Productive and Healthy Computer-based Work*. Mahwah, NJ: Lawrence Erlbaum Associates.

Virtanen, M., Kurvinen, T., Terho, K., Oksanen, T., Peltonen, R. Vahtera, J., Routamaa, M., Elovainio, M. and Kivimäki, M. 2009. Work hours, work stress, and collaboration among ward staff in relation to risk of hospital-associated infection among patients. *Medical Care*, 47, 310–18.

Waterson, P. 2009. A critical review of the systems approach within patient safety research. *Ergonomics*, 52(10), 1185–95.

White, D. 1995. Application of systems thinking to risk management: A review of the literature. *Management Decision*, 33(10), 35–45.

Xu, W. 2007. Identifying problems and generating recommendations for enhancing complex systems: Applying the abstraction hierarchy framework as an analytical tool. *Human Factors*, 49, 975–94.

Yeung, K. and Dixon-Woods, M. 2010. Design-based regulation and patient safety: A regulatory studies perspective. *Social Science and Medicine*, 71, 502–9.

Zohar, D. 2010. Thirty years of safety climate research: Reflections and future directions. *Accident Analysis and Prevention*, 42, 1517–22.

Chapter 6

Creating a Safety Culture:
Learning from Theory and Practice

Ruth McDonald and Justin Waring

Introduction

It is widely held that an organisation's performance is a reflection of its underlying culture. Although the concept of a 'culture' can be difficult to define, it is associated with the shared attitudes, values, norms and behaviours that shape individual, group and organisational activities (Schein 1988). These ideas have become particularly influential in many high-risk settings, where unsafe behaviours and systems are associated with catastrophic failures (International Atomic Energy Agency IAEA 1991). As such, policy-makers, regulators and safety-experts have called for high-risk organisations to foster shared attitudes, values and behaviours that continually assure operational quality and safety (Health and Safety Commission HSC 1993; IAEA 1991). Such calls are now evident in the healthcare sector, where evidence of service failure, iatrogenic illness and patient harm have necessitated a change in how clinicians and service leaders think about, respond to and prioritise quality and safety (Department of Health 2000). Quality improvement programmes, for example, routinely call for the creation of a new philosophy, mindset or culture that promotes quality and safety in clinical behaviours and care processes. This is reflected in the idea of a 'patient safety culture' that instils awareness of safety issues, openness about the potential threats to safety and shared values and behaviours that continually promote safety and learning (e.g. NPSA 2003).

Despite the apparent association between a safety culture and safe operational performance, there remains considerable debate both about the nature of a 'safety culture' and whether it is feasible to create or foster such a culture (Rowley and Waring 2012). This chapter explores the challenges of making culture and behaviour change in healthcare organisations. It provides a short review and critical commentary on the safety culture concept and deconstructs the idea of culture change by exploring different approaches to staff engagement and behaviour change. To elaborate its argument, the chapter focuses on the use of guidelines and incentive structures as two prominent examples of where policies seek to promote culture and behaviour change amongst healthcare professionals. Culture change is often necessary and, when successful, can be effective in promoting safety (Dixon-Woods et al. 2011). However, it remains a challenging process, especially in the healthcare context, given that this remains a highly complex, differentiated and political organisational setting.

Creating a Safety Culture

A prominent aspect of the patient safety movement has been to change how people think about and behave in relation to healthcare quality and safety. First, policies and research have worked to change how we think about safety and, importantly, the threats to patient safety (Department of Health 2000; Reason 2000). Drawing on theories of human error, human factors and ergonomic and safety science, it has consistently been argued that threats to safety are brought about, not by individual or group behaviours alone, but by the influence of upstream or latent factors that enable or condition the opportunities for active error. As such, efforts to improve patient safety should focus on the causal relationships between system-wide or root cause factors and front line clinical behaviours. This 'system thinking' has significant implications for how we should learn about and tackle the threats to patient safety through addressing, not individual behaviours at the sharp-end, but rather the influence of underlying root causes or systemic factors that are arguably experienced across many clinical and organisational settings. This 'systems thinking' is also important for challenging the 'blame culture' that is said to inhibit openness and communication about safety events. Specifically, clinicians too often feel they will be blamed for patient harm due to the lack of appreciation of upstream systemic factors which show how clinicians at the 'sharp-end' are often the victims of unsafe systems.

Second, policies have introduced procedures to enhance organisational learning and to control for potentially unsafe practices (NPSA 2003). These are often associated with the introduction of formal risk or knowledge management procedures that a) routinely capture information about threats to safety through forms of incident reporting, b) provide opportunities to investigate and identify the latent sources of these risks and c) inform the development and implementation of recommendations that promote system-wide safety improvement. Such procedures are clearly based upon the type of 'systems' thinking outlined above. This is important for challenging the 'blame culture' so that clinicians feel confident that, in reporting their experiences of unsafe care, these will engender learning and improvement, rather than disciplinary actions. This systems thinking is also integral to the investigation of safety events where service leaders undertake techniques – such as 'root cause analysis' – that are explicitly directed towards detecting the latent sources of risk and informing the implementation of system-wide safety improvements. Equally, when safety interventions are introduced, staff should recognise that these are concerned with optimising the safety of clinical systems and procedures, rather than questioning the inherent safety of an individual's practice. Here we find particular support for interventions that promote safer and more integrated team work, such as the Safe Surgery Saves Lives Checklist, or operational policies that simplify or standardise clinical communications or decision-making processes.

Third, and of primary interest in this chapter, these linked aspirations are often bound together in the idea of culture change or, more specifically, the creation of a 'safety culture' which is believed to be essential for guiding safer and more reflexive clinical behaviours (NPSA 2003; Parker et al. 2005; Pidgeon and O'Leary 2000). Clinicians are sometimes portrayed as defending their individual reputations, hoarding knowledge within occupational silos, adhering to professional norms around collegiality or self-regulation and advancing the interests of their profession over those of patients or the need for learning (Department of Health 2000). Changing how clinicians think about and act in relation to safety events, i.e. recognising the importance of systemic factors and participating in safety enhancing activities, is often therefore presented as a radical departure from more customary and institutionalised ways of responding to safety events. As with other conceptual and theoretical developments within the patient safety movement, the elaboration of the safety culture similar reflects advances in other industries.

Safety Culture – Conceptualisations and Critiques

The idea of a safety culture is much debated both within the wider field of safety science and in the specific context of healthcare. Without unnecessarily repeating other contributions in this collection, the concept of a safety culture is described in relation to a number of shared attributes and behaviours that are commonly associated with what are termed high reliability organisations, such as nuclear energy and petrochemicals (Flin et al. 2000; Parker et al. 2008; Pidgeon and O'Leary 2000). In broad terms, it is elaborated in relation to the values, attitudes and behaviours that characterise individual or group behaviour in relation to safety, especially their commitment to ensuring optimally safe operating practices. As such, organisations with a positive safety culture are associated with enhanced awareness and attention to safety issues, openness and communication about the potential threats to safety, shared values about the importance of safety and behaviours to promote learning and embrace safety enhancing interventions. These broad ideas are often elaborated, for example with reference to:

- 'situational awareness' or 'mindfulness to danger' – where staff are attentive to the potential risks in the work place
- individual and collective 'sense-making' – whereby staff recognise and understand the influence of upstream and system level factors, rather than individual fault alone
- 'communication and openness' – where staff feel confident and recognise the need to share their experiences of unsafe practice to promote problem-solving and learning

- 'valuing and prioritising' safety – where staff share the belief that safety is essential to operational practice
- 'learning and change' – where experience of unsafe practice provides the foundations for reflection, analysis and the development of safety enhancing interventions with tailored learning procedures
- 'proactive leadership' and 'organisational commitment' – so that the importance of safety is articulated in organisational strategy, performance expectations and practices.

In the context of patient safety these are often elaborated along a number of lines, although it is interesting to note that the different frameworks used by policy and research communities emphasise slightly different dimensions. For example, the Manchester Patient Safety Framework builds on theories that emphasise the importance of communication. According to the Agency for Healthcare Research and Quality, a patient safety culture is associated with a) acknowledging the high risk nature of the healthcare environment and having the determination to achieve consistently safe levels of performance, b) a blame-free environment that encourages practitioners to report their experiences of unsafe practice without fear of blame or disciplinary action, c) collaboration and teamwork across disciplinary and hierarchical boundaries that seek to develop robust and integrated safety solutions and d) organisational and leadership commitment to safety.

What we find therefore is that the idea of a safety culture is somewhat broad and arguably poorly defined (Guldenmund 2000, see also Chapter 2). In various ways, these frameworks and models emphasise the importance of cognition and interpretation, i.e. how we perceive and make sense of safety; norms and values, i.e. how we prioritise and show commitment to safety; and behaviours and practice, i.e. how we respond to safety events by reporting or embracing new safety interventions. Interestingly, when we look at how 'safety culture' has been operationalised in the form of safety surveys or safety tools we find an emphasis on espoused attitudes, beliefs and self-reported behaviours (Flin et al. 2000).

Without detracting from our main argument, a number of conceptual points of debate might be highlighted. The first is that most approaches see cultures as something that an organisation or workplace 'has' rather than something that they 'are'. In other words culture is developed as some ontologically distinct property of an organisation. The second, and linked to this, is that the attributes of a safety culture can be measured or assessed in the form of espoused attitudes and beliefs. This is reflected in the widespread use of surveys and psychometrics to measure the extent of a safety culture. The third is that these attitudes and, in turn, cultures can be changed along some form of linear spectrum of safer-less safe.

Another way to think about safety culture might be to start by setting aside, at least for a short time, the very idea of a safety culture and begin instead with the idea that complex organisations are characterised by a multitude of shared meanings, norms, values and beliefs that are manifest in and recursively recreated in various form of symbolic practice (Schein 1988). From this perspective it is difficult to see cultures as a resource or property of an organisation but rather they are (re-)constituted within day-to-day organisational life. They are integrally bound up with not only what we do, or how we see the world, but who we are and how we see ourselves, whether as individuals or occupations. Whilst cultures are open to investigation, analysis and indeed measurement – including espoused attitudes or beliefs, they are also manifest through a much broader range of tacit, taken for granted or non-cognitive actions, artefacts, materials and rituals. A rich history of anthropological and ethnographic research highlights how much of what characterises the underlying beliefs and meanings of a culture or community is manifest in symbolic practice, ceremonies and rituals that are often poorly articulated in the spoken word. As such, it becomes difficult to isolate and measure all those aspects of culture that shape the safety or quality of what people do or do not do. Many aspects of these cultures will certainly relate to what we think of as safety or safe behaviours, but there is arguably no such single 'thing' as a safety culture. Rather communities possess a range of different values, beliefs, meanings and behaviours, that researchers may associated with 'safety' but that members associate with 'being'. This then casts doubt on the ease to which cultures and the behaviours they inform can be changed. Furthermore, cultures do not exist in a vacuum but are closely linked to our sense of self, group belonging and the wider socio-political context of organisational life. This highlights how cultures can be resistant to processes of change or modification, especially where change is driven by outside groups. In other words questions of resource and politics, or who has power and influence, are closely bound up with culture. Bearing in mind these points we now examine some of the prominent drivers of culture and behaviour change in relation to clinical quality and safety.

Changing and Controlling Behaviours

As suggested above, safety related cultures and behaviours can be difficult to conceptualise and understand which, in turn, can make culture or behaviour change difficult. By way of an introduction, and with the explicit intention of stimulating debate around the idea of behaviour change, we first turn to the seminal work of Etzioni (1961). In thinking about the dynamic influence of organisational power, Etzioni delineated three base forms of influence:

- Coercive – force individuals to comply, including the use of punishments (sticks)
- Remunerative – reward and pay individuals to comply (carrot when new money or stick when withholding part of existing payment)
- Normative – use symbolic values and norms to encourage individuals to accept and believe in compliance (ideas).

Interestingly, these broad approaches have different effects upon individual or staff engagement:

- Coercive change promotes limited commitment and fosters alienation and detachment
- Remunerative change promotes calculative commitment where engagement is only maintained where rewards outweigh costs
- Normative approaches promote high levels of moral and ideological alignment and commitment.

This well-established framework highlights some of the benefits and pitfalls of seeking to change behaviours. Building on these ideas, we identify five common approaches to changing behaviours and cultures within the healthcare context.

'Hard' Technological and Design Interventions

One way to change or alter behaviours is through the use of 'hard' or 'technology' interventions that force or encourage certain forms of practice. These are often developed through identifying recurrent and routine behaviours that directly or indirectly undermine safety and then developing appropriate interventions in advance of these behaviours that limit or restrict the potential for unsafe behaviour. Classical examples are computer interface prompts that require confirmation before proceeding, or automatically prevent behaviours that are known to result in harm. In the healthcare context, this might be a warning on a prescribing doctor's computer that identifies potential medication-related errors for particular patient types, i.e. those with penicillin allergies. In other instances hard technological or design solutions can simply make it impossible for unsafe practices to occur, such as the design of needles and syringes for different types of drugs to prevent the inappropriate administration of potential toxic medicine. Less extreme examples include alerts or alarms that indicate the potential for danger, thereby giving the operator sufficient time to adjust behaviours. Although these can be highly effective for promoting safety, it is arguably the case that their impact on behaviour change is rather shallow or limited. By this we mean, they rarely engage with individual's deep understanding or values about why certain behaviours are unsafe, but simply prevent or discourage unsafe practices from occurring. As such, the extent of change might only be limited to a particular intervention or behaviour, without fostering deeper culture change. Moreover, there is growing evidence of ways in which users find 'work-arounds' for behaviour limiting technologies or routinely over-ride safety alerts.

Regulatory, Rule-based and Prescriptive Interventions

Another way to direct and change behaviours is through prescriptive rules or instructions. Here our understanding of rules is relatively broad, and includes formal guidelines and requirements, but also more taken-for-granted conventions and customs that shape social behaviour. Changing the rules or expectations that inform behaviour clearly necessitates some forms of appreciation of the rewards and penalties that motivate compliance, which are discussed below, but rule-based interventions have become prominent in their own right and often with little appreciation of incentives. In the area of clinical quality and safety these are particular evident in the form of evidence-based guidelines and standard operating procedures or checklists. Over the last two decades, these have become the *modus operandi* for promoting higher and more consistent levels of clinical practice (Timmermans and Berg 2003) and exemplify attempts, especially in the area of patient safety, to standardise and simplify high-risk activities. Gawande's Checklist Manifesto is a prominent example of this trend and, with the Safe Surgery Saves Life Checklist, demonstrates how standardised and directive checking procedures can direct and, over time, recast group behaviours. Again, these slightly softer behaviour change interventions rarely engage with underlying motives and beliefs about safety, but rather aim to prescribe behaviours through step-by-step or formulaic instruction or checking. As above, there is growing evidence that such guidelines are difficult to implement because they run counter to the prevailing customs, cultures and identities of clinicians and result in limited or superficial compliance (McDonald et al. 2006).

Incentives and Calculative Interventions

Linking financial incentives to policy goals (such as improving safety and quality) is becoming increasingly popular in health settings. Much of the traditional neoclassical economics literature on incentives suggest that those on the receiving end of incentives ('agents') derive their utility solely from the money income attached to incentives and disutility from the effort exerted on behalf of the payer ('principal'). More recently, some economists have approached the issue of incentives from a perspective which recognises that agents may get their utility from things other than money (e.g. agents may trade off potential gains in income for more leisure time). There is a growing literature examining public service motivation as opposed to motivation more generally (Perry and Hondeghem 2008), looking at the importance of non-monetary incentives, i.e. where agents share 'some idealistic or ethical purpose' (Dixit 2002) served by the organisation in which they work. For example, Julian Le Grand's (2003) theory of public motivation classifies behaviours and responses to incentives as either knavish or knightly. The former concerns the pursuit of self-interest. Knaves may be motivated to help others, but only if it serves their own interests. In contrast, knights are motivated to help others, without the prospect of personal gain and may even do so in circumstances where this is to the detriment of their personal interests. Le Grand's analysis emphasises the need to design incentive structures that 'align knightly and knavish motivations in a fashion that directs the individuals concerned towards producing the desired outcomes' (Le Grand 2003: 155).

Economic theory also suggests that the recipient of an incentive must be compensated for the additional cost of undertaking the extra work required to hit a target. Whilst, this might imply that money is a key motivator, it would seem reasonable to assume that if clinicians incur additional costs as a result of undertaking extra work, they will seek reimbursement of these costs. The extent to which clinicians undertake calculations of incremental costs and benefits when responding to incentives is uncertain. However, evidence suggests that they do not necessarily do so (Spooner et al. 2001). Additionally, the literature on loss aversion (Kahneman 2011; Kahneman and Tversky 1979; Tversky and Kahneman 1991) suggests that people prefer avoiding losses to making gains, which might suggest that clinicians are keen to avoid a loss, even if they are required to work harder to do so. There is also a growing psychological literature concerning intrinsic motivation, or the desire to do something for its own sake, as opposed to responding to some external incentive. This evidence suggests that the provision of performance-contingent rewards may undermine or 'crowd-out' intrinsic motivation (Deci, Koestner and Ryan 1999). Whilst individuals may respond to financial incentives in the short-run, these may be negative reinforcers in the long-run, since they may conflict with intrinsic motivation (the individual's desire to perform a task for its own sake) by signalling to the individual that they are not trusted to perform in the absence of inducements. Feelings of competence and autonomy appear to be important for intrinsic motivation. Feedback is also important. Positive feedback is seen as facilitating intrinsic motivation by promoting a sense of competence. Negative feedback, can reduce individuals' perceptions of their competence, leaving them feeling demotivated and reducing both intrinsic and extrinsic motivation (Gagne and Deci 2005). The fact that the evidence supporting the use of financial incentives in healthcare settings is at best equivocal and the potential for changes in incentive structures to have unintended consequences means that great care should be exercised when designing and implementing such initiatives.

Normative Engagement

Our final approaches relate to efforts to promote what we see as the deeper and softer forms of normative engagement. These efforts at culture and behaviour change seek to address the deeply held values, beliefs and interpretative frameworks that influence shared meanings and behaviours. Arguably these are the most intractable aspects of culture to both understand and change but, where change is made, it is likely to be most lasting and significant. In the context of patient safety, these types of change are often associated with raising awareness and understanding about the nature and causes of the threats to patient safety. In particular, media campaigns, professional publications and expert reports illustrate ways in which new ways of thinking have been communicated and legitimised in professional communities. However, the reach and relevance of these professional beliefs and behaviours remains unclear. Other, more direct, examples include forms of training and staff engagement. These often follow the examples of Crew Resource Management training developed in the aviation sector seeking to raise awareness about the threats to safety and challenge potentially unsafe behaviours. Similarly, culture tools such as the Manchester Patient Safety Framework are ostensibly aimed at fostering reflection, inter-group communication, shared learning and culture change amongst clinical teams, rather than simply providing a basis for measuring the extent of a safety culture. More recently, efforts to promote changed thinking and behaviours in relation to safety have been developed at the level of professional training and socialisation, so the clinicians acquire and learn safer behaviours from their early years of training. These approaches promise a deeper and more lasting change in culture and behaviours, but arguably are more difficult to enact and evidence. In particular, the process of making change at this level can take many years of intervention and sharing practices as change becomes institutionalised.

Changing Culture – Rules and Guidelines in the Operating Theatre

Despite the benefits of normative engagement, the rule-based approach to patient safety has achieved the status of accepted wisdom in policy circles. Our research, carried out in a hospital operating theatre setting, found little support for such approaches from the doctors in our study (McDonald et al. 2006). Whilst managers placed great faith in written rules and guidelines, these threatened surgeons' identities since the latter defined professionalism as the ability to practise without guidelines. Rather than creating a 'safety culture', attempts to apply rules and standardisation were resisted by surgeons who saw these as likely to undermine their autonomy and ability to practise medicine according to their own individual view of what works best. The surgeons in our study did not question variations in practice; indeed they saw this as the norm. Their narratives stressed variability and unpredictability and rejected the idea of routinised behaviour specified in guidelines. Although they complained about inadequate resources and disruption caused by non-medical staff, they were prepared to work in these circumstances, even though this increased the risk to the patient. To refuse to work because of disruption to the list or inadequate resources might suggest that an individual did not have the 'right stuff' to be a doctor (Bosk 1979). Indeed, one of the unwritten rules appears to be a willingness to react to whatever is thrown at one, although it is perfectly acceptable to complain about 'external factors' in interviews. Indeed, the existence of less than perfect (or even unsafe) circumstances means that the doctors can claim even greater credit for a successful outcome, since this is testimony to their technical skills and ability to cope under pressure.

Looking at the situated behaviours and practices of surgeons, rather than their espoused attitudes, further reveals the underlying meanings and values attached to the uncertainty and complexity of surgical practice. In particular, we found surgeons often responded to potential risk or uncertainty in three shared and largely tacit ways (Waring et al. 2007). First, they would *tolerate* a degree of uncertainty and risk given that, as outlined above, it is an inevitable feature of surgical practice. Second, they would *accommodate* certain risks through making slight alterations to normal practice so as to absorb the risk. Third they would *innovate* in their practices through trying out new approaches or solutions to maintain patient safety whilst managing the risk. These ritualistic behaviours show that, for doctors, there are often unwritten conventions and rules that guide and, importantly, reinforce what are believed to be 'professional' standards and behaviours, especially the importance of coping in a crisis.

With regard to written rules, doctors' accounts suggested that the rule is 'there are no rules'. Pope (2002) suggests that the 'contingent' nature of surgery makes it difficult to codify and specify in advance but that this may have been specific to the particular and non-routine aspect of the incontinence surgery she studied. Our study suggests that doctors from a range of specialties describe their work as being non-routine. At the same time, the importance of routine and the suggestion that, over time, experienced doctors internalise this routine is used as a rationale for not following guidelines. However, if each doctor develops their own routine and if each operation is unpredictable, this may mean that medical staff are not convinced of the potential to learn from other people's mistakes. The fact that previous mistakes have occurred elsewhere and in different sets of circumstances, together with the acceptance that mistakes are inevitable, may also mean that medical staff are unwilling to report errors. If mistakes are seen as inevitable and as a matter of bad luck arising from a particular set of circumstances, then this implies that attempting to learn from other people's mistakes (and by implication reporting those mistakes) is not regarded as a valuable exercise.

Changing Culture – Financial Incentives in English Primary Medical Care

The Quality and Outcomes Framework (QOF) introduced in 2004 in the NHS combines guideline driven care with financial incentives, technology and software developed specifically to make QOF service delivery and data collection relatively easy (Roland 2004). The fact that performance data for each primary medical care practice are in the public domain can be seen as an attempt to influence norms within the GP community. QOF was part of a package of reforms which resulted in increased investment in primary medical care. Although GP partners benefitted from a large pay increase as a result of participating in QOF, there was also increased staffing in general medical practices. This, combined with other factors such as the technology developed specifically to facilitate QOF related care delivery, meant that the organisational context enabled staff to respond in the desired way (McDonald et al. 2010).

QOF was implemented in a context where guidelines and electronic medical records had increasingly been part of the landscape (Checkland 2004). Furthermore, in contrast to the surgeons we studied, GPs are generalists whose work covers a broad range of medical conditions. These factors meant that QOF was not seen as a threat to values and identities and the negotiation of the content of QOF with members of the profession, combined with the fact that QOF was voluntary and its introduction proceeded only after support from a majority of GPs in a national ballot, all suggest that QOF was seen as legitimate. Most, but not all QOF indicators built on existing practice. One exception to this related to mental health screening and could be interpreted as threatening GP identities. The formalisation of screening and the requirement to use a specific screening instrument seems to suggest that the informal approaches used by GPs to diagnose and manage depression were deficient (Dowrick et al. 2009). However, since this was one small element of QOF, for the most part, GPs have complied. The fact that patients support the use of this measure (Dowrick et al. 2009) may add to its legitimacy. In addition, the widespread acceptance of QOF and an increasing acceptance that practice legitimacy is linked to high QOF scores (McDonald et al. 2010) means that practices may comply, even where practitioners dislike this indicator.

QOF is also underpinned by staff engagement since, although indicators are nationally specified, practices are free to decide how to organise and deliver care to meet them. Giving team members a voice, combined with regular feedback on performance, has served to motivate clinicians. Furthermore, the practice owns the data which are used to assess performance, which means that they are familiar with it and produce the data set as part of their everyday practice. However, the distribution of rewards is an important factor here and some staff are demotivated by what they perceive as a mismatch between effort and reward. Although they may see the indicators as fair, they are less happy with remuneration arrangements which have afforded GP partners large pay increases whilst nurses and salaried doctors do not share similar benefits (McDonald et al. 2010).

A key component of the 2004 policy involved a move away from contracts with individual doctors to a shared contract (and collective responsibility) which requires practice staff to work in a more 'joined up' way. Indications are that QOF has changed the culture of general practice but whether these changes are wholly for the better is the matter of some debate. Furthermore, QOF has come at a substantial financial cost and it is not clear that the benefits are commensurate with these costs (National Audit Office 2008).

Assessing the Likely Impact of Quality and Safety Initiatives

Whilst it is important to assess the costs of initiatives to improve quality and safety, assessment of cost effectiveness may be difficult if 'safety culture' is a proxy for a panoply of assumed benefits. Economic models have been used to estimate benefits from quality and safety improvement programmes (Walker et al. 2010), and these can tend to 'black box' explanations of how intervention 'A' results in impact 'B'. In the English NHS in recent years the requirement to produce Impact Assessments as part of the policy development process is a welcome development. Impact Assessments specify how policies are expected to work. The IA for CQUIN, a scheme which is intended to support a 'cultural shift by embedding quality improvement and innovation as part of the commissioner-provider discussion everywhere' (Department of Health 2008) suggests that this is to be achieved by increasing the focus of healthcare organisations' boards on quality and safety by linking these to financial rewards, as well as engaging care providers by involving them in discussions about CQUIN goals.

These goals relate primarily to safety, effectiveness, experience or innovation. The combination of rewards, measurement and a greater focus on quality, as opposed to costs and activity, is intended to enhance relationships between commissioners and providers and to improve care provision. The Department of Health considered but rejected a nationally mandated scheme content, preferring to let local decision makers and clinical teams agree goals. This desire to harness the enthusiasm of local clinical teams can be seen as an attempt at normative engagement. In theory at least, this should help to ensure legitimacy but evidence suggests that, in practice, this is not working as intended.

An emphasis on 'stretch' targets (or goals) means that new CQUIN schemes are negotiated each year in an attempt to extract improvements from fixed resources. This contributes to a heavy workload in the run up to the next financial year which leaves little time for ongoing meaningful engagement with staff. In some cases CQUIN goals are not agreed until several months after the year in which they take effect has commenced. Furthermore, far from being engaged, staff are often handed a set of CQUIN goals and asked to comply (McDonald et al. 2013).

In contrast to QOF, there has not been a substantial investment in CQUIN rewards. Instead staff members feel that they are being asked to do more to earn what would previously have been an automatic entitlement to income for the organisation. There is also the issue that, once financial incentives are withdrawn, behaviour change does not endure (McMurran et al. 2013) highlighting the fact that, unlike the impact of QOF, changes in behaviour are not becoming institutionalised.

Since the scheme was introduced in 2009, the number of national goals has increased (for 2012/13 the NHS Safety Thermometer was included) and the magnitude of reward has also increased (up from 1.5 percent to 2.5 percent of total contract value). These developments may be interpreted as indicative that policy makers are keen to see more central direction and that an increase in the size of rewards (or penalties) will improve performance. It is difficult to see how such measures will overcome the problems of engagement highlighted by our research and they appear to clash with the theory of change outlined in the IA.

The extent to which attempts to improve quality and safety are successful is likely to depend on a number of factors. Consideration of these in the context of theories of change should enable assessment of the likely impacts of policies intended to improve quality and safety. One of these factors is the extent to which the messages are clear and targeted properly. Awareness as a precondition for action may seem an obvious point but our research in the USA and England (McDonald 2012; McDonald et al. 2013) suggests that front line clinicians are often unaware of, or at best ill-informed about, the content of culture change initiatives which target their behaviour and attitudes. Furthermore, adopting an 'enlightenment' approach which treats individuals as empty vessels to be filled with knowledge, is unlikely to succeed, as our operating theatres study illustrates. This is because responses to quality and safety initiatives are also likely to depend on the extent to which the initiatives are seen as fair and legitimate and, linked to this, whether they are seen as threatening cherished norms, values and identities.

Whilst much has been written about professional resistance to initiatives (much of it by ourselves), it would be wrong to assume that overcoming this or aligning initiatives to existing norms will guarantee successful implementation. Other important factors here include the extent to which professionals are able to respond in the desired way and, linked to this, the extent to which the organisational setting provides the capacity to respond as intended. Incentivising quality and safety improvements in Californian primary medical care did not lead to the introduction of a safety culture, in part because the organisational setting made it difficult for doctors to participate. Unlike their English counterparts, the US practices did not routinely have access to computerised medical records, guidelines and patient registers. Combined with poor communication about the content of good practice guidelines and related quality and safety targets, this made it difficult for doctors to respond in the way required. Similarly, due to budget constraints, the surgeons we studied were required to use outdated and relatively unsafe equipment which made it impossible to conform to best practice standards.

As the context in which 'culture change' initiatives are introduced is often replete with other initiatives, consideration must also be given to the 'fit' with these existing attempts to change behaviours and attitudes. We found that initiatives to reduce hospital acquired infections clashed with a policy to respond in a timely and systematic manner to pneumonia patients, since the latter placed great emphasis on blood cultures to provide information on diagnosis and subsequent treatment (McDonald et al. 2012). These problems were not insurmountable, but could have been minimised if thought had been given to these issues prior to implementation.

A further factor concerns the extent to which the desired result is subject to significant influence by the individuals or groups whose behaviour is targeted. The US doctors we studied were unable to exclude from performance calculations patients who refused treatment. Making doctors accountable for circumstances beyond their control in this way contrasts with the QOF, which permits English GPs to use their clinical judgement to remove inappropriate patients from achievement calculations, a process known as 'exception reporting'. Furthermore, the fact that practice incomes have not significantly increased in California (unlike in England) following the introduction of the financial incentive scheme may also account for the differences in attitudes between the two groups.

Finally, since evaluation of progress and promotion of learning requires data collection and reporting, the extent to which measures are perceived as valid and data as accurate is likely to influence professional responses with regard to both data recording and behaviour change. In California, the use of data from third parties (as opposed to the medical practices' own performance data) contributed to disengagement. Combined with the refusal to allow exception reporting, this conveys to doctors that they are not trusted, which is likely to undermine their engagement further. Data accuracy is linked to perceptions of legitimacy but also informs feedback, which is a key aspect of culture change initiatives (Dixon-Woods et al. 2011). Praise may increase motivation but the relationship between feedback and performance is complex and feedback can actually reduce performance (Kluger and DeNisi 1996). Praise and verbal feedback have been found to be less effective as 'reinforcers' than computerised feedback that was task focused, feedback that created a clear feedback-standard discrepancy at the task level and feedback that supplied the correct solution at the task level (such as computerised prompts for QOF actions). All of which highlights the importance of data accuracy and validity prior to implementing 'culture change' policies.

Conclusion

Our chapter intentionally offers a critical commentary on the concept of a safety culture and, more specifically, the processes of making culture and behaviour change. Through the examples of safety-enhancing guidelines and quality-based incentives systems, we show that changing behaviours and cultures is often difficult, because cultures are themselves complex and deeply embedded within cognitive, normative and regulatory institutions. Whilst it might be possible to change the 'rules of the game' or 'limit behaviours', this does not mean that underlying meanings, values and tacit assumptions will be changed. Classic approaches to organisational behaviour and control show that coercive and calculative techniques often struggle to engage with the underlying ideologies and norms that shape collective practices.

Furthermore, this chapter draws on studies which employ in-depth qualitative analysis: the fact that such studies provide (implicit) critiques of simplistic assumptions about 'culture change' highlights the importance of going beyond broad surveys. These often fail to engage with the day to day messy and complex realities of organisational life. Placing reliance on their findings can be likened to getting a 'feel for' a country's ways of life by flying over it in an aeroplane.

We are not suggesting that culture change is impossible or should not be undertaken; only that we need to think more deeply about what this involves. As research shows, attempts to change culture can be effective (Dixon-Woods et al. 2011), but this is not always the case. Since such initiatives can be costly (in financial and emotional terms), care needs to be taken to ensure that they are designed and implemented after careful elaboration of theory and assessment of likely impact. This is particularly important given the potential unintended consequences of attempts at 'culture change', which may result in damage to relationships and morale (McDonald et al. 2005) and harm to patients.

References

Bosk, C. 1979. *Forgive and Remember*. Chicago: University of Chicago Press.
Checkland, K. 2004. National Service Frameworks and UK general practitioners: Street-level bureaucrats at work? *Sociology of Health & Illness*, 26(7), 951–75.
Deci, E.L., Koestner, R. and Ryan, R.M. 1999. A meta-analytic review of experiments examining the effects of extrinsic rewards on intrinsic motivation. *Psychological Bulletin*, 125(6), 627–68.
Department of Health. 2000. *An Organisation with a Memory*. London: TSO.
Department of Health. 2008. *Impact Assessment of the Commissioning for Quality and Innovation Payment Framework*. Available at: http://www.dh.gov.uk/prod_consum_dh/groups/dh_digitalassets/@dh/@en/documents/digitalasset/dh_091690.pdf (last accessed on 28 July 2012).
Dixit, A. 2002. Incentives and organizations in the public sector: An interpretative review. *Journal of Human Resources*, 37, 696–727.
Dixon-Woods, M., Bosk, C., Aveling, E.L., Goeschel, C.A. and Provonost, P.J. 2011. Explaining Michigan: Developing an ex post theory of a quality improvement program. *Milbank Quarterly*, 89, 167–205.
Dowrick, C., Leydon, G., McBride, A., Howe, A., Burgess, H., Clarke, P., Maisey, S. and Kendrick, T. 2009. Patients' and doctors' views on depression severity questionnaires incentivised in UK quality and outcomes framework: Qualitative study. *British Medical Journal*, 338, b663.
Etzioni, A. 1961. *Comparative Analysis of Complex Organizations*. New York; Free Press.
Flin, R., Mearns, K., O'Conner, P. and Bryden, R. 2000. Measuring safety climate: Identifying the common features. *Safety Science*, 34, 177–92.
Gagné, M. and Deci, E. 2005. Self determination theory and work motivation. *Journal of Organizational Behaviour*, 26, 331–62.
Gawande, A. 2010. *The Checklist Manifesto*. London: Penguin.
Guldenmund, F.W. 2000. The nature of safety culture: A review of theory and research. *Safety Science*, 34, 215–57.
HSC (Health and Safety Commission) 1993. *Third Report: Organizing for Safety*. ACSNI Study Group on Human Factors. London: HMSO.
IAEA 1991. *Safety Culture* (Safety Series No. 75-INSAG-4). International Atomic Energy Agency.
Kahneman, D. 2011. *Thinking Fast and Slow*. London: Allen Lane.

Kahneman, D. and Tversky, A. 1979. Prospect theory: An analysis of decision under risk. *Econometrica*, 47, 263–91.

Kluger, A. and Denisi, A. 1996. The effects of feedback interventions on performance: A historical review, a meta-analysis, and a preliminary feedback intervention theory. *Psychological Bulletin*, 119, 254–84.

Le Grand, J. 2003. *Motivation, Agency, and Public Policy: Of Knights and Knaves, Pawns and Queens.* Oxford: Oxford University Press.

McDonald, R. 2012. Restratification revisited: The changing landscape of primary medical care in England and California. *Current Sociology*, 60, 441–55.

McDonald, R., Cheraghi-Sohi, S., Tickle, M., Roland, M., Doran, T. and Campbell, S. 2010. *The Impact of Incentives on the Behaviour and Performance of Primary Care Professionals.* Report for the National Institute for Health Research Service Delivery and Organisation programme. Available at: http://www.netscc.ac.uk/ hsdr/files/project/SDO_FR_08-1618-158_V06.pdf (last accessed on 17 April 2013).

McDonald, R., Sutton, M., Boaden, R., Lester, H. and Roland, M. 2012. *Evaluation of the Advancing Quality Pay for Performance Programme in the NHS North West.* Report for the National Institute for Health Research Service Delivery and Organisation programme. Available at: http://www.nets.nihr.ac.uk/ projects/hsdr/081809250 (last accessed on 11 April 2014).

McDonald, R., Sutton, M., Kristensen, S., Zaidi, S., Todd, S., Koneth, F., Hussein, K. and Brown, S. 2013. *The Implementation and Impact of the Commissioning for Quality and Innovation (CQUIN) Payment Framework: Final Report.* Available at: http://www.nottingham.ac.uk/business/ documents/news-documents/evaluation-of-the-commissioning-for-quality- and-innovation-framework---final-report---feb-2013 (last accessed on 17 April 2013).

McDonald, R., Waring, J. and Harrison, S. 2006. Rules, safety and the narrativization of identity: A hospital operating theatre case study. *Sociology of Health & Illness*, 28(2), 178–202.

McDonald, R., Waring, J., Harrison, S., Walshe, K. and Boaden, R. 2005. Rules and guidelines in clinical practice: A qualitative study in operating theatres of doctors' and nurses' views. *Quality and Safety in Health Care*, 14, 290–94.

McMurran, M., Robertson, E., Coffey, F. and Miller, P. 2013. The profile of risky single occasion drinkers presenting at an emergency department. *Journal of Substance Misuse*, 18(6), 484–91.

National Audit Office 2008. *NHS Pay Modernisation: New Contracts for General Practice Services in England.* London: The Stationery Office.

National Patient Safety Agency. 2003. *Seven Steps to Patient Safety.* London: NPSA.

Parker, D., Lawrie, M. and Hudson, P. 2005. A framework for understanding the development of organisational safety culture. *Safety Science*, 44, 551–62.

Perry, J. and Hondeghem, A. 2008. *Motivation in Public Management.* Oxford: Oxford University Press.

Pidgeon, N. and O'Leary, M. 2000. Man-made disasters: Why technology and organizations (sometimes) fail. *Safety Science*, 34, 15–30.

Pope, C. 2002. Contingency in surgical work and some implications for evidence based surgery. *Sociology of Health and Illness*, 24(4), 369–84.

Reason, J. 1998. Achieving a safe culture: Theory and practice. *Work and Stress*, 12, 293–306.

Reason, J. 2000. Human error: Models and management. *British Medical Journal*, 320, 728–70.

Reason, J., Parker, D. and Lawton, R. 1998. Organizational controls and safety: The varieties of rule-related behaviour. *Journal of Occupational and Organizational Psychology*, 71, 289–304.

Roland, M. 2004. Linking physician pay to quality of care: A major experiment in the UK. *New England Journal of Medicine*, 351, 1448–54.

Rowley, E. and Waring, J. 2012. *A Socio-cultural Perspective on Patient Safety*. Farnham: Ashgate.

Schein, E.H. 1988. *Organizational Culture*. Cambridge: MIT Press.

Spooner, A., Chapple, A. and Roland, M. 2001. What makes British general practitioners take part in a quality improvement scheme? *Journal of Health Services Research and Policy*, 6, 145–50.

Timmermans, S. and Berg, M. 2003. *The Gold Standard*. London: Temple.

Tversky, A. and Kahneman, D. 1991. Loss aversion in riskless choice: A reference dependent model. *Quarterly Journal of Economics*, 106, 1039–61.

Walker, S., Mason, A.R., Claxton, K., Cookson, R., Fenwick, E., Fleetcroft, R. and Sculpher, M. 2010. Value for money and the quality and outcomes framework in primary care in the UK NHS. *British Journal of General Practice*, 60(574), 352–7.

Waring, J., Harrison, S. and McDonald, R. 2007. A culture of safety or coping: Ritualistic behaviours in the operating theatre. *Journal of Health Services Research and Policy*, 12, Suppl. 1, 3–9.

Williamson, A., Feyer, A.-M., Cairns, D. and Biancotti, D. 1997. The development of a measure of safety climate: The role of safety perception and attitudes. *Safety Science*, 25, 15–27.

PART II
Methods and Tools

Chapter 7
Improving Safety Culture in Healthcare Organisations

Jane Carthey

Introduction

Much has been written about the need for healthcare organisations to create a positive safety culture (Carthey and Clarke 2009; Department of Health 2000, 2001; National Patient Safety Agency 2004; Reason 2000). In essence, when national agencies like the former UK National Patient Safety Agency and the new National Health Service (NHS) Commissioning Board describe wanting to achieve a positive safety culture, they are aiming to create organisations that are open, just and informed, in which reporting and learning when things go wrong is the norm. This chapter describes how three types of safety tools have been adapted and implemented in healthcare (culture surveys, safety walk rounds and checklists), and discusses the barriers to improving safety culture in healthcare.

Safety Culture Tools and Approaches

Various safety culture tools and approaches have been adapted from other industries and tailored to measure safety culture in healthcare organisations. These include safety culture measurement tools, safety walk rounds and checklists which aim to standardise patient care and improve reliability. So how have safety culture tools been adapted and implemented in healthcare organisations? What lessons have we learnt so far that could inform future work in this area?

Safety Climate Surveys

Safety climate surveys are well embedded as measures of safety culture in industry and have also been translated and applied in healthcare (Abdullah et al. 2009; Cox and Cheyne 2000; Cox and Cox 1991; Coyle et al. 1995; Flin et al. 2006; Helmreich and Merritt 1998; Mearns et al. 1998, 2003; Modak et al. 2007; Nieva and Sorra 2003; Smits et al. 2008; Sorra and Nieva 2004). Safety climate is regarded as the surface features of the underlying safety culture (Flin et al. 2000). Surveys typically assess workforce perceptions of procedures and behaviours in the work environment that indicate the priority given to safety, relative to other organisational goals (Clarke 1998, 2006; Cox and Cheyne 2000; Cox and Cox 1991; Helmreich and Merritt 1998). Survey data is collected from an individual level, and then aggregated to a higher level. The resulting data offer managers insights into worker perceptions of safety and can also be used for benchmarking purposes and for analysing trends.

Numerous studies have shown that safety climate survey results predict safety-related outcomes (Yule et al. 2007), for example, accidents and injuries (Huang et al. 2006), safety performance (Nahrgang et al. 2011; Shannon and Norman 2009) and workers' safety behaviour (Griffin and Neal 2000). Previous studies on the relationship between positive safety climate and lower accident rates demonstrated that employees with a 'positive safety attitude' were less likely to be involved in accidents (Barling et al. 2002; Hofmann and Stetzer 1996; and Lee 1998).

Translation of Safety Climate Surveys in Healthcare

Flin et al. (2006) reviewed the psychometric properties of safety climate questionnaires used in healthcare (i.e. construct validity, criterion validity, reliability and factor structure). They identified a lack of explicit theoretical underpinning in most questionnaires and some instruments that did not report standard psychometric criteria. Where information on psychometric properties was available, many surveys had limitations. Several of the instruments reviewed by Flin et al. (2006) had been developed from measures used in other industries (for example, aviation, oil and nuclear). These authors argued that considerable care needs to be taken when adapting measures from these very proceduralised, high risk industries. In healthcare the nature of the work carried out is unpredictable and complex. The managerial relationships are also more ambiguous, varying between different professional groups (e.g. doctors and nurses). Additionally, whereas safety climate studies in industry all focus on the worker, in healthcare the focus is on patient harm. Hence there are important differences between industry and healthcare in terms of what a safety measure is aiming to measure (Flin et al. 2006).

Flin et al's findings are supported by other research in both healthcare and other industries. Another meta-analysis of nine patient safety climate surveys identified considerable variability across surveys and concluded that more effort should be expended on understanding the relationship between measures of patient safety climate and patient outcomes (Colla et al. 2005). Furthermore, in industry Cooper and Philips (2004) noted that much of the research demonstrating a relationship between safety climate survey results and safety is based on accident rates. Some researchers have attempted to assess concurrent validity (i.e. safety performance at the time of distribution) or predictive validity (i.e. forecast future safety performance) by correlating the scale or factor scores against actual accident rates (e.g. Lee and Harrison 2000; Mearns et al. 2003; Niskanen 1994; O'Toole 2002; Silva et al. 2004; Varonen and Mattila 2000; Vredenburgh 2002; Zohar 2000), expert ratings (Arboleda et al. 2003; Diaz and Cabrera 1997;

Zohar 1980), human error analysis (Glendon and Stanton 2000), ratings of behavioural compliance (Garavan and O'Brien 2001), and actual safety behaviour (Glendon and Litherland 2000).

The limitations with repeated use of climate survey tools as a safety measure have also been identified. In one study an analysis of a large sample of safety climate survey data (n=110,014), collected over 10 years from U.S. Naval aircrew using the Command Safety Assessment Survey (CSAS), was carried out. Results showed that there was substantial non-random response bias associated with the data (the reverse worded items had a unique pattern of responses; there was an increasing tendency over time to provide only a modal response; the responses to the same item towards the beginning and end of the questionnaire did not correlate as highly as might be expected; the faster the questionnaire was completed the higher the frequency of modal responses). These findings occurred because of a combination of questionnaire design, lack of a belief in the importance of the response, participant fatigue, and questionnaire administration (O'Connor et al. 2011).

Safety Climate Tools – Applications and Misapplications in Healthcare Settings

The Safety Attitudes Questionnaire (Sexton et al. 2006) and Hospital Safety Culture Survey (Nieva and Sorra 2003; Sorra and Nieva 2004) are the two most commonly used safety climate surveys in healthcare. Like surveys applied in other industries, each tool is structured to include core dimensions of safety culture. The Safety Attitudes Questionnaire, for example, focuses on: teamwork climate, safety climate, perceptions of management, job satisfaction, working conditions and stress recognition. It has been used to measure safety climate in operating theatres and critical care and pharmacy teams (Nordén-Hägg et al. 2010; Sexton et al. 2011), amongst others. Sexton et al. (2011) used the Safety Attitude Questionnaire (SAQ) to evaluate the impact of a comprehensive unit-based safety programme on safety climate in a large cohort of intensive care units participating in the Keystone intensive care unit project. The Safety Attitudes Questionnaire was administered to ICU staff at baseline in 2004 and after two years of exposure to the safety program to assess improvement. Results showed significant improvements in overall mean safety climate scores in a large cohort of 71 intensive care units that had participated in the programme.

Research has been carried out to test the psychometric properties of the Safety Attitudes Questionnaire and AHRQ Hospital Safety Culture Survey (Nieva and Sorra 2003; Sexton et al. 2006; Sorra and Nieva 2004). However, there have been important misapplications of survey tools in healthcare. A proliferation of climate surveys now exist, including many where the factor structure and construct validity have not been tested. Many hospitals have developed their own bespoke survey tools and these are poorly designed because basic survey design rules have not been followed. One common issue is surveys that do not counter-balance positive and negative statements, thus increasing the risk of response set bias where the respondent simply ticks the same response without reading each individual item. Sampling is another area where misapplication has occurred. Hospitals have drawn inferences from the findings of climate surveys even when the response rate has been low and certain staff groups (for example, consultants) are under-represented. Quite often design and sampling errors occur because there is a lack of understanding that in order for the results to be meaningful, survey design has to be robust.

Matrix-based Maturity Models

Safety climate has also been measured using safety culture maturity matrixes (Ashcroft et al. 2005; Dingsdag et al. 2006, 2008; Hayes et al. 2007, 2008; Kirk et al. 2007). Safety culture maturity matrices display a set of key indicators on the y axis and an evolutionary measure of cultural maturity on the x axis. Originally developed as part of the 'Hearts and Minds' project for Shell plc, one such matrix, the Manchester Patient Safety Framework (MaPSaF) was adapted for use in primary care settings and has since been developed into versions for mental health, acute and ambulance trusts (Kirk et al. 2007).

The Manchester Patient Safety Framework uses critical dimensions of patient safety and, for each of these, describes five levels of increasingly mature organisational safety culture. The dimensions relate to areas where attitudes, values and behaviours about patient safety are likely to be reflected in the organisation's working practices (see Table 7.1), for example, how patient safety incidents are investigated, staff education and training in risk management. The levels of maturity, based on a model originally put forward by Westrum (1992) and modified by Parker and Hudson (2001), show the journey from a pathological, reactive, bureaucratic, proactive to a generative organisational typology.

Table 7.1   Manchester patient safety framework risk dimensions

Manchester Patient Safety Framework Risk Dimensions
1. Commitment to overall continuous improvement
2. Priority given to safety
3. System errors and individual responsibility
4. Recording incidents and best practice
5. Evaluating incidents and best practice
6. Learning and effecting change
7. Communication about safety issues
8. Personnel management and safety issues
9. Staff training and education
10. Team working

Information is elicited by setting up a series of focus groups where healthcare staff rate the maturity of their local team and organisation. MaPSaF is a flexible tool that can be used to encourage healthcare teams to reflect on their safety culture, identify strengths and weaknesses and reveal differences in perception between staff groups and to evaluate safety interventions (by measuring safety culture before and post-implementation. The case study shown in Box 7.1 provides an example of how one hospital used MaPSaF and how the information elicited informed the organisation about its safety culture. Chapter 16 describes an NHS manager's experience of using the tool.

Box 7.1   Using the Manchester Patient Safety Framework to evaluate hospital safety culture

---

Hospital A, an acute NHS Foundation Trust, set up 10 focus groups with a cross-section of its staff. The organisation wanted to take a baseline measure of its safety culture. Focus groups were facilitated by an external facilitator. The focus groups comprised pharmacists, doctors (including consultants, registrars and senior house officers), nurses of all grades, radiographers, nursing assistants and ward clerks.
The key weaknesses identified with the organisation's safety culture were:

- Lack of feedback from the incident reporting system prevented organisation-wide learning.
- An excessive number of policies and procedures and too little emphasis on reviewing whether people are following policies and procedures.
- Lessons learnt from clinical audit were not fed back to frontline healthcare teams.
- There was a poor risk assessment culture. Risk assessments were viewed by staff as a 'tick box' exercise.
- Disconnects in terms of how well safety solutions were disseminated across the entire organisation.
- Recommendations from incident investigations were often not implemented.
- Senior managers had a top down approach to patient safety and did not engage staff in the development of safety solutions. Sustainability of solutions would be achieved if frontline staff were engaged in developing solutions, rather than having them imposed from above.

Key quotes from the focus groups give insights into translating safety attitudes into safety behaviours:

'All staff would say patient safety is a priority, but they do not always behave in ways that put patient safety at the centre of clinical care.' Infection Control Nurse

'Within the department there is a mixture of blame and support when an incident occurs. Some line managers practice what they preach and are supportive of staff involved in incidents whereas others are not.' Radiographer

---

The case study demonstrates that MaPSaF elicits useful information about safety culture that can be fed into improvement work. On the whole, healthcare teams like the tool and find participating in focus groups constructive. There are some practical challenges with using the tool however: firstly, some healthcare staff are members of more than one team. For example, an infection control nurse is part of the hospital's infection control team and may also be allocated to a specific clinical team on a ward. Facilitators need to clearly introduce how ratings should be carried out, making it clear what 'team' and 'organisation' means in the context of the focus group. Secondly, the usual issues with team hierarchies and social conformity apply. Good facilitation is essential. Setting the ground rules at the start of the focus group and structuring the workshops so that each participant independently rates each risk dimension first is important. Poor use of the framework inevitably colours the results. One notable example was a facilitator who asked the Deputy Chief Executive in an open forum to give ratings of risk dimensions first, then asked more junior staff share their own ratings on each risk dimension. The approach increased the likelihood that the junior team members would give the same ratings as the Deputy Chief Executive because of perceived pressure to conform with a senior manager's viewpoint in a public situation.

Approaches to Improving Safety Culture in Healthcare

Other methods that have been adapted and applied to measure and improve safety culture in healthcare organisations include safety walk-rounds and checklists designed to standardise patient care (for example, the World Health Organisation Surgical Safety Checklist).

Safety Walk-rounds

Safety walk-rounds enable operational staff to discuss safety issues with senior managers directly. The benefits of safety walk-rounds are well described in the industrial safety research literature (Mears 2009; Packard 1995; Peters and Austin 1985). Walk rounds provide an important source of 'real-world safety intelligence' and increase open communication between senior managers and the workforce. As a safety performance measure, safety walk-rounds are used as a visible indicator of senior management commitment to safety which has been identified in many studies as a key element of a good safety culture (O'Dea and Flin 2001; Reason 1997; Zohar 2000). Safety walk-rounds have been adapted and implemented successfully in healthcare organisations (Frankel et al. 2003, 2006, 2008; National Patient Safety Agency 2009, 2011). Research has shown that they can have a positive impact on safety culture when implemented as part of broader safety improvement programmes (Frankel et al. 2008).

Some important lessons can be learnt about how safety walk rounds have been implemented in healthcare. Most notable are the measurement paradox and the purpose paradox:

The measurement paradox   The measurement paradox relates to how safety walk-round performance is measured. The walk-round measures used in the National Patient Safety First Campaign in 2009 involved using run charts to measure:

- The number of executive safety walk-rounds per month (with the aim of carrying out two per month).
- The number of actions identified on walk rounds completed every month.

The first measure has, in some organisations, led to a focus on counting the number of walk-rounds carried out, thus putting too much emphasis on quantity rather than the quality of the walk-rounds. It has encouraged a 'we carry out the target number of walk-rounds per month' culture, rather than one that is truly grounded in using walk rounds to gather safety intelligence from the coal-face. The second measure has also posed problems for healthcare organisations and has highlighted the issue of poor human factors integration in the design of patient pathways and medical devices, and the allocation of staffing and resources on clinical units. This issue is illustrated in Box 7.2:

Box 7.2   Hospital B's safety walk-round story

Hospital B has carried out over 150 Executive Patient Safety Walk rounds in the last three years. Patient safety walk-rounds are led by one of the executive directors of the hospital. High priority actions identified on walk rounds are allocated to a named Executive Patient Safety Walk-round team member to follow up and resolve within one month.

A thematic analysis of the issues identified on walk-rounds is routinely carried out to measure the completion rate. Trust data showed that the clear-up rate of high priority issues has improved year on year; in 2011, 50 percent of actions were resolved and 26 percent were partly resolved.

Thematic analysis also showed that the most frequently identified issues are environmental – the physical space and design of the hospital accounted for just over 25 percent of all issues identified, followed by equipment (23 percent) and clinical processes (20 percent). Many of these issues were unresolvable within the one month time frame set because they related to the environment in which healthcare teams work. They are therefore not easily fixed.

Hospital B's experience is not unique. The net effect of the completion rate measure has been that it produces data which show that many actions are not completed within a specified time frame. This has, in some hospitals led senior managers to call into question the value of the walk rounds or to drive behaviour to focus on problems that can be fixed easily. Thus the lack of integration of human factors into hospital and medical device design and the allocation of staffing and resources gets put into an 'unresolvable box' and remain latent conditions that continue to create error traps, inefficiencies and poor patient experience.

Unlike other domains, healthcare has not recognised the importance of 'participative management' (O'Dea and Flin 2001) as a crucial indicator of an organisation's safety culture. That is to say, it is not just management participation and involvement in safety activities which is important, but the extent to which management encourages and devolves responsibility and accountability to the workforce by encouraging them to become actively involved in developing safety interventions and safety policy, rather than playing a passive role of recipient (Niskanen 1994; O'Dea and Flin 2001; Simard and Marchand 1995; Williamson et al. 1997).

The purpose paradox   The second paradox relates to the lack of understanding of the purpose of walk-rounds in healthcare. Whereas some hospitals use walk-rounds as a performance management tool, others under-utilise their potential value by reducing them to being informal chats between senior healthcare managers and frontline healthcare staff or 'royal visits' from the senior management team (Frankel et al. 2006). There has been, in some healthcare organisations, a lack of understanding that walk rounds are a source of safety information and that the 'soft intelligence' they provide needs to be integrated into the formal quality and safety structures of an organisation. One leading proponent of safety walk rounds has commented:

> … many organizations mistakenly think the key component is leadership walking around, and that WalkRounds is an informal conversation between leadership and providers. In fact, the real power is that these conversations elicit useful information within a formal structure, the information is then documented and analysed, combined with relevant information from root cause analyses and other reporting systems, and regularly discussed in meetings involving the Clinical chairs, chiefs, and senior leaders. These leaders of the organization accept and have clear responsibility for actions to resolve identified problems. Learning around these issues and the actions to be taken then becomes part of the operations-committee agenda. (Frankel et al. 2006)

The measurement and purpose paradoxes are in themselves indicators of cultural maturity. Healthcare organisations which do not exhibit these paradoxes may be more mature than those that do.

The WHO Surgical Safety Checklist

In England and Wales, 129,419 incidents relating to surgical specialties were reported to the National Reporting and Learning System in 2007 with a range of degrees of harm, including 271 deaths (NPSA 2009). The WHO Surgical Safety Checklist aims to improve the safety of surgery by reducing deaths and complications. This includes improving anaesthetic safety practices, ensuring correct site surgery, avoiding surgical site infections and improving communication within the team. The checklist comprises a core set of safety checks for use in any operating theatre environment including (amongst others) 'is the surgical site marked?', 'has the patient confirmed his/her identity/site/consent?' and 'is all essential imaging displayed?'

Launched as a national safety improvement initiative by the National Patient Safety Agency in 2009, the wide variation in approaches to its implementation provide a good case study about how safety cultures vary across different hospitals. Implementation approaches for the WHO Surgical Safety Checklist have ranged from the 'please use this checklist in your operating theatre from next Monday' to, at the other end of the spectrum, well designed implementation plans (Box 7.3).

Box 7.3   It's not about the checklist

Curran (2010) has described how the WHO Surgical Safety Checklist was implemented at UCLH NHS Hospitals Foundation Trust.

At UCLH, operating theatre teams led the adaptation and implementation of the checklist. Most notably, implementation did not focus on the checklist per se, but on resolving issues with poor communication, teamwork and culture in the operating theatres. Whilst the checklist provides a platform for structuring team communication, the hospital had a clear goal that bigger cultural issues needed to be addressed in order to improve patient safety.

Together with the Trust's Education Team, theatre staff developed multi-disciplinary team training focusing on safe surgery. Senior management commitment for the training was evidenced by releasing theatre teams who worked together to train together, thus giving a clear message from the Board that safety was being prioritised over efficiency and targets. This was strengthened by a letter to all theatre staff by the Corporate Medical Director, who introduced the aims and purpose of the theatre team training, and who made sure his team was one of the first to participate in the training.

Training was delivered in an operating theatre simulator using a mock operating list. Each team completed a briefing, WHO Surgical Safety checklist for each patient and a debriefing. After each simulation, a facilitated debrief was held which explored human factors issues and how the checks in the checklist could act as an aide memoire.

Training was not a one off event. Following the simulation training, checklist mentors worked with theatre teams during cases to resolve issues with the checklist and to provide real time feedback.

Story-telling was used to spread the word about successful interventions where, for example, a scrub nurse had constructively challenged a consultant and thwarted a potential incident. A newsletter was produced called 'It's complicated' which shared good practice and enabled theatre teams to learn from each other.

Like in all hospitals, on-going monitoring of levels of compliance with using the checklist was introduced and regular auditing is carried out. The audits provide a further form of feedback to theatre teams about their performance. Whilst the training ensured that the implementation approach was grounded at the outset in winning hearts and minds, the compliance data and feedback from theatre teams identified staff who did not engage with using the checklist even after the investment in training. The momentum created by the training meant that these people found themselves marginalised.

Whereas the training programme and on-the-job mentoring created 'horizontal peer pressure' whereby using the checklist became the norm, the audit data supported the 'appropriate use of hard edges (Dixon-Woods et al. 2011), by identifying those theatre teams who were not engaging.

Implementation is viewed as a continuous process: in Safe Surgery Week 2012, the Trust carried out a series of workshops about the usability of the checklist with a view to understanding how its design and content could be improved.

The WHO Surgical Safety Checklist provides useful insights into the limitations with rolling out national 'top down' initiatives that aim to improve patient safety. Securing 'bottom up' engagement from doctors, nurses and allied healthcare professionals is more challenging when a safety solution is part of a national directive because they are not active participants in its design and development. Many hospitals have struggled to communicate the purpose of the checklist to operating theatre teams. The pursuit of standardisation at hospital level, whereby one version of the checklist is used in all operating theatres, has at times led to lack of engagement by consultant surgeons and increased the risk of non-compliance.

Two other problems have arisen during the local adaptation of the checklist by NHS Trusts: The NPSA's patient safety alert stated that *'This checklist contains the core content but can be adapted locally or for specific specialties through usual clinical governance procedures'* (NPSA 2009). The statement aimed to support local adaptation so that it could be widely applied to different surgical domains. In practice, local adaptation has sometimes resulted in important safety checks being removed. For example, checking consent prior to anaesthetic induction has been removed because local Trust consent policies dictate that consent checking should be carried out as part of the surgical admissions process. Furthermore, at times, managers have used the checklist as a forum for collecting additional types of information (for example, a patient's CJD status). The net effect of local adaptation has at times been to increase the complexity and length of the checklist. This in turn has reduced its usability and disenfranchised theatre teams.

There is also a danger of checklists being seen as a panacea for all patient safety problems: in short, healthcare is suffering from an epidemic of 'checklistitis.' Checklistitis involves not understanding the clinical situations in which checklists act as a layer of defence and a platform for conversation amongst healthcare teams. The proliferation of checklists to cover all clinical situations – like handovers, ward rounds, admissions processes, medication checking etc – creates a 'tick box' culture and fosters 'involuntary automaticity' (Toft and Mascie-Taylor 2005). Just as safety walk rounds have exposed the ramifications of poor human factors in hospital and device design, healthcare's experience with the WHO Surgical Safety Checklist demonstrates a need for further education about when checklists are appropriate and the need for checklist development using a whole systems approach.

Barriers to Improving Safety Culture in Healthcare

The description of how safety climate tools, walk-rounds and the WHO Surgical Safety Checklist have been implemented in healthcare has identified some of the challenges to measuring and improving culture in healthcare teams and organisations. Other important barriers to improving safety culture exist:

Measuring the Measurable

Firstly, the culture in healthcare is to measure the easily measurable. Unlike other improvement goals, measuring cultural change can feel amorphous and challenging. One NHS Trust Chairman described trying to understand his organisation's safety culture as like '… grasping at smoke.' Measures of safety culture don't always fit easily onto a hospital's safety dashboard. Cultural change takes time to effect. So there is not a comfortable fit between the current drive to demonstrate measurable improvement quickly and the slower evolutionary process that drives an organisation to become more mature.

The focus on measuring the measurable quite often translates into a reductionist approach and the desire to identify a single indicator which will measure the health of an organisation. The recent introduction of the Friends and Family test in NHS organisations is a good example of reductionism. The test elicits patient feedback on a single question 'would you recommend this service to a friend or family?' The question itself is ambiguous and is therefore likely to be interpreted by different patients in different ways: patients may respond by rating the quality of care by nursing staff, their overall experience, the cleanliness of the ward or unit on which they were treated or the politeness of the healthcare team who treated them (Lynn 2013).

Data from the Friends and Family test is used to compare the proportion of respondents answering that they would be 'likely' or 'extremely likely' to recommend the department/ward. Comparing simple proportions in this way takes no account of differences between the trusts or wards in terms of the patient case mix, the reasons for admission and the outcomes of treatment (Lynn 2013). In essence, the NHS may be using an over-simplistic measure to compare apples and pears and, as a result, drawing conclusions about patient experience that are meaningless.

Regulatory Approach

The nature of regulation in the NHS is another barrier to improving safety culture. Many regulators assess the performance of healthcare organisations on patient safety improvement initiatives. The current structure of some regulatory assessments drives hospitals towards a 'bureaucratic safety culture' (Parker and Hudson 2001; Westrum 1992). That is to say, there is a tick box approach to safety. All of the necessary mechanisms are in place to prove to the regulator that the organisation is safe but the focus is on passing the exam, rather than on improving the safety culture from Board to ward.

The current regulatory complexity of healthcare and the approach of some (but not all) regulators has in itself become a barrier to developing more mature safety cultures. For example, around 20 national bodies and regulators produce guidance and/or policies on safe anaesthesia (see Box 7.4) (Carthey et al. 2011). This guidance has to be translated into local policies and guidelines. Healthcare organisations have to demonstrate that they are compliant with the guidance. Although the guidance aims to support safe practice (and often does), fragmentation across different professional bodies and regulators increases the volume and complexity of information. This in turn creates a high workload and leads to resources being subsumed by ensuring that local policies and guidelines exist to demonstrate that local practice is compliant.

Box 7.4   Professional bodies and national agencies who publish guidelines for anaesthetists

---

Association of Anaesthetists of Great Britain and Ireland
Association of Medical Royal Colleges
Association of Cardiac Anaesthetists
British Association of Day Surgery
British National Formulary
British Pain Society
Department of Health
Difficult Airway Society
European Society of Anaesthesiology
Faculty of Pain Medicine
General Medical Council
Health and Safety Executive
Intensive Care Society
Medicines and Healthcare Products Regulation Authority
National Patient Safety Agency
National Institute for Health and Clinical Excellence
Obstetric Anaesthetists Association
Resuscitation Council (UK)
Royal College of Anaesthetists
Scottish Intercollegiate Guidelines Network

At its worst, the fragmentation of regulation in healthcare can delay the response when problems are identified. The Francis Inquiry report into Mid-Staffordshire NHS Foundation Trust shows how different regulators had information that indicated serious safety culture issues but, because the information was fragmented across many different regulatory bodies, it was not assimilated and acted upon in a timely way.

'Biomedical Culture' versus 'Quality of Care' Culture

The biomedical model of healthcare delivery is still very dominant. In the biomedical model, poor quality care is often seen as an inevitable side effect of being cured. Pressure ulcers and hospital acquired infections were traditionally viewed by many managers, doctors and nurses as complications of treatment rather than indicators of poor quality care. As a result, lessons were not learnt to prevent other patients being harmed. Although notable work on quality and safety improvement has improved safety culture in some teams and organisations, there is still a long way to go before the 'quality of care' culture replaces the biomedical culture. Targets for reducing the incidence of MRSA and C Diff infections and Hospital Acquired Pressure Ulcers (HAPUs) have improved performance and, in the case of HAPUs, changed attitudes about what is an acceptable complication of treatment. But universal attitudinal and behavioural change has not yet been achieved; in community settings the attitude that pressure ulcers are unavoidable in patients with long term conditions being treated at home still predominates, as it does in some acute hospitals.

Training and Education

Another key barrier to improving team cultures in healthcare is the uni-professional education model. Doctors and nurses train in different systems and are accountable to different professional bodies. Each has their own distinct set of values and attitudes. Multi-disciplinary training opportunities are rare (although simulation is being used increasingly), so multi-disciplinary teamwork skills are learnt on the job. Current training and education models focus mainly on technical skills, even though incident investigation findings often identify failures in non-technical skills like teamwork, leadership, communication and situational awareness. For example, several serious incidents in the operating theatre, including the Elaine Bromiley case, show how steep authority gradients mean that theatre nurses are reluctant to challenge consultants, even when they can see an incident scenario unfolding.

In summary, current training and education create another barrier to improving safety culture because they do not support the development of shared norms and behaviours across different professional groups. Rather they create professional silos. The net effect of the current training approach is that doctors and nurses cannot 'walk in each other's shoes.' Handovers, where multi-disciplinary team input is needed to communicate safety information across healthcare interfaces, illustrate this issue very well. All too often, nurses seek 'nursing information' and doctors want to hand over a clinical management plan in which key information required by the nurses is not included.

The Focus on Tools

Much of the work on improving safety culture in healthcare has focused on developing tools that healthcare organisations can take off the shelf and use locally. Tools and approaches are necessary but, used singly, they are insufficient to improve safety culture. The discussion, earlier in this chapter, of how different tools have been adopted, and sometimes misapplied, supports this viewpoint. Healthcare organisations lack experts who can work within their organisations to help them to effect cultural change. Very few NHS Trusts employ an appropriately qualified human factors expert. Going forward, there is a need to strike a better balance between developing tools which healthcare organisations sometimes struggle to use locally because of the lack of internal human factors expertise.

Conclusion

This chapter has discussed some applications and misapplications of safety culture tools in healthcare. Lessons learnt from the early adoption of safety culture surveys, safety walk-rounds and the WHO Surgical Safety Checklist are very informative. By learning these lessons and removing some of the barriers we have discussed, healthcare organisations can move forward and improve patient safety.

References

Abdullah, N., Spickett, J.T., Rumchevc, K.B. and Dhaliwald, S.S. 2009. Validity and reliability of the safety climate measurement in Malaysia. *International Review of Business Research Papers*, 5(3), 111–41.

Arboleda, A., Morrow, P.C., Crum, M.R. and Shelley II, M.C. 2003. Management practices as antecedents of safety culture within the trucking industry: Similarities and differences by hierarchical level. *Journal of Safety Research*, 34, 189–97.

Ashcroft, D.M., Morecroft, C., Parker, D. and Noyce, P. 2005. Safety culture assessment in community pharmacy: Development, face validity and feasibility of the Manchester patient safety assessment framework. *Quality and Safety in Health Care*, 14(6), 417–21.

Barling, J., Loughlin, C. and Kelloway, E. 2002. Development and test of a model linking safety-specific transformational leadership and occupational safety. *Journal of Applied Psychology*, 87(3), 488–96.

Carthey, J. and Clarke, J. 2009. *How to Guide to Implementing Human Factors in Healthcare*. Published as part of the National Patient Safety First campaign. Available at: http://www.patientsafetyfirst.nhs.uk/ashx/Asset.ashx?path=/Intervention-support/Human%20Factors%20How-to%20Guide%20v1.2.pdf (last accessed on 10 February 2014).

Carthey, J., Walker, S., Deelchand, V., Vincent, C. and Griffiths, W.H. 2011. Breaking the rules: Understanding non-compliance with policies and guidelines. *British Medical Journal*, 343, d5283.

Clarke, S. 1998. Safety culture on the UK railway network. *Work and Stress*, 12(3), 285–92.

Clarke, S. 2006. Safety climate in an automobile manufacturing plant: The effects of work environment, job communication and safety attitudes on accidents and unsafe behavior. *Personnel Review*, 35(4), 413–30.

Colla, J.B., Bracken, A.C., Kinney, L.M. and Weeks, W.B. 2005. Measuring patient safety climate: A review of surveys. *Quality and Safety in Health Care*, 14, 364–6.

Cooper, M.D. and Phillips, R.A. 2004. Exploratory analysis of the safety climate and safety behaviour relationship. *Journal of Safety Research*, 35(5), 497–512.

Cox, S. and Cheyne, A. 2000. Assessing safety culture in offshore environments. *Safety Science*, 34, 111–29.

Cox, S. and Cox, T. 1991. The structure of employee attitudes to safety: A European example. *Work and Stress*, 5, 93–106.

Coyle, I.R., Sleeman, S.D. and Adams, N. 1995. Safety climate. *Journal of Safety Research*, 26(4), 247–54.

Curran, N. 2010. Implementing the WHO surgical safety checklist at University College London NHS Foundation Trust. Personal communication.

Department of Health 2000. *Organisation with a Memory*. The Stationery Office, London. UK. Available at: http://webarchive.nationalarchives.gov.uk/+/www.dh.gov.uk/en/Aboutus/MinistersandDepartmentLeaders/ChiefMedicalOfficer/ProgressOnPolicy/ProgressBrowsableDocument/DH_5016613 (last accessed on 16 April 2014).

Department of Health 2001. *Building a Safer NHS for Patients*. Department of Health, London. UK. Available at: http://webarchive.nationalarchives.gov.uk/+/www.dh.gov.uk/en/publicationsandstatistics/publications/publicationspolicyandguidance/browsable/DH_4097460 (last accessed on 16 April 2014).

Diaz, R.I. and Cabrera, D.D. 1997. Safety climate and attitude as evaluation measures of organizational safety. *Accident Analysis and Prevention*, 29(5), 643–50.

Dingsdag, D.P., Biggs, H.C., Sheahan, V.L. and Cipolla, D.J. 2006. *A Construction Safety Competency Framework: Improving OH&S Performance by Creating and Maintaining a Safety Culture*. Brisbane: Cooperative Research Centre for Construction Innovation, Icon.net Pty Ltd.

Dingsdag, D.P., Biggs, H.C. and Cipolla, D. 2008. Safety effectiveness indicators (SEIs): Measuring construction industry safety performance, clients driving innovation: Benefiting from innovation. *Third International Conference, CRC for Construction Innovation*, 12–14 March 2008 Surfers Paradise Marriott Resort & Spa, Gold Coast, Queensland, Australia.

Dixon-Woods, M., Bosk, C.L., Aveling, E.L., Goeschel, C. and Pronovost, P.J. 2011. Explaining Michigan: Developing an ex post theory of a quality improvement program. *Milbank Quarterly*, 89(2), 167–205.

Flin, R., Mearns, K., O'Connor, P. and Bryden, R. 2000. Measuring safety climate: Identifying the common features. *Safety Science*, 34, 177–92.

Flin, R., Burns, C., Mearns, K., Yule, S. and Robertson, E.M. 2006. Measuring safety climate in health care. *Quality and Safety in Health Care*, 15(2), 109–15.

Francis, R. 2013. The Mid Staffordshire NHS Foundation Trust Public Inquiry. Report of the Mid Staffordshire NHS Foundation Trust Public Inquiry. Available at: http://www.midstaffspublicinquiry.com/report (last accessed on 16 April 2014).

Frankel, A., Graydon-Baker, C., Neppl, T., Simmonds, M., Gustafson, M. and Gandhi, T.K. 2003. Patient safety leadership walkrounds. *Joint Commission Journal on Quality and Safety*, 29(1), 16–26.

Frankel, A.S., Leonard, M.W. and Denham, C. 2006. Fair and just culture, team behavior, and leadership engagement: The tools to achieve high reliability. *Health Services Research*, 41, 1690–709.

Frankel, A., Pratt Grillo, S., Pittman, M., Thomas, E.J., Horowitz, L., Page, M. and Sexton, B. 2008. Revealing and resolving patient safety defects: The impact of leadership walk rounds on frontline caregiver assessments of patient safety. *Health Services Research*, 43, 2050–66.

Garavan, T.N. and O'Brien, F. 2001. An investigation into the relationship between safety climate and safety behaviours in Irish organizations. *Irish Journal of Management*, 22, 141–70.

Glendon, A.I. and Litherland, D.K. 2000. Safety climate factors, group differences and safety behaviour in road construction. *Safety Science*, 39, 157–88.

Glendon, A.I. and Stanton, N. 2000. Perspectives on safety culture. *Safety Science*, 34, 193–214.

Griffin, M.A. and Neal, A. 2006. Perceptions of safety at work: A framework for linking safety climate to safety performance, knowledge, and motivation. *Journal of Occupational Health Psychology*, 5, 347–58.

Hayes, A., Lardner, R., Medina, Z. and Smith, J. 2007. *Personalising Safety Culture: What Does it Mean for Me?* Paper presented at Loss Prevention conference, 22–4 May, Edinburgh, UK.

Hayes, A., Novatsis, E. and Lardner, R. 2008. Our safety culture: Our behaviour is the key. Paper presented at Society of Petroleum Engineers HSE Conference, 15–17 April, Nice, France.

Helmreich, R.L. and Merritt, A.C. 1998. Organizational culture. In R.L. Helmreich and A.C. Merritt (eds), *Culture at Work in Aviation and Medicine*. Brookfield, VT: Ashgate Publishing, 107–74.

Hofmann, D.A. and Stezer, A. 1996. A cross-level investigation of factors influencing unsafe behaviours and accidents. *Personnel Psychology*, 49, 307–39.

Huang, Y.H., Ho, M., Smith, G.S. and Chen, P.Y. 2006. Safety climate and self-reported injury: Assessing the mediating role of employee safety control. *Accident Analysis & Prevention*, 38(3), 425–33.

Kirk, S., Parker, D., Claridge, T., Esmail, A. and Marshall, M.L. 2007. Patient safety culture in primary care: Developing a theoretical framework for practical use. *Quality and Safety in Health Care*, 16, 313–20.

Lee, T. 1998. Assessment of safety culture at a nuclear reprocessing plant. *Work & Stress*, 12, 217–37.

Lee, T. and Harrison, K. 2000. Assessing safety culture in nuclear power stations. *Safety Science*, 30, 61–97.

Lynn, P. 2013. The friends and family test is unfit for purpose. *Guardian Professional*, 9 April 2013. Available at: http://www.guardian.co.uk/healthcare-network/2013/apr/09/friends-family-test-unfit-for-purpose (last accessed on 16 April 2014).

Mearns, K., Flin, R., Gordon, R. and Fleming, M. 1998. Measuring safety climate in offshore installations. *Work & Stress*, 12, 238–54.

Mearns, K., Whitaker, S.M. and Flin, R. 2003. Safety climate, safety management practice and safety performance in offshore environments. *Safety Science*, 41, 641–80.

Mears, M. 2009. *Leadership Elements: A Guide to Building Trust*. Bloomington, IN: Universe Ltd.

Modak, I., Sexton, J.B., Lux, T.R., Helmreich, R.L. and Thomas, E.J. 2007. Measuring safety culture in the ambulatory setting: The safety attitudes questionnaire–ambulatory version. *Journal of General Internal Medicine*, 22(1), 1–5.

Nahrgang, J., Morgenson, F. and Hofman, D. 2011. Safety at work: A meta-analytic investigation of the link between job demands, job resources, burnout, engagement, and safety outcomes. *Journal of Applied Psychology*, 96, 71–94.

National Patient Safety Agency 2004. *Seven Steps to Patient Safety*. Crown Publishing, Department of Health. London. Available at: http://www.nrls.npsa. nhs.uk/resources/collections/seven-steps-to-patient-safety/ (last accessed on 16 April 2014).

National Patient Safety Agency. 2009. *Patient Safety Alert. WHO Surgical Safety Checklist*. Issued 26 January 2009. Available at: http://www.nrls.npsa.nhs.uk/resources/?entryid45=59860 (last accessed on 16 April 2014).

National Patient Safety Agency 2011. *National Patient Safety First Campaign, 2008–2010*. The campaign review. Available at: http://www.patientsafetyfirst. nhs.uk/ashx/Asset.ashx?path=/Patient%20Safety%20First%20-%20the%20 campaign%20review.pdf (last accessed on 16 April 2014).

National Patient Safety First Campaign 2009. *How to Guide for Leadership for Safety*. Available at: http://www.patientsafetyfirst.nhs.uk/ashx/Asset. ashx?path=/How-to-guides-2008-09-19/Leadership%201.1_17Sept08.pdf (last accessed on 16 April 2014).

Nieva, V.F. and Sorra, J. 2003. Safety culture assessment: A tool for improving patient safety in healthcare organizations. *Quality and Safety in Health Care*, 12, ii17–ii23. doi: 10.1136/qhc.12.suppl_2.ii17.

Niskanen, T. 1994. Safety climate in the road administration. *Safety Science*, 7, 237–55.

Nordén-Hägg, A., Sexton, J.B., Kälvemark-Sporrong, S., Ring, L. and Kettis-Lindblad, A. 2010. Assessing safety culture in pharmacies: The psychometric validation of the Safety Attitudes Questionnaire (SAQ) in a national sample of community pharmacies in Sweden. *Biomed Central Clinical Pharmacology*, 10, 1–40.

O'Connor, P., Buttrey, S.E., O'Dea, A. and Kennedy, Q. 2011. Identifying and addressing the limitations of safety climate surveys. *Journal of Safety Research*, 42(4), 259–65.

O'Dea, A. and Flin, R. 2001. Site managers and safety leadership in the offshore oil and gas industry. *Safety Science*, 37, 39–57.

O'Toole, M. 2002. The relationship between employees' perceptions of safety and organizational culture. *Journal of Safety Research*, 33, 231–43.

Packard, D. 1995. *The HP Way: How Bill Hewlett and I Built our Company*. New York, NY: Harper Business.

Parker, D. and Hudson, P. 2001. *Understanding your Culture*. Shell International Exploration and Production.

Peters, T. and Austin, N. 1985. *A Passion for Excellence: The Leadership Difference*. London: Collins.

Reason, J.T. 1990. *Human Error*. Cambridge, UK: Cambridge University Press.

Reason, J.T. 1997. *Managing the Risks of Organisational Accidents*. Aldershot: Ashgate.

Reason, J.T. 2000. Human error. Models and management. *British Medical Journal*, 320.

Sexton, J.B., Berenholtz, S.M., Goeschel, C.A., Watson, S.R., Holzmueller, C.G., Thompson, D.A., Hyzy, R.C., Marsteller, J.A., Schumacher, K. and Pronovost, P.J. 2011. Assessing and improving safety climate in a large cohort of intensive care units. *Critical Care Medicine*, 39(5), 934–9.

Sexton, J.B., Helmreich, R.L., Neilands, T.B., Rowan, K., Vella, K., Boyden J., Roberts, P.R. and Thomas, E.J. 2006. The safety attitudes questionnaire: Psychometric properties, benchmarking data, and emerging research. *BMC Health Services Research*, 6, 44.

Shannon, H.S. and Norman, G.R. 2009. Deriving the factor structure of safety climate scales. *Safety Science*, 47(3), 327–9.

Silva, S., Lima, M.L. and Baptista, C. 2004. OSCI: An organisational and safety climate inventory. *Safety Science*, 42, 205–20.

Simard, M. and Marchand, A. 1995. A multilevel analysis of organisational factors related to the taking of safety initiatives by workgroups. *Safety Science*, 21, 113–29.

Smits, M., Christiaans-Dingelhoff, I., Wagner, C., van der Wal, G. and Groenewegen, P.P. 2008. The psychometric properties of the 'Hospital Survey on Patient Safety Culture' in Dutch hospitals. *Biomed Central Health Services Research*, 8, 230.

Sorra, J.S. and Nieva, V.F. 2004. *Hospital Survey on Patient Safety Culture. Rockville, MD*: Agency for healthcare research and quality. Available at: www.ahrq.gov/qual/hospculture/hospcult.pdf (last accessed on 16 April 2014).

Toft, B. and Mascie-Taylor, H. 2005. Involuntary automaticity: A work-system induced risk to safe health care. *Health Services Management Research*, 18(4), 211–16.

Varonen, U. and Mattila, M. 2000. The safety climate and its relationship to safety practices, safety of the work environment and occupational accidents in eight wood-processing companies. *Accident Analysis and Prevention*, 21, 761–9.

Vredenburgh, A.G. 2002. Organizational safety: Which management practices are most effective in reducing employee injury rates? *Journal of Safety Research*, 33, 259–76.

Westrum, R. 1992. Cultures with requisite imagination, in J. Wise, D. Hopkin and P. Stager (eds), *Verification and Validation of Complex Systems: Human Factors Issues*. Berlin: Springer-Verlag, 401–16.

Williamson, A.M., Feyer, A., Cairns, D. and Biancotti, D. 1997. The development of a measure of safety climate: The role of safety perceptions and attitudes. *Safety Science*, 25(1–3), 15–27.

Yule, S., Flin, R. and Murdy, A. 2007. The role of management and safety climate in preventing risk-taking at work. *International Journal of Risk Assessment and Management*, 7(2), 137–51.

Zohar, D. 1980. Safety climate in industrial organizations: Theoretical and applied implications. *Journal of Applied Psychology*, 65, 96–102.

Zohar, D. 2000. A group-level model of safety climate: Testing the effect of group climate on microaccidents in manufacturing jobs. *Journal of Applied Psychology*, 85, 587–96.

Chapter 8

Assessing Psychometric Scale Properties of Patient Safety Culture

Jeanette Jackson and Theresa Kline

Introduction

The measurement of patient safety culture is a key topic of interest to researchers in the field of patient safety. However, there is little agreement about how to define and assess patient safety culture and a range of different measurement tools has been developed. There does seem to be consensus, though, that dimensions of patient safety culture need to include management commitment, teamwork, communication and incident reporting. The Hospital Survey on Patient Safety Culture (HSOPSC), designed for the Agency for Healthcare Research and Quality (AHRQ; Sorra and Nieva 2006), is especially popular and is now the recommended tool of choice for some reviewers (European Society for Quality in Healthcare 2010). The HSOPSC has been extensively used by different research teams across different countries (see Chapters 10 and 14) and includes 42 items measuring 10 dimensions of safety culture and four outcome dimensions, including two 'single item outcomes', namely *'Patient Safety Grade'* and *'Number of Events Reported'*. Its implementation in practice is illustrated in Chapters 10, 11, 12 and 14.

Psychometric criteria (such as validity and reliability) have not been reported for many safety culture questionnaires, raising concerns about their utility to inform healthcare managers who are designing effective safety management systems and interventions (Flin et al. 2006). For example, when initiating change, a psychometrically sound instrument is needed to detect changes in culture resulting from the initiative. However, the validity and reliability of a scale are important considerations in both research and applied settings when judging the quality of measurement instruments and ensuring that patient safety improvements will be effective and sustainable.

This chapter provides an overview of the psychometric properties of the HSOPSC using both classical test theory (CTT) and the modern approach, often referred to as Item Response Theory (IRT). To enhance the understanding and importance of IRT, the basic principles will first be introduced. In particular, three fundamental outcomes of the IRT approach will be highlighted: (1) item characteristic curves, (2) measurement information, and (3) invariance. Moreover, this chapter will present data that have been previously analysed and published, using the classical approaches of exploratory and confirmatory factor analysis by Waterson and colleagues (2010) in order to contrast and discuss IRT results with findings based on the factor analytical approach. Finally, practical implications will be highlighted to encourage future healthcare research and healthcare service evaluations to apply and implement IRT findings when measuring patient safety culture.

Construct Validity

Construct validity has many definitions. However, the most appropriate is the degree to which inferences can be made about the 'scores' obtained on a test. For example, if a person gets a score at the 95th percentile on the Medical College Admissions Test, this suggests that person will do well in Medical School. Thus, construct validity also implies that the measure will act in ways that are theoretically meaningful and appropriate. One piece of evidence that suggests the test items are behaving as expected is to identify the underlying factor structure, using latent variable theory applied through exploratory and confirmatory factor analyses (EFA and CFA) which both belong to the common factor model (Thurstone 1947). The underlying principle of the common factor model is that shared variability among observable variables (e.g. scales on a safety culture survey) can be attributed to the presence of a smaller set of common but unobserved (i.e. latent) variables. Using this classical approach, some studies have validated the original 12-factor structure (e.g. Chen and Li 2010; Hellings et al. 2007; Olsen 2008; Sarac et al. 2011), while others have reported contradictory results (e.g. Bodur and Filiz 2010; Pfeiffer and Manser 2010; Smits et al. 2008; Waterson et al. 2010). Establishing the veracity of a factor analytical model is accomplished by ascertaining the goodness of fit of latent variables within a given sample dataset. It is not surprising that various factor models have been uncovered, given that different studies use different samples with unique characteristics. Regardless of the discrepancies in findings, these approaches are used within the framework of CTT. Although many scales have been developed and validated using the analyses prescribed by CTT, a more comprehensive tactic includes the addition of IRT.

Both CTT and IRT can, and should, be used to assess the psychometric properties of a scale. Each proposes ways of conceptualising dimensions of safety culture and specifies how they relate to observed scores. The central focus of CTT is the true score on the test, which is the expected value of the observed test score over an infinite number of trials. A primary goal of test developers is to minimise random error around the true score by aggregating scores across several related items. If such a set of items is successfully identified by EFA and CFA, correlations between items are consistent and moderately high.

The longer the test (i.e. the greater the number of such items), the higher the internal consistency of the scale will be.

In IRT, on the other hand, a latent trait, such as 'patient safety culture', is an unobserved construct measured by a set of observed items. The unobserved construct, in this case 'patient safety culture', has a causal relationship to the observed scores'. Thus, CTT tends to focus more on total test scores, whereas IRT focuses on the items and their characteristics (Borsboom 2005). CTT is useful when the test score is of central interest, while IRT is most useful in examining the individual building blocks (items) of the total test scores. Both have their strengths and both are needed to form a complete picture of the properties of a set of items.

CTT has been the dominant approach in psychometric theory for the past 100 years. Although IRT has been around since the late 1960s (Lord and Novick 1968), until computer programmes were accessible to researchers it was not widely adopted. Its use is now reported in various literatures including healthcare (Chan et al. 2004), education (Kane 1987) and clinical psychology (Cooke and Michie 1997), to name a few. It is important, then, to pursue using such an approach in the field of assessment of patient safety culture as this technology becomes an expected standard in psychometric evaluation.

Fundamentals of Item Response Theory

Table 8.1 provides a glossary that summarises and explains some common terms used when carrying out IRT in order to support the reader's understanding of the IRT.

Table 8.1   Glossary of terms used with Item Response Theory

Term	Explanation
Item Characteristic Curve (ICC)	IRT is a group of measurement models that describe the mathematical relationship between underlying latent construct of safety culture and the individual's performance (observed item responses) as a logistic function called an item characteristic curve (ICC). Each single item in a 5-point Likert-type response format is characterised by an ICC defined by the estimated slope parameter and 4 item location parameters representing the movement from the lower options to the higher options of the Likert scale.
Item Discrimination (Slope Parameter)	The item discrimination is determined by the steepness of the logistic function. Slope parameters greater than 1.35/1.70 are considered to be 'high'/'very high' in discriminating between individuals with different safety culture levels. The item measures most precisely at the steepest part of the logistic curve,
Item Difficulty (Location Parameter)	Each HSOPSC item has five possible response options and thus four location parameters. These are estimates indicating at what point along the 'patient safety construct' that the respondent will shift from one option (say a '1') to the next option (say a '2') on the Likert scale. This location, then, determines the point where the item's severity in assessing safety culture equals the individual respondent's level of safety culture.
Person Parameter	Based on estimated HSOPSC item characteristics (i.e. slope parameter and four location parameters), the person parameter of theta (θ) can be estimated (i.e. IRT scores) for each respondent. Each individual responds to the items differently and each item has different characteristics. Based on these item and response differences, unique IRT-based scores can be found for each HSOPSC dimension. These IRT scale scores can be compared with the original mean scale scoring (i.e. sum of ordinal raw scores on items ranging from '1 = Strongly Disagree' to '5 = Strongly Agree' divided by the number of items comprising each of the HSOPSC dimensions), based on CTT.

Item Characteristic Curves

The relationship between the construct of safety culture (x-axis) and the probability of choosing different response options for every item (y-axis) is graphically illustrated by item characteristic curves (ICCs) as shown in Figure 8.1. The underlying safety culture construct has a mean of 0, a standard deviation of 1 and is normally distributed. For practical considerations the plotted values range from -3 to +3 (Baker 2001). Figure 8.1 displays the slope (*a*) and location parameters (*b*s) of the different response options for a hypothetical example item (cf. Table 8.1). Assume, for interpretive purposes, that the item is: 'I feel patient safety is a priority issue in my organisation' and that respondents can choose from 5 Likert-type response options, 1 = strongly disagree and 5 = strongly agree. Figure 8.1 displays four ICCs that denote differences between those individuals who choose adjacent response options. That is, the ICC furthest on the left represents the likelihood function of moving from choosing 'Strongly Disagree' to 'Disagree', whereas the ICC furthest on the right hand side represents the likelihood function of moving from the 'Agree' to 'Strongly Agree'. If the respondent feels that there is a low priority put on patient safety, then the person would select '1' on the Likert scale, corresponding to 'Strongly Disagree'. If the respondent feels less negatively, then they are expected to move to option '2' ('Disagree'). The respondent will select '1' over '2' if they are lower than -1.5 standard deviations below the mean in terms of his/her level of the latent trait, 'patient safety culture'. The respondent moves from '1' to '2' when his/her 'patient safety culture' trait is more than -1.5 standard deviations below the mean. If the respondent feels that there is even more of a priority put on patient safety in their organisation – his/her latent trait on 'patient safety culture' is -0.5 standard deviations below the mean – the individual would be expected to move to option '3' ('Neither Agree or Disagree'). If the respondent feels that there is even more of a priority put on patient safety in their organisation – their 'patient safety culture' is +0.5 standard deviations above the mean they are expected to move to option '4' ('Agree'). Finally, if they really believe that their organisation prioritises patient safety culture, then they are likely to select option '5' ('Strongly Agree') over '4'. They will only do this if their perception of the patient safety culture is +1.5 or higher standard deviations above the mean.

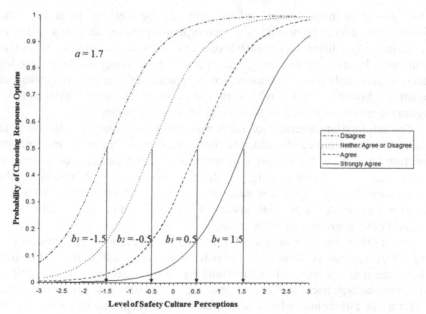

Figure 8.1   Hypothetical Item Characteristic Curves (ICCs) based on estimated item discrimination *a* (slope parameter) and item difficulty *b* (four location parameters)

Thus, one fundamental of IRT is that it highlights at what point along the underlying 'safety culture' construct dimension (ranging from very low to very high for each individual test-taker) each item measures most sensitively and reliably (Kline 2005). Each item can be characterised by the degree to which individuals with varying 'safety culture' perceptions are likely to endorse the different response options (such as 'Strongly Agree' or 'Strongly Disagree').

The slope of the responses (discrimination parameter) represents the precision with which the item measures different levels of the underlying construct of safety culture. Slope parameters larger than 1.35/1.70 are considered to be 'high'/'very high' (Baker 2001). Thus, items with high/very high discrimination power are able to differentiate between individuals on the construct of patient safety culture. Each HSOPSC item has five possible response options, from 'Strongly Disagree' through to 'Strongly Agree' as in the sample item described in Figure 8.1. Using the IRT approach can increase the reliability of the HSOPSC by identifying items with high discrimination power. At the same time construct validity can be ensured by identifying HSOPSC items with response options endorsed by individuals located throughout the entire scale. This means that the four estimated response option parameters should ideally be spread across the full range of the underlying safety culture scale.

Measurement Information

The amount of measurement information can be derived from the item characteristic curves by providing a number that represents an item's ability to assess individual differences at each level of patient safety culture. In particular, items with location parameters spread across the full range of the underlying safety culture scale provide measurement information throughout the measured construct. Moreover, items with higher discrimination power provide more measurement information than those with lower slope parameters.

The information functions of items measuring the separate dimensions of patient safety culture can be added to form category or dimension information functions. Similarly, measurement information for each category or dimension can be added to provide insights into the measurement precision of the entire scale. Item, category and scale measurement information functions are analogous to item and test reliability in CTT. However, an important difference is that, in IRT measurement information (precision) can vary at different levels of patient safety culture, whereas in CTT reliability (precision) is the same for all individuals regardless of their level of patient safety culture (Reise et al. 2005). Thus, measurement information functions are a very important characteristic of IRT methodology because it shows how each item, dimension, or entire scale performs in estimating individual respondents' perceptions of patient safety culture throughout the whole range of possible levels. Because information functions are directly related to the standard error of measurement (*SE*) and reliability (*r*), this allows for the determination of where the underlying patient safety culture scale measures most reliably. Across all HSOPSC dimensions the amount of measurement information should be high in order to accurately assess individuals presenting various levels of perceptions of patient safety culture.

Invariance

A third fundamental of IRT is related to invariance, which has two implications. Firstly, the person parameter Θ (cf. Table 8.1) can be estimated from responses by an individual to any set of items with known item characteristic curves (Reise et al. 2005). Secondly, the estimated item slope and location parameters are independent of the sample under study (Reise et al. 2005). This invariance is displayed below, where individuals with low levels (Figure 8.2a) and with high levels (Figure 8.2b) of safety culture perceptions responded to the same item.

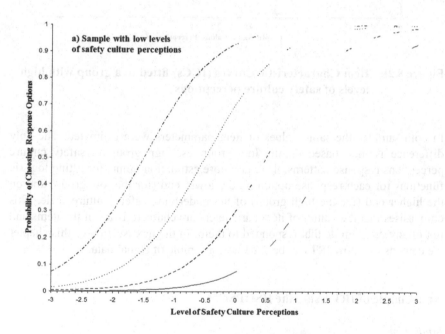

Figure 8.2a   Item Characteristic Curves (ICCs) fitted to a group with low levels of safety culture perceptions

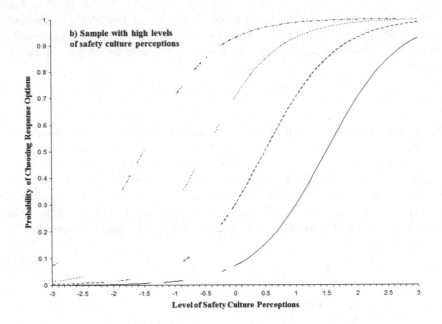

**Figure 8.2b Item Characteristic Curves (ICCs) fitted to a group with high
levels of safety culture perceptions**

In both samples the same values of item parameters were estimated. The only
difference is that, based on the low group vs. high group in safety culture
perceptions response patterns, there are more estimation points for fitting logistic
functions for each response option on the lower end (for the low group) and on
the higher end (for the high group) of the underlying safety culture scale. This
emphasises that the values of item parameters are characteristics of the items and
not of any one sample that responded to them. In the next section of this chapter
we demonstrate how IRT can be used with a sample of actual data.

An Example of IRT Using the HSOPSC

Study Aim

The purpose of this study was to probe deeper into the psychometric properties
of the HSOPSC using the IRT approach. In particular, using IRT will allow a
determination of how well items can discriminate between individuals who have
varying levels of perceptions of patient safety culture. This can be used to judge the
quality of single items and thus, enhance the HSOPSC's ability to provide valid and
reliable measurement information independently of the sample under study.

The Hospital Survey on Patient Safety Culture

The HSOPSC was originally designed for the US healthcare context to provide a means of measuring safety culture within hospitals and to facilitate comparisons between departments, wards and sub-units within them, involving a range of clinical and non-clinical staff. More recently, the US Agency for Healthcare Research and Quality has designed a comparative database containing the HSOPSC results from 1,128 US hospitals (Agency for Healthcare Research and Quality). Findings from using the HSOPSC have been reported from a number of countries within Europe, Asia and the Middle-East (Waterson and Hutchinson 2011).

The HSOPSC questionnaire is based on a set of pilot studies carried out in 21 different hospitals, involving 1,437 hospital staff across the US (Sorra and Nieva 2006). As a result of a series of item and content, reliability and exploratory and confirmatory factor analyses, the resultant 42-item scale has been grouped into 14 dimensions: four outcome dimensions and ten safety perception dimensions (seven related to unit level perceptions and three related to hospital level perceptions). For each item there are five possible response categories and the anchoring of these categories varies across dimensions. Of the 42 items, 17 have a 'negative' valence and are subsequently reverse-scored when analysing the data or creating subscale or total scores.

In order to tailor the HSOPSC to a UK healthcare setting, a number of changes were made to the original version of the instrument. Specifically, in 11 items, the expressions 'area' and 'unit' were changed to 'ward' and 'department'; for five items 'error' and 'mistake' were replaced by 'adverse outcome'; 'over and over' was replaced by 'repeatedly' for one item; two items were excluded based on feedback from hospital management and change of item meaning (see Waterson et al. 2010, for more details).

Participants and Response Rates

Three hospitals within a large NHS Acute Trust in the East Midlands were contacted and invited to participate in the study. At the time of the study the NHS Trust was planning a set of initiatives designed to improve patient safety within acute care. Part of this involved assessing levels of patient safety in hospitals. The chair of the local ethics committee, as well as the research and development department, approved the present study. Ward management staff distributed 4,000 questionnaires to clinical and non-clinical staff across the three hospitals. One thousand four hundred and seventy-one questionnaires (60 percent nursing staff, 21 percent allied health professionals, 11 percent management and administrative staff and 8 percent medical staff) were returned by post in an envelope provided (a 37 percent response rate). Specifically, the highest response rates were obtained among healthcare professionals (50 percent) and nursing staff (27 percent). In contrast, the lowest response rates were obtained in junior doctors (2 percent), registrars (4 percent), administrative staff (4 percent), and management (11 percent). These response rates are comparable with other studies using the HSOPSC (Bodur and Filiz 2010; Chen and Li 2010; Hellings et al. 2007; Olsen 2008; Pfeiffer and Manser 2010; Sarac et al. 2011; Smits et al. 2008; Waterson et al. 2010). It should be noted that they are likely to be underestimated because, while the number of questionnaires sent to each hospital was known, it was not known how many were actually distributed within each hospital unit.

Statistical Analysis

The MULTILOG 7 software program (Thissen 1991) was used to fit a graded IRT response model (Samejima 1969) estimating item discrimination and location parameters (Embretson and Reise 2000) for each HSOPSC item. SPSS software (PASW 18.0) was used to calculate correlations between mean scale scores and corresponding IRT scale scores.

Results

Table 8.2 displays HSOPSC discrimination (a), as well as the four location parameters (b_1, b_2, b_3, and b_4) for each item. Note that HSOPSC items labelled with an 'r' are reverse-scored. Items with poor discrimination (i.e. $a < 1.35$) are highlighted. These items (B3r, B4r, A9, A11, A5r, A7r, F6r and A15) match those that were dropped, based on the results of an exploratory factor analysis conducted on data for the present study (Waterson et al. 2010), which highlights that the slope parameter is analogous to a factor loading (item-factor correlation) in factor analysis. In addition, item F11r *'Shift changes are problematic for patients in this hospital'* also had low discriminating power.

Table 8.2   Item parameters for each HSOPSC dimension and outcome measure (correlation between IRT-scoring and original mean-scale-scoring)

HSOPSC Item	a	b₁	b₂	b₃	b₄
Individual Perceptions of Unit Level Dimensions					
1. Supervisor Expectations (.80**)					
B1 My supervisor manager provides positive feedback when he/she sees a job done according to established patient safety procedure	2.24	-6.6	-2.0	-0.8	-0.1
B2 My supervisor manager seriously considers staff suggestions for improving patient safety	3.43	-5.8	-2.1	-1.3	-0.6
B3r Whenever pressure builds up, my supervisor/ manager wants us to work faster, even if it means taking shortcuts	1.24	-9.4	-3.4	-1.7	-0.6
B4r My supervisor/manager overlooks patient safety problems that happen repeatedly	0.73	-13.6	-4.6	-2.3	-1.0
2. Organisational Learning (.88**)					
A6 We are actively doing things to improve patient safety	2.56	-6.2	-3.0	-1.9	-0.8
A9 Mistakes have led to positive changes here	1.19	-9.5	-3.8	-2.2	-0.6
A13 After we make changes to improve patient safety we evaluate their effectiveness	1.67	-7.5	-3.3	-1.5	-0.2
3. Teamwork Within Units (.82**)					
A1 People support one another in this ward/ department	3.60	-5.6	-2.6	-1.5	-1.0
A3 When a lot of work needs to be done quickly we work together as a team to get the work done	2.18	-6.6	-2.8	-1.7	-1.1
A4 In this ward/ department people treat each other with respect	2.53	-6.2	-2.4	-1.3	-0.7
A11 When one area in this ward department gets busy others help out	0.92	-11.6	-3.1	-0.9	-0.1
4. Communication Openness (.78**)					
C2 Staff will freely speak up if they see something that may negatively affect patient care	2.36	-6.2	-3.0	-1.9	-0.5
C4 Staff feel free to question the decisions or actions of those with more authority	1.84	-7.0	-2.4	-1.0	0.4
C6r Staff are afraid to ask questions when something does not seem right	1.45	-8.3	-3.1	-1.8	-0.1
5. Feedback and Communication About Error (.83**)					
C1 We are given feedback about changes put into place based on event reports	2.14	-6.5	-2.4	-1.1	0.3
C3 We are informed about events that happen in this ward/ department	2.62	-6.0	-2.4	-1.2	-0.1
C5 In this ward/department we discuss ways to prevent events from happening again	1.88	-6.9	-2.8	-1.5	-0.1

Table 8.2   Item parameters for each HSOPSC dimension and outcome measure (correlation between IRT-scoring and original mean-scale-scoring) (*continued*)

6. Nonpunitive Response to Error (.92**)					
A8r Staff feel like their mistakes are held against them	1.95	-6.9	-2.3	-0.8	0.1
A16r Staff worry that mistakes they make are kept in their personal files	2.02	-6.8	-2.1	-0.4	0.7
7. Staffing (.85**)					
A2 We have enough staff to handle the workload	1.48	-8.2	-1.5	0.2	0.8
A5r Staff in this ward/ department work longer hours than is best for patient care	0.94	-11.3	-2.8	-0.8	0.8
A7r We use more agency/temporary staff than is best for patient care	0.56	-15.8	-6.1	-0.7	-0.4
A14r We often work in crisis mode trying to do too much too quickly	2.82	-5.8	-1.3	0.1	0.8
Individual Perceptions of Hospital Level Dimensions					
8. Hospital Management Support (.88**)					
F8 The actions of hospital management show that patient safety is top priority	3.45	-6.3	-1.9	-0.7	0.2
F9r Hospital management seems interested in patient safety only after an adverse event happens	1.89	-7.2	-1.9	-0.1	0.7
9. Teamwork Across Hospital Units (.91**)					
F2r Hospital Wards Departments do not co-ordinate well with each other	2.08	-6.8	-1.4	0.2	1.1
F4 There is good co-operation among hospital wards/ departments that need to work together	2.31	-6.4	-2.3	-0.7	0.3
F6r It is often unpleasant to work with staff from other hospital wards/departments	0.81	-12.7	-4.9	-2.3	-0.1
F10 Hospital wards/departments work well together to provide the best care for patients	2.10	-6.6	-2.8	-1.2	0.0
10. Hospital Handoffs and Transitions (.95**)					
F3r Things fall between the cracks when transferring patients from one ward/department to another	1.89	-7.0	-1.4	0.3	1.3
F5r Important patient care information is often lost during shift changes	2.28	-6.4	-1.9	-0.5	0.4
F7r Problems often occur in the exchange of information across hospital wards/departments	2.39	-6.3	-1.8	0.1	1.0
F11r Shift changes are problematic for patients in this hospital	1.28	-9.1	-3.0	-0.9	0.7
Outcomes					
11. Overall Perception of Safety (.84**)					
A10r It is just by chance that more serious mistakes don't happen around here	1.60	-7.9	-2.3	-0.7	0.1
A15 Patient safety is never sacrificed to get more work done	1.10	-10.3	-3.1	-1.1	-0.1

Table 8.2 **Item parameters for each HSOPSC dimension and outcome measure (correlation between IRT-scoring and original mean-scale-scoring) (*concluded*)**

A17r We have patient safety problems in this ward/ department	1.85	-7.2	-2.7	-1.0	-0.1
A18 Our procedures and systems are good at preventing errors from happening	1.45	-8.4	-3.3	-1.8	-0.5
12. Frequency of Event Reporting (.72**)					
D1 When an event occurs but is caught and identified before affecting the patient how often is this reported	1.91	-6.9	-3.0	-1.5	-0.2
D2 When an event occurs but has no adverse outcome to the patient how often is this reported	4.86	-6.3	-2.4	-1.2	-0.3
D3 When an event occurs that could have an adverse outcome to the patient but does not, how often is this reported	2.90	-5.9	-3.0	-1.6	-0.5
13. Patient Safety Grade (.74**)					
E Please give your Ward/ Department in this hospital an overall grade on patient safety	1.62	-8.2	-3.9	-2.6	-0.5
14. Number of Events Reported (.86**)					
G1 In the past 12 months how many incident reports have you filled out and submitted	1.44	-9.0	-0.8	0.7	1.9/ 3.0#

Note: Items labelled with an 'r' were reverse-scored. ** Correlation between the original mean scale scoring (i.e. sum of ordinal raw scores on items ranging from '1 = Strongly Disagree' to '5 = Strongly Agree' divided by the number of items comprising each of the HSOPSC dimensions) and corresponding IRT scale scores for each HSOPSC dimension is statistically significant at the $p < .001$ level (2-tailed). # Threshold parameter b_5 (item G1 had 6 response options and thus 5 threshold parameters). Items shadowed in grey indicate items with poor discrimination (< 1.35).

As described in the introduction, measurement information can be derived from the item characteristic curves (ICCs) by providing a number that represents an item's, category's or scale's ability to measure individual differences at each level of patient safety culture. To illustrate this, the three graphs on the left in Figure 8.3a display ICC and item information for the HSOPSC item measuring 'Teamwork within Units' that had the highest discrimination power, as well as the total amount of measurement information provided by all four items within the dimension 'Teamwork within Units'. The three graphs on the right in Figure 8.3b display ICC and item information for the HSOPSC item measuring 'Staffing' that had the lowest discrimination power, as well as the category information provided by all four items measuring 'Staffing'.

Comparing the ICCs for both items shows that the response options of the HSOPSC item in Figure 8.3a are sensitive ($a > 1.35$) and will discriminate between individuals with different levels of safety culture perceptions, except for those who choose 'Strongly Disagree' versus 'Disagree' because the estimated difficulty parameter b_1 is located below -3 standard deviations. In contrast, the response options of the HSOPSC item in Figure 8.3b are not sensitive ($a < 1.35$).

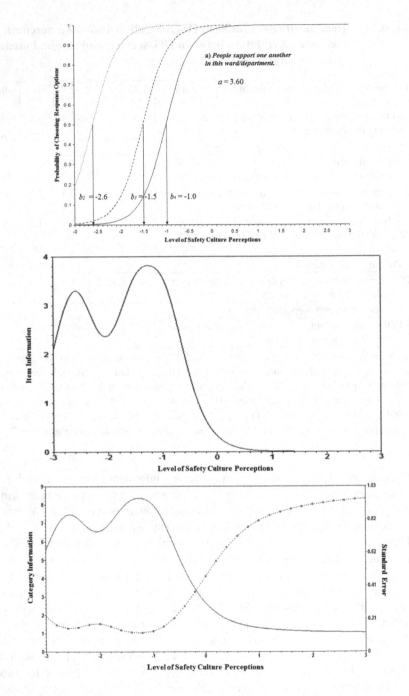

Figure 8.3a   Item Characteristic Curves (ICCs), Item Information, and 'teamwork within units' Category Information

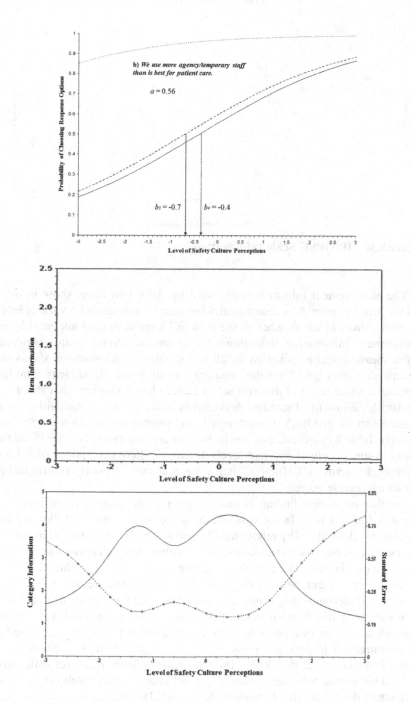

b) *We use more agency/temporary staff than is best for patient care.*

$a = 0.56$

$b_3 = -0.7$ $b_4 = -0.4$

Figure 8.3b   Item Characteristic Curves (ICCs), Item Information, and 'staffing' Category Information

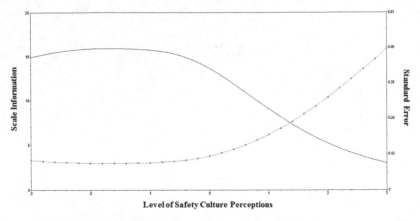

Level of Safety Culture Perceptions

Figure 8.3c   HSOPSC Scale Information

The measurement information provided by these two items show in detail that the item in Figure 8.3a discriminates between individuals with various levels of safety culture below 0, whereas the item in Figure 8.3b does not provide any measurement information throughout the entire underlying scale. However, adding measurement information for all items within each dimension shows that 'Teamwork within Units' reliably measures levels below 0, whereas 'Staffing' measures a wider range of different safety culture levels ranging from about -1.5 standard deviations to +1 standard deviation as indicated by low standard errors of measurement (dotted line). The estimated item parameters for the four 'Staffing' items (cf. Table 8.2) indicate that two items measure sensitively ($a < 1.35$) and two items measure poorly at different levels of safety culture perceptions ($a > 1.35$). Interestingly, across all HSOPSC items the majority of poorly discriminating items were reverse-scored.

Another interesting finding is that the prior study using a factor analytical approach found a nine factor structure to be optimal, with the following two dimensions deleted: 'Organisational Learning' and 'Hospital Management Support'. However, four out of the five items within these categories showed good discrimination in our IRT analyses, suggesting that these items should be kept, at least for the present, because they provide valuable measurement information. These two dimensions might not have been supported by factor analysis since there were only two items in each of these two dimensions. However, the IRT approach allows an evaluation of item and dimension performance, as well as the measurement information provided by the entire HSOPSC instrument. As shown in Figure 8.3c, the entire HSOPSC scale reliably measures differences between individuals with perceptions of patient safety culture levels ranging from -3 standard deviations to +1 standard deviations. However, the instrument fails to differentiate well among individuals with high levels of patient safety culture perceptions.

The amount of measurement information is influenced by the discrimination power and the range of measurement information is influenced by location parameters spread across the entire underlying safety culture scale. Therefore, the following paragraphs report in more detail the estimated location parameters for each HSOPSC item. The first location parameter, b_1, discriminates between those people who 'Strongly Disagree' and those who 'Disagree'. As can be seen in Table 8.2, b_1 is located below -3 standard deviations across all HSOPSC items, indicating that the likelihood of respondents selecting 'Strongly Disagree' or 'Disagree' for any of the items is negligible.

The second location parameter, b_2, discriminates between those people who 'Disagree' and those who 'Neither Agree/Disagree'. Estimates for b_2 ranged from -3.9 to -1.3 for items with slope parameters above 1.35. These results imply that even people who are about 3 standard deviations below the mean on the construct of patient safety culture are likely to respond at least 'Neither Agree/Disagree' on a 5-point Likert scale. For the *Number of Events Reported* the b_2 parameter was -0.8 (almost 1 standard deviation below the mean on patient safety culture) indicating that at this point respondents were likely to report at least three to five incidents (option '2' in the Likert-type scale).

The third location parameter, b_3, discriminates between those people who 'Neither Agree/Disagree' and those who 'Agree'. Estimates for b_3 ranged from -2.6 to +0.3 for HSOPSC items with slope parameters above 1.35. This implies that even people who are about 3 standard deviations below the mean are likely to at least 'Agree' on a 5-point Likert scale. For the *Number of Events Reported* the b_3 parameter was +0.7 (almost 1 standard deviation above the mean on patient safety culture) indicating that at this point respondents were likely to report at least six to ten incidents.

The fourth location parameter, b_4, discriminates between those people who 'Agree' and those who 'Strongly Agree'. Estimates for b_4 ranged from -1.1 to +0.3. This suggests that respondents at the average level of patient safety culture are likely to select the highest point on the Likert scale. For the *Number of Events Reported* the b_4 parameter was +1.9 (almost two standard deviations above the mean on patient safety culture) indicating that at this point respondents were likely to report at least 11 to 20 incidents. Finally, because the *Number of Events Reported* item had six response options, a fifth location parameter was estimated (b_5) to be +3.0 (three standard deviations above the mean on patient safety culture), indicating that at this point respondents were likely to report more than 20 incidents.

Table 8.2 also displays Pearson's correlation coefficients between the original mean scale scoring as often reported in studies using the HSOPSC (i.e. sum of ordinal raw scores on items ranging from '1 = Strongly Disagree' to '5 = Strongly Agree' divided by the number of items comprising each of the HSOPSC dimensions) and corresponding IRT-based person scores for each HSOPSC dimension. Each correlation was strong and positive as well as statistically significant at the $p < .0001$ level and ranged from $r = .72$ (*Frequency of Event Reporting*) to $r = .95$ (*Hospital Handoffs and Transitions*), indicating that that the rank orders of individuals are similar using both methods for estimating test scores.

Discussion

The present study provides insights into the discrimination power of individual HSOPSC items. The majority of the poorly discriminating items were also those that were dropped when an alternative factor structure was proposed (Waterson et al. 2010). Moreover, the majority of poorly discriminating items were those with a 'negative' valence (reverse-scored). One possible explanation might be that respondents tend to agree or strongly agree with positive items but do not disagree or strongly disagree with a negatively worded item. In addition, this phenomenon of poor performance by negative valence (reverse-coded) items has been documented elsewhere (DeVellis 2003; Kline 2005; Netemeyer et al. 2003). While it is a laudable goal to ensure that respondents 'pay attention' by changing the valence of some items, this has an unintended consequence of confusing respondents who are pressed for time and/or are not used to completing questionnaires. Respondents in this study were a field sample of busy healthcare system workers and thus may not have been used to confronting negative valence items. Our findings in this area support the contention that use of negative valence items is not appropriate when using non-university student samples.

The present study also provides insight into the appropriateness of the choice of response options for each HSOPSC item. The HSOPSC items did not discriminate between people who 'Strongly Disagreed' versus 'Disagreed' indicating that those response options could be collapsed. Most importantly, the low values of the location parameters, as well as the scale measurement information function (cf. Figure 8.3c), indicate that the HSOPSC discriminates among individuals with patient safety culture levels ranging from -3 standard deviations to +1 standard deviation. Given these results, perhaps changing the response options to reflect more precise levels of agreement could increase measurement information throughout the underlying safety culture scale, including +1 to +3 standard deviations. For example, response options could be changed to a frequency-based anchoring system such as 'Never agree', 'Rarely agree', 'Sometimes agree', 'Often agree', or 'Always agree'.

In addition, it was found that the 'neutral' response option includes important measurement information which has implications for studies reporting positive percentage scores and ignoring 'neutral' response options (Jackson et al. 2010). For example, 24 to 36 percent of respondents chose the 'neutral' option when rating the four items measuring the dimension 'Hospital Handoffs and Transitions'. Similarly, high frequencies of 'neutral' responses were shown for items measuring 'Teamwork Across Hospital Units', 'Hospital Management Support', 'Organisational Learning', 'Staffing', and 'Nonpunitive Response to Error'.

Person scores based on original mean scoring and IRT scoring correlated highly, as has been shown in previous studies (see for example MacDonald and Paunonen 2002). Thus, choices between these methods may seem inconsequential because the rank ordering of individuals remains about the same across both scoring methods. However, findings from this study provide evidence for strongly non-linear associations between scores that explicitly take into account ordinal item distributions (i.e. IRT scores) and scores that violate assumptions about underlying item distributions (i.e. original mean scoring). This is in line with the finding that the original mean scoring within the CTT framework is biased at the ends of score distributions (Wright 1999). Moreover, there are reasons for the detrimental consequences of ignoring the non-linear associations between both scoring methods. First, it has been demonstrated that both scoring methods can lead to different conclusions when analysing change over time (Seltzer et al. 1994) or interaction effects between constructs (Kank and Waller 2005). An interaction might be masked if a respondent scores high on one dimension and low on another while, together, the dimensions add to a 'neutral' overall score. Secondly, summed scale scores, but not IRT-based scores, tend to overestimate relationships between variables (Yan et al. 2010) – an issue that would have a detrimental effect on assessments of criterion-related validity.

The present study has limitations. IRT scoring is complex and there is no consensus on model fit (Embretson and Reise 2000). Practical problems arise from IRT software complexity (Hays et al. 2000) and computational power requirements (Hembleton and Swaminathan 1995), as well as lack of agreement in presentation of IRT results. However, the increasing interest in IRT methodology in many disciplines suggest this method will become common with the introduction of more user-friendly software and associated literature around issues of model fit and item selection. Also, it is not known if studies conducted in other countries would produce similar results to those found in the present research. Future research is necessary to determine the generalizability of these findings.

Implications and Future Directions

Theoretical and Practical Implications

The aim of this chapter was to summarise the modern approach used to provide evidence for psychometric scale properties of the HSOPSC and contrast them with the classical approach. Basic concepts (i.e. item characteristics, measurement information, and the group invariance principle) of Modern Test Theory were introduced in order to then contrast IRT findings with findings based on the more frequently used classical factor analytical approach. One major difference between approaches is that factor analysis accounts for the covariance between test items and IRT models account for individual item responses (Reise et al. 1993). However, IRT slope and location parameters are analogous to factor loadings and intercepts in factor analyses (Stark et al. 2006). Thus, the primary approach to address measurement issues of safety culture using factor analytical approaches is studying the covariance patterns of item-factor relations (Windle et al. 1988), whereas IRT focuses on the level of safety culture perceptions presumed to underlie and cause variation in the observed item response (Reise et al. 1993). The advantage of IRT is that the response patterns of each single respondent are analysed and related to the underlying factorial structure, whereas CFA identifies item-factor relations which are not the same for every single participant (Borsboom 2005). In particular, the IRT approach allowed further investigation of the functioning of individual items comprising the HSOPSC scale and results showed that patient safety culture perceptions are accurately measured by the HSOPSC instrument at the low and moderate levels. However, IRT findings suggest that the scale validity and reliability could be improved to differentiate more precisely those at higher levels of safety culture perceptions. In terms of validity, the HSOPSC can be improved by discarding items with poor discrimination power. In terms of reliability, items need to be added which sensitively measure high levels of safety culture. Alternatively, the response options for each HSOPSC item might be changed so that they reflect different levels of agreement rather than ranging from 'Strongly Disagree' to 'Strongly Agree'. Thus, the present item analysis showed that the HSOPSC can be adjusted to enhance the ability of providing credible and accurate information consistently.

Moreover, precise definition and measurement of patient safety culture is important for correctly identifying improvement areas, as well as correctly measuring effectiveness of interventions. Researchers interested in introducing innovations into clinical practice and organisational management need to assess culture so that they can design interventions that will have a high chance of adoption and success. Accurate measurement tools and an understanding of measured dimensions of patient safety culture allow tracking over time to monitor progress in healthcare organisations. The beauty of IRT is that item banks can be created by calibrating items with high measurement information from different patient safety culture instruments, since items can be scaled and compared based on their item characteristics. Item banks can facilitate accurate and precise measurement using computerised adaptive testing, where individuals start to respond to items which have the highest measurement precision at medium levels of patient safety culture. Depending on their level of (dis)agreement, the next presented item would be high in measurement information at (low) high levels of patient safety culture. With each item responded to, the computer estimate of the individual's level of patient safety culture becomes more precise, using fewer items compared to traditional approaches. This is especially important when monitoring patient safety culture on a regular basis, given the workload of clinical staff in their daily practice. In addition, different dimensions of patient safety culture might be important depending on the nature of the intervention or implemented changes into clinical practice. For example, when the health organisation changes from a 3 shift pattern into a 2 change pattern, it might be important to monitor unit level dimensions related to teamwork, communication and staffing whereas, when implementing leadership walkrounds in different clinical areas, hospital level dimensions such as management and leadership support, as well as teamwork and transitions across hospital units might be of interest in assessing patient safety culture perceptions.

Future Directions and Conclusions

Item banks can be created across different patient safety culture instruments and also across different countries or healthcare settings. Based on estimated item parameters, items can be identified with the same and different characteristics across healthcare systems, which will in turn allow a better understanding of similarities and unique attributes depending on the contexts within which healthcare is delivered.

In conclusion, this chapter demonstrates that the IRT approach can provide additional insights to psychometric properties of the HSOPSC. Both, the classical and modern approaches, are needed to form a complete picture of the properties of a set of items. Understanding the basic principles of IRT will hopefully foster its use more widely within the field of patient safety culture assessment. This will ultimately enhance our ability to measure this important construct accurately.

References

Agency for Healthcare Research and Quality (AHRQ 2012). 2012 User Comparative Database Report: Hospital Survey on Patient Safety Culture. AHRQ, Rockville, MD. Available at: http://www.ahrq.gov/qual/hospsurvey12/ (last accessed on 14 June 2013).

Baker, F. 2001. *The Basics of Item Response Theory.* College Park, MD: ERIC Clearinghouse on Assessment and Evaluation. Available at: http://edres.org/irt (last accessed on 9 March 2006).

Bodur, S. and Filiz, E. 2010. Validity and reliability of Turkish version of 'Hospital Survey on Patient Safety Culture' and perception of patient safety in public hospitals in Turkey. *BMC Health Service Research*, 10, 28.

Borsboom, D. 2005. *Measuring the Mind: Conceptual Issues in Contemporary Psychometrics.* Cambridge: Cambridge University Press.

Chan, K.S., Orlando, M., Ghosh-Dastidar, B., Duan, N. and Sherbourne, C.D. 2004. The interview mode effect on the Center for Epidemiology Studies Depression (CES-D) scale: An item response theory analysis. *Medical Care*, 42, 281–9.

Chen, I.-C. and Li, H.-H. 2010. Measuring patient safety culture in Taiwan using the Hospital Survey on Patient Safety Culture (HSOPSC). *BMC Health Service Research*, 10, 152.

Cooke, D.J. and Michie, C. 1997. An item response analysis of the Hare Psychopathy Checklist-Revised. *Psychological Assessment*, 9, 3–14.

DeVellis, R.F. 2003. *Scale Development: Theory and Applications* (Second edition). Thousand Oaks, CA: Sage.

Embretson, S.E. and Reise, S.P. 2000. *Item Response Theory for Psychologists.* Erlbaum: Mahmwah.

European Society for Quality in Healthcare 2010. *Use of Patient Safety Culture Instruments and Recommendations.* EUNetPas Project Report, Aarhus, Denmark.

Flin, R., Burns, C., Mearns, K., Yule, S. and Robertson, E.M. 2006. Measuring safety climate in health care. *Quality and Safety in Health Care*, 15(2), 109–15.

Hays, R.D., Morales, L.S. and Reise, S.P. 2000. Item response theory and health outcomes measurement in the 21st century. *Medical Care*, 38(9), 28–42.

Hellings, J., Schrooten, W., Klazinga, N. and Vleugels, A. 2007. Challenging patient safety culture: Survey results. *International Journal of Health Care Quality Assurance*, 20, 620–32.

Hembleton, R.K. and Swaminathan, H.R. 1995. *Item Response Theory: Principles and Application.* Boston: Kluwer.

Jackson, J., Sarac, C. and Flin, R. 2010. Hospital safety climate surveys: Measurement issues. *Current Opinion in Critical Care*, 16, 632–8.

Kane, M.T. 1987. On the use of IRT models with judgmental standard setting procedures. *Journal of Educational Measurement*, 24, 333–45.

Kank, S.M. and Waller, N.G. 2005. Moderated multiple regression, spurious interaction effects, and IRT. *Application of Psychological Measures*, 29, 87–105.

Kline, T.J.B. 2005. *Psychological Testing: A Practical Approach to Design and Evaluation.* Vistaar Publications: New Delhi.

Lord, F.N. and Novick, M.R. 1968. *Statistical Theories of Mental Test Scores.* Reading, MA: Addison-Wesley.

MacDonald, P. and Paunonen, S.V. 2002. A Monte Carlo comparison of item and person statistics based on item response theory versus classical test theory. *Educational and Psychological Measurement*, 62, 921–43.

Netemeyer, R.G., Bearden, W.O. and Sharma, S. 2003. *Scaling Procedures: Issues and Applications.* Thousand Oaks, CA: Sage.

Olsen, E. 2008. Reliability and validity of the Hospital Survey on Patient Safety Culture at a Norwegian hospital. In J. Øvretveit and P. Sousa (eds), *Quality and Safety Improvement Research: Methods and Research Practice from the International Quality Improvement Research Network (QIRN).* Lisbon: Escola Nacional de Saúde Pública, 173–86.

Pfeiffer, Y. and Manser, T. 2010. Development of the German version of the Hospital Survey on Patient Safety Culture: Dimensionality and psychometric properties. *Safety Science*, 48(10), 1452–62.

Reise, S.P., Ainsworth, A.T. and Haviland, M.G. 2005. Item response theory: Fundamentals, applications, and promise in psychological research. *Current Directions in Psychological Science*, 14(2), 95–101.

Reise, S.P., Widaman, K.F. and Pugh, R.H. 1993. Confirmatory factor analysis and item response theory: Two approaches for exploring measurement invariance. *Psychological Bulletin*, 114, 552–6.

Samejima, F. 1969. Estimation of latent ability using a pattern of graded scores. *Psychometrika Monograph*, 34(4), Pt. 2.

Sarac, C., Flin, R., Mearns, K. and Jackson, J. 2011. Hospital Survey on Patient Safety Culture: Psychometric analysis on a Scottish sample. *BMJ Quality & Safety*, 20, 842–8.

Seltzer, M.H., Frank, K.A. and Bryk, A.S. 1994. The metric matters: The sensitivity of conclusions about growth in student achievement to choice of metric. *Education and Evaluation Policy Annals*, 16, 41–9.

Smits, M., Christiaans-Dingelhoff, I., Wagner, C., van der Wal, G. and Groenewegen, P.P. 2008. The psychometric properties of the 'Hospital Survey on Patient Safety Culture' in Dutch hospitals. *BMC Health Service Research*, 8, 230.

Sorra, J. and Nieva, V. 2006. *Reliability and Validity of the Hospital Survey on Patient Safety*. Rockville, MD: Westat. Ref Type: Report.

Stark, S., Chernyshenko, O.S. and Drasgow, F. 2006. Detecting differential item functioning with confirmatory factor analysis and item response theory: Toward a unified strategy. *Journal of Applied Psychology*, 91(6), 1292–306.

Thissen, D. 1991. *Multilog*. Mooresville, IN: Scientific Software.

Thurstone, L.L. 1947. *Multiple-factor Analysis*. Chicago: University of Chicago Press.

Waterson, P.E., Griffiths, P., Stride, C., Murphy, J. and Hignett, S. 2010. Psychometric properties of the hospital survey on patient safety: Findings from the UK. *Quality and Safety in Health Care*, Published Online First: 8 March 2010. doi:10.1136/qshc.2008.031625 (last accessed on 30 November 2011).

Waterson, P.E. and Hutchinson, A. 2011. Use of the Hospital Survey of Patient Safety Culture – A review of the current evidence base, in S. Albolino, S. Bagnara, T. Bellandi, J. Llaneza, G. Rosal-Lopez and R. Tartaglia (eds), *Healthcare Economics and Patient Safety (HEPS 2011)*, 22–4 June 2011, Oviedo, Spain. Taylor & Francis: CRC Press, 473–4.

Windle, M., Iwawaki, S. and Lerner, R.M. 1988. Cross-cultural comparability of temperament among Japanese and American preschool children. *International Journal of Psychology*, 23, 547–67.

Wright, B.D. 1999. Fundamental measurement of psychology. In S.E. Embretson and S.L. Hershberger (eds), *The New Rules of Measurement: What Every Psychologist and Educator Should Know*. Mahwah, NJ: Erlbaum, 65–104.

Yan, C., Nay, S. and Hoyle, R.H. 2010. Three approaches to using lengthy ordinal scales in structural equation models: Parcelling, latent scoring, and shortening scales. *Applied Psychological Measurement*, 34(2), 122–42.

Chapter 9

Patient Safety Indicators as Tools for Proactive Safety Management and Safety Culture Improvement

Teemu Reiman and Elina Pietikäinen

Introduction

The effort to improve patient safety has increasingly focused on the management of the healthcare organisation. It has been realised that management creates the preconditions for safe care. Concepts, practices and tools from other safety-critical industries and from quality management have been adopted to manage patient safety in hospitals and other healthcare organisations. Among them is the systematic use of safety indicators to measure, monitor and develop patient safety.

If used correctly, patient safety indicators can help leaders and other agents working in the healthcare system to manage safety in a proactive manner, that is, they can be used as tools for patient safety culture improvement. However, safety indicator systems can also become extra bureaucracy for the organisation and, in the worst case, narrow or misdirect people's attention and blindfold the organisation to its actual hazards. Thus, indicators have to be used in a considerate manner based on a sound model of safety management. Safety indicators can contribute to a good patient safety culture but, if misused, they can create hazards instead of detecting them.

In this article, we discuss the use of patient safety indicators in proactive patient safety management and safety culture improvement. We aim to illustrate how indicators can be used in healthcare organisations to facilitate and support a strong organisational patient safety culture. In light of recent safety theories, we start by discussing what patient safety management in a complex healthcare organisation should look like. We continue by describing how safety indicators have been used in healthcare and other safety-critical industries. We then propose a model of patient safety indicators that is in line with recent theories of system safety and give examples of possible indicators to be used as part of an organisation's patient safety management. Finally, we point out key issues to be considered when using patient safety indicators effectively to improve organisational safety culture.

Patient Safety Management

The way safety is managed in an organisation depends heavily on the beliefs and assumptions managers have concerning organisational behaviour and safety (Reiman and Rollenhagen 2011). Traditionally, safety management in safety-critical industries (such as nuclear power, aviation and oil) has been based on the idea of safety as a lack of incidents and accidents. It has also assumed the future to be relatively knowable and the activities in the organisation well-defined, regular and stable. Procedures and instructions have been considered the appropriate tools to define and regulate behaviour and its intended outcomes. The role of management in supervising and directing organisational behaviour has been emphasised. Good safety culture has been characterised by, for example, a clear division of responsibilities, upper management safety commitment, awareness of risks and adherence to rules and procedures. In line with these assumptions, safety management has focused on identifying the possible ways in which things can go wrong and then seeking to prevent these deviations by implementing barriers, emphasising procedural adherence, creating redundancy, supervising work and making the distribution of responsibilities clear. The numbers of accidents and other negative events, such as breakdowns, adverse events and process leaks have been used as indicators of safety. No doubt developments in the traditional safety management paradigm have very often contributed to safety and will continue to do so. The effects have not always been as positive as predicted, however, and unexpected negative events have continued to occur. Human error has been found to be too narrow a concept to explain adverse events, much less patient safety as a phenomenon. The phenomenon of safety itself has recently been reconceptualised in a broader way, in part because the traditional safety management paradigm has proven somewhat ineffective in preventing harm.

Contrary to the traditional view, safety can be seen not only as the absence of something negative but also as the presence of something positive. Instead of aiming to predict and control individual acts, proactive safety management is putting more focus on increasing the potential of the organisation to cope with the work on a daily basis. Safety management has been seen as requiring an understanding of various normal organisational phenomena, coupled with a shared model of safety that encourages and endorses a holistic view of the dynamics of the sociotechnical system in question.

The management of safety has been conceptualised as culminating in the problem of system control (Rasmussen 1997; Reiman and Oedewald 2009). Hollnagel and Woods (2006: 348) summarised 'in order to be in control it is necessary to know what has happened (the past), what happens (the present) and what may happen (the future), as well as knowing what to do and having the required resources to do it.' However, as healthcare organisations are complex adaptive systems, composed of various semi-independent agents, understanding what is really going on in them (the present) is challenging for the leaders and other agents in the organisation. Since all the agents that act in the system have the ability to create their own future, it is also practically impossible to know exactly what will happen in the organisation in the future. In fact, instead of talking about the future, it would be more beneficial to talk about several possible futures. Patient safety indicators should help in understanding all three of these aspects – the past, the present and the potential future – not just the past, as outcome indicators such as adverse event counts do. Furthermore, the indicators should help managers guide the system towards the envisioned future and away from other imagined – and unwanted – futures.

New proactive safety management approaches often view the healthcare system as a complex adaptive system that cannot be controlled in the traditional sense of the word (e.g. De Savigny and Taghreed 2009; Dekker 2011; Plsek and Greenhalgh 2001; Rouse 2007). Phenomena such as self-organisation and emergence have been seen as inherent characteristics of healthcare organisations. Activities in healthcare organisations involve a variety of rather independent – semi-autonomous – agents (such as patients and their family members, nurses, medical doctors and managers) who all have the ability to process information, change their activity accordingly and exchange information with other agents, and who can thus affect the functioning of the entire system. New structures and forms of behaviour to deal with the challenges at hand emerge spontaneously and nonlinearly from the interactions between different agents. Furthermore, there are always tensions between the aspirations, goals and conceptions of different agents. These tensions affect the performance of the system and may thus have safety impacts. Patient safety management in complex adaptive healthcare organisations should in fact be seen as something that takes place in the interactions between the different agents, not something done by an individual leader. In practice, many of the activities of managing and leading safety take place in everyday collaboration and discussions.

As healthcare organisations comprise several semi-autonomous agents that all take part in the activity of managing safety, it is not enough to rely only on traditional, centralised control strategies. To manage patient safety in a complex adaptive system effectively and proactively, diverse – and paradoxically somewhat contradictory – principles need to be applied (e.g. see Pietikäinen et al. 2012), that is, it is not enough to provide boundaries and limitations for the system in the form of rules, barriers and prohibitions for the people working in the organisation. Proactive safety management is also about supporting the agents' capability to self-organise in a safe manner. This means giving permission to cross and redefine boundaries and roles, as well as to adjust and interpret rules according to situational requirements. This also means that leaders, as well as other agents of the system, need to maintain mindfulness towards potential emerging risks (McDaniel and Driebe 2001) and sustain tension in the system to allow for innovation, development and change (Lichtenstein et al. 2006). Slack resources and time for reflection are also critical to the ability to organise situationally (McDaniel and Driebe 2001). In addition to boundaries and the capacity to self-organise, the system needs shared guiding principles for its activities. Leaders need constantly to orient the organisational activity towards the core task of the organisation and incorporate safety considerations into the core task. To facilitate information flow and collaboration, an open climate is needed. This means an environment in which people trust each other, respect each other's competences and are willing to share information and learn from each other (McDaniel and Driebe 2001). Finally, leaders need to steer the organisation in a goal-oriented manner. They need to prioritise and select areas where they will focus their effort and emphasise some connections and persons over others, depending on their potential contributions to organisational goals. Safety indicators can support prioritisation and also be used to facilitate interaction, define boundaries, enforce safety values and encourage innovation and self-organising.

Indicating Safety

Safety Performance Indicators

Wreathall (2009: 494) defines safety indicators as 'proxy measures for items identified as important in the underlying model(s) of safety'. When defining safety indicators, we are making explicit and systematising the issues we understand to be important in creating safety and on which we want people to focus. Conversely – whether or not we want to – at the same time, we are defining what is not so important and what does not need so much attention. The influence of the underlying safety model can be explicit when a certain model is acknowledged as being the background model of the indicator system or implicit when the indicators are chosen based on standards or regulations, or on taking what everyone else is using. Naturally, having an explicit safety model is a prerequisite for the proactive use of indicators in safety management.

Safety indicators have commonly been used in safety-critical industries such as aviation (Herrera 2012), nuclear (EPRI 2001, www.wano.org) and chemical (HSE 2006). In these industries, safety indicators are often categorised into two types – 'lead' and 'lag'. Lead and lag indicators are typically considered on a time-scale on which lead indicators precede unwanted events or incidents and lag indicators follow the unwanted events. Lag indicators show the safety performance in terms of measures of past performance, e.g. infection rates, adverse events and medication errors. Lead indicators target the processes and activities, creating the safety performance outcomes. However, this distinction is not clear-cut. Some safety scientists and practitioners have described them more as a continuum than as two separate entities and have even suggested that the distinction between leading and lagging is not all that important (Hale 2009; Hopkins 2009; Wreathall 2009; see also Reiman and Pietikäinen 2012). As recent safety theories view safety as an emergent property of the functioning of the entire system (e.g. Hollnagel 2004; Dekker 2005), indicators should not only indicate the abnormal and negative, they should also indicate how the system functions normally and be sensitive to changes before those changes actualise as harm (Kjellén 2009: 486). For the indicator to be sensitive to changes in the organisational risk control system that predate the rise of the risk level, it cannot focus on 'failings', 'holes' or even 'near misses' or 'deviations'. The indicator has to provide information on the organisational activities and the organisational means of controlling risk. Thus, leading indicators should measure things that may one day become precursors to harm or cause a precursor to harm, as well as things that contribute to safety in positive terms. Since lead/lag distinction is ultimately a question of a reference point on a timeline, if we are indicating safety (and not adverse events as is usually done), 'lag' then implies those indicators that measure past changes in safety levels whereas lead indicators measure (future) potential for safety. However, it can be argued that the true lead indicator of future safety is impossible to derive, and all indicators are thus by necessity lag indicators. Thus, instead of speaking about lead and lag indicators it is more beneficial to speak about monitor, drive and outcome indicators (Reiman and Pietikäinen 2012). We will return to this argument later in the article.

Patient Safety Indicators

In recent years, much international effort has been put into the healthcare domain for selecting valid and feasible patient safety indicators (e.g. EUNetPas 2010; Kristensen et al. 2007; Millar and Mattke 2004). However, the purpose of using patient safety indicators has not always been clearly articulated. Patient safety indicators are often considered national or international tools for defining large-scale political goals and following whether these goals are met. Safety indicators can also be seen as a way to communicate safety issues to the public, thus increasing transparency and accountability of care. In both of these cases, indicators should be comparable. They should be such that they can be used in several diverse contexts and compared in a reliable way between organisations, or even countries. In this chapter, however, we consider patient safety indicators an integral part of a healthcare organisation's safety management. We conceptualise them as tools that can provide information on the performance of a specific organisation from a safety perspective, motivate people to work on safety and ultimately increase the organisational potential for safety. In this case, inter-organisational comparability is not necessarily the only or first criterion for selecting indicators. Indicators should be such that they contribute towards better learning and sense-making in the organisation.

Patient safety indicators have typically been understood as measures that produce information on the number of unsafe outcomes in healthcare. The indicator data typically collected and followed in healthcare deal with infections associated with healthcare, such as urinary infections from catheters and a central line due to bloodstream infections (EUNetPas 2010). In the light of recent system safety theories, this use of indicators seems narrow since it reflects the idea of patient safety as a lack of adverse events.

If patient safety is understood as something more than the negation of risk and absence of adverse events, indicators should also focus on this positive side of safety – on the presence of something (Hollnagel 2008: 75; Rollenhagen 2010). Patient safety can be considered in a positive light as an emergent property of everyday interactions of patients, their relatives, individual professionals, technology, units and organisations. Thus, safety indicators should measure the presence of organisational attributes that enable safe everyday interaction. In this way, the safety indicators can be used proactively as part of patient safety management in the given healthcare organisation.

A focus on the positive side of safety does not mean that anything negative cannot or should not be measured. Negative indicators, such as the number of medication errors or patient safety incidents, are most likely to alert the public and personnel and they can be useful in safety management if used correctly. However, they do not directly tell us why problems occur. Even more importantly, they do not tell us about why certain problems did not occur and why certain activities have been successful most of the time. Thus, these indicators alone do not give us the whole picture of what is going on in the organisation's daily activities. Nor do they alone enable us to steer the organisation towards better patient safety. A more proactive approach is needed in order to develop organisational capability to deliver high quality care.

Framework for Patient Safety Indicators

We have been developing a theoretical framework for using safety performance indicators in safety-critical organisations that incorporates three types of safety performance indicators – outcome, monitor and drive (Reiman and Pietikäinen 2012). Underlying the model is a conception of a healthcare organisation as a complex adaptive system (CAS) and patient safety as an emergent property of that system. The framework is based on the idea that indicators should measure the presence of certain organisational attributes necessary for safety instead of only indicating harm.

Figure 9.1 illustrates the influence of a safety model on using indicators. The safety model in Figure 9.1 refers to the underlying, often implicit, ideas of what safety is and how it is achieved in an organisational context. The safety model includes the hazards that are perceived as threatening the organisation, personnel or clients. Hazards that are not acknowledged or foreseen in the model will not be transformed into safety interventions and corresponding drive indicators. Thus, the safety model influences what kind of data on safety are gathered and how the data will be interpreted. The conception of current safety level in Figure 9.1 refers to how leaders perceive their organisation in terms of safety: how well it is doing and what the strengths and weaknesses are in terms of safety. These interpretations are influenced by how the leaders understand safety (safety model). The conception of safety level in turn influences the goals that are set for safety development and thus which issues are selected as drive indicators.

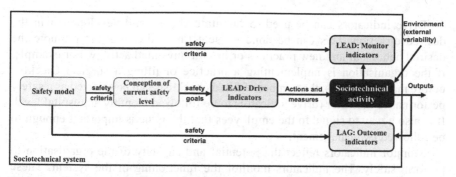

Figure 9.1   System model illustrating the different types of indicators
Source: Adapted from Reiman and Pietikäinen 2012.

Outcome indicators are among those typically called 'lag indicators' in safety science literature, as the outcomes always follow something; they are the consequences arising from multiple other situational and contextual factors. Outcome indicators measure the outcomes of the sociotechnical system. However, it has to be remembered that safety is not an outcome and thus cannot be measured with outcome indicators. Safety is a dynamic non-event (Weick 1987) and non-events cannot be characterised or counted. Thus, we have to look at the term 'dynamic' and seek to identify the way the non-event is created and acknowledge that we cannot ever capture the non-event itself.

What we call monitor and drive indicators are typically called 'lead indicators' in the safety science literature. However, as we discussed above, the distinction between lead and lag loses some of its relevance when we aim to manage the safety potential of the nonlinearly functioning complex adaptive system and use indicators as one of the tools of safety management.

The main function of the drive indicators is to direct the sociotechnical activity in the organisation by motivating certain safety-related activities. Monitor indicators on the other hand provide a view on the dynamics of the organisation: the practices, abilities, skills and motivation of the personnel – the organisational potential for safety. An important distinction is thus made here between two types of 'leading' indicators: those that monitor safety and those that drive safety. Monitor indicators indicate the potential of the organisation to achieve safety. They do not directly predict the safety-related outcomes of the sociotechnical system since these are also affected by numerous other factors such as external circumstances, situational variables and chance. Drive indicators in turn indicate development activities aimed at improving safety – the chosen priority areas of the organisational safety activity. The topic areas of the drive indicators are turned into control measures (or control measures are turned into drive indicators) that are used to manage the system: to change, maintain, reinforce or reduce something.

Drive indicators can be used to facilitate change and development in the desired direction. This can be done by selecting indicators that promote the desired behaviour and new practices or inhibit unwanted activity. For example, if the organisation is implementing a practice of filling a surgical checklist before major surgeries, the number of filled checklists can be selected as a safety performance indicator (drive indicator) to be followed monthly at hospital level. It is also a way to signal to the employees that the issue is important enough to be followed by management.

Monitor indicators reflect the potential and capacity of the organisation to perform safely. The indicators monitor the functioning of the system. These indicators seek to measure the internal dynamics of the sociotechnical system and provide information on the activities of the system. Information should be provided not only about safety management initiatives but also about the personnel, work processes, structures and technology in the system. Cultural issues such as norms, values and shared practices should also be incorporated into monitor indicators. Figure 9.2 illustrates the ways in which safety indicators can be used to provide information to support safety management.

In practice, outcome indicators alone are often used to define safety priorities or to draw conclusions about the level of patient safety. This is not a correct use of outcome indicators. Outcome indicators provide information about what has happened in the past. Thus, the values of the outcome indicators – however well the indicators themselves have been selected – do not have a direct relation to the current level of patient safety in the organisation. However, outcome indicators can provide information on the functioning and failure of the safety management efforts carried out in the organisation if the effects of the efforts have been predicted by the underlying safety model. This means that they can be used in fine-tuning and adjusting safety management activities within the boundaries of the current safety model.

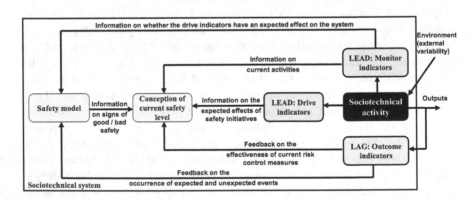

Figure 9.2   The extended system model showing the feedback from the indicators

Source: Modified from Reiman and Pietikäinen 2012.

For example, if the number and severity of adverse events do not decrease when safety management developments have been initiated in a given hospital, the content and implementation of the development actions can be reconsidered. Changes in the outcome indicators' values can also motivate the organisation to inspect further its monitor indicators and other means for monitoring performance (such as auditing) for possible changes in the organisation's capabilities for ensuring safety both now and in the future.

In conclusion, proactive safety indicators either influence safety management priorities and the chosen actions for safety improvement or provide information about the dynamics of the sociotechnical system (not merely about the functioning of safety barriers and absence of harm). These proactive indicators are respectively labelled drive indicators and monitor indicators in this chapter. Safety indicators should be capable of measuring (monitor indicators) or facilitating (drive indicators) the presence of organisational attributes necessary for ensuring adequate patient safety.

Defining the Indicators and Gathering Data

In order for the drive, monitor and outcome indicators to support proactive safety management, the indicators should be meaningful to the organisation in question. Healthcare organisations comprise several independent and self-organising agents (such as patients, medical doctors, nurses) who can all be considered to be taking part in the activity of organisational safety management. All of these actors can also be involved in defining meaningful safety indicators for the organisation and ideating ways for gathering reliable data on them. As a minimum, all the actors should understand why certain issues are considered important in the organisation.

Several authors have connected the concept of leading safety indicators to the concept of safety culture and proposed the use of safety culture or climate measures as safety indicators (cf. Grabowski et al. 2007; Mearns 2009; see also Zwetsloot 2009: 495). Safety culture researchers have tried to identify the key characteristics of safely functioning organisations and developed ways to measure those reliably. Safety culture research thus provides one starting point for considering what should be indicated in order to understand and manage patient safety in healthcare organisations in a proactive manner.

We have argued that the essence of safety culture is the ability and willingness of the organisation to understand safety, hazards and means of preventing them, as well as the ability and willingness to act safely, prevent hazards from actualising and promote safety (Reiman et al. 2010, 2012). Safety culture refers to the potential of the organisation to perform safely in specific situations. In specific situations, the preconditions created by a safety culture actualise in a manner that is dependent on many other task- and situation-specific variables. These include the nature of the patient's illness or injury and the clinician's competence, personal (work orientation, personality) and situational (fatigue, stress tolerance) characteristics.

Table 9.1 describes our conceptualisation of the criteria for a good safety culture and Table 9.2 the necessary organisational functions to achieve the criteria.

Table 9.1   Our conceptualisation of criteria for a good safety culture

Safety culture criteria
1. Safety is genuinely valued and the members of the organisation are motivated to put effort into achieving high levels of safety.
2. It is understood that safety is a complex phenomenon. Safety is understood as a property of an entire system and not just the absence of incidents.
3. People feel personally responsible for the safety of the entire system; they feel that they can have an effect on safety.
4. The organisation aims to understand the hazards and anticipate the risks in their activities.
5. The organisation is alert to the possibility of an unanticipated event; it is mindful in its practices.
6. There are good prerequisites for carrying out daily work; the organisation is controllable.

Table 9.2   Organisational control functions required to achieve a good safety culture

Organisational control functions
Strategic management Refers to how the organisation addresses long-term preconditions for work (workforce availability, environment, investments), sets long-term goals for development and guarantees financial viability.
Work conditions management Structuring the work in terms of the constraints and requirements it puts on the workers: management of the physical conditions (e.g. workspace, lighting), structural means necessary for carrying out the work (e.g. tools, instructions) and human resources.
Work process management How cooperation, communication and information flow are managed in the organisation.
Safety leadership How safety considerations are included in management decision-making. This function involves gathering feedback and information, making expectations clear, communicating on safety issues and ensuring that management is up to date on the way work is conducted in the field.
Supervisory support for safety Supervisors are organising the work in the immediate work environment in such a manner that it can be safely accomplished, providing positive feedback on the safety-conscious behaviour of personnel, treating subordinates fairly and monitoring subordinates' coping skills, stress, fatigue levels and skills.

Table 9.2   Organisational control functions required to achieve a good safety culture (*continued*)

Proactive safety development Includes using 'operating' experience and leading safety indicators, as well as continuous development of practices and constant vigilance for weak signals. This function deals with how learning takes place and it supports the ability of the organisation to recognise the boundaries of safe performance.
Hazard control How known risks are prevented from actualising. This function deals with the provision and implementation of barriers (e.g. quality assurance, back-up systems, checklists and physical barriers) to prevent unwanted human and technical variance.
Competence management How competence needs are identified and the skills and knowledge of personnel developed and maintained. This function also includes the training and socialisation of newcomers and transfer of knowledge from the experienced personnel to newcomers.
Change management Handling of changes in organisational structures, practices and technology; planning, implementation, as well as follow up on changes already implemented. Change management should also take into account incremental changes in the organisation.
Management of third parties How contractors and leased employees are selected and trained in safety-related issues, and how their expertise in the field of interest is ensured. This function also concerns practices to facilitate organisational learning from contractors as well as contractors' own learning.

The safety culture criteria depicted in Table 9.1 can be understood as the goals or ideal end states of patient safety management and they can be used as a frame to define monitor indicators. Organisational activities can be viewed from the point of view of how they contribute to these six ideal states. The control functions in Table 9.2 on the other hand can be used for selecting drive indicators that facilitate change in the right direction.

Examples of possible patient safety indicators derived from the criteria and the control functions are presented in Tables 9.3 and 9.4.

Table 9.3 **Examples of monitor indicators measuring the fulfilment of the six safety culture criteria**

	Monitor
Safety as a shared value	-Employee perceptions of management safety commitment -Management perceptions of employee safety commitment -Personnel's reported safety attitudes in safety culture surveys -Management raising safety issues in their talks
Systemic understanding	-The extent to which patients understand the purpose and possible hazards relating to their treatment -The extent to which personnel perceive competence as a system-level issue instead of an individual attribute -The extent to which personnel consider mistakes system properties -The amount of individual blaming after near misses and events (NEG)
Responsibility	-The extent to which personnel feel personal responsibility for patient safety and the quality of care in their organisation (questionnaires or interviews) -The extent to which personnel take initiatives to improve organisational practices or report problems to management -The percentage of investigated major adverse events
Understanding of hazards	-The extent to which personnel understand the hazards that are connected to their work -The extent to which personnel understand the inherent trade-offs and goal conflicts in their work -The understanding of hazards prevalent in surgery teams
Mindfulness	-The ratio of near miss reports versus adverse events in the voluntary incident reporting system -The extent to which external audits provide results that are in accordance with findings in internal audits or the prevalent conceptions of personnel
Controllability	-Employees' reported sense of control in safety culture surveys -The extent to which personnel perceive the organisation as supporting their work instead of hindering it -Average working hours per week of various occupational groups (NEG)
All the criteria	-Employees feeling worried about patient safety in their organisation -Uncovered systemic factors underlying successes and failures in care

One means of gathering data for monitor indicators is patient safety culture questionnaires. We have developed a patient safety culture questionnaire, TUKU, that measures employees' perceptions of the organisational functions depicted in Table 9.4 as well as employees' psychological states, such as sense of control and worry about patient safety (Reiman et al. 2013). In one hospital, the results of the questionnaire were compared with the ratio of patient safety incidents at the hospital's 40 units 16 months after the safety culture questionnaire was administered. The results, which must be treated with caution due to the small sample size, indicated that perceptions of 'work process management', 'work conditions management', 'safety management' and 'management of third parties' had a statistically significant relation to future events (correlation ranging from -.28 to -.41, with the negative sign meaning that the higher the unit level score on a given scale, the fewer adverse events the unit reported in the next 16 months) (Reiman et al. 2013). Although the results are based on a limited sample, they give a promising foundation for clarifying how a decline in employees' perceptions of how well the organisation currently functions does in fact predict a future decline in objective performance as well – if the management does not take action before that happens.

Table 9.4   Examples of drive indicators facilitating the implementation of the 10 control functions

Control function	Examples of potential drive indicators
Strategic management	-Long-term personnel availability is considered annually -Upcoming health service needs of the region are scanned annually
Work conditions management	-Instructions have been written for key tasks -Shifts are fully manned
Work process management	-Process descriptions are up to date -Division of responsibilities in inter-departmental tasks has been agreed upon
Safety leadership	-Number of lessons from safety incidents communicated to personnel -Time devoted to safety in management meetings -Weekly management safety walks (visits) -Potential risks are systematically considered when making management decisions
Supervisory support for safety	-Positive feedback from supervisors to employees on safety-conscious behaviour -Development discussions have been kept
Proactive safety development	-Number of new safety developments initiated during the year -Number of internal and external audits -Number of employee safety initiatives -Response rate to safety culture surveys
Hazard control	-Typical task hazards have been identified -Surgical checklist is used in surgeries (%)
Competence management	-Number of employees going through a formal induction -Number of employees having an in-service training plan
Change management	-Risk analyses are conducted before technical and organisational changes -An implementation plan is written for all technical and organisational changes -There is a follow-up evaluation for all technical and organisational changes
Management of third parties	-Safety issues are used as criteria when purchasing outside services -Work requirements and changes in organisation's practices are communicated to the service provider

Table 9.5 depicts possible outcome indicators. These indicators can be seen as the result of the organisation's safety culture and the effects of situational and environmental variability, such as the nature of the patient's illness or injury or the fatigue of the clinician. The aggregation of these outputs into an outcome indicator can provide the organisation with clues about possible changes to their safety potential.

Table 9.5   Examples of outcome indicators

Outcome indicators
• Sick leave of personnel
• Adverse events in surgeries
• Hospital-acquired infections
• Number and trend of adverse events reported during past years
• Number and trend of near misses reported during past years
• Staff turnover
• Medical errors
• Patient falls
• Operative and post-operative complications

As the construction of safety is a complex activity, it is often challenging to obtain reliable data on the things that it is really important to indicate in terms of safety. Nonetheless, the selection of indicators should not be based too much on what is easy to measure. There are several possible ways to gather data on the indicators (Figure 9.3). Sometimes it may be necessary to use several types of data to obtain a reliable idea of the status of a specific indicator. For example, if we seek to measure the patient safety worry experienced by personnel, we could combine information from a survey with interviews of personnel. Unfortunately, it is rare for organisations to use more than one source of information when deducing the value of a specific indicator.

Patient safety indicators have almost always been considered quantitative by nature (e.g. EUNetPas 2010: 33; Kristensen et al. 2007: 7). However, qualitative indicators can also provide important information on patient safety. For example, personnel's expressions of worry about patient safety issues in the organisation or quality audits that provide a written summary evaluation provide important information about patient safety – often more important than the quantitative indicators do. The problem of the quantification of, for example, adverse events and near misses is the loss of context, i.e. the loss of the story behind the incidents (Dekker 2011; Reiman and Rollenhagen 2011). Instead of data analysis it is merely a classification of data (Dekker 2011: 143). Too often this classification is taken as a sufficient basis for managerial decision-making in situations that should require more in-depth analysis of the individual and organisational dynamics underlying the numbers. Condensing complex information into numbers may oversimplify reality. Too strong an emphasis on quantitative safety indicators may also misdirect personnel and leaders to manage the numbers instead of the underlying phenomena and increase the unpredictability of the organisation by pointing the management focus in the wrong direction. An extreme example would be putting a reduced number of harmful patient safety incidents as a merit pay criterion, thus, as a side effect, prompting personnel to under-report incidents.

Using Indicators in Proactive Patient Safety Management

Indicators as Samples of Organisational Activities

Indicators can be used at different levels of the organisation. They can be used by the top management as well as at the unit level to acquire an understanding of where the organisation is going in terms of safety and in steering it in a desired direction. It is good to keep all the members of the organisation (the self-organising agents) as informed as possible on what indicators say about the safety of their organisation in order for indicators genuinely to steer their everyday work in a safer direction. The values of specific indicators are not so important here and it may not be sensible to communicate all of these values to personnel. The things that have to be communicated are management's interpretations of what the values say about patient safety.

Sometimes safety indicators are used as part of an incentive scheme (merit pay scheme). However, monitor indicators should never be used that way and outcome indicators only after careful consideration. One of the main practical differences between monitor and drive indicators is that managers should enforce and reward the fulfilment of the performance measured by the drive indicator but never the fulfilment of performance measured by the monitor indicator. This is due to the fact that personnel pay attention to things the management measures and often try to optimise their performance on those, and only those, measures. When it comes to drive indicators that do not matter, the point of selecting something as a drive indicator is to focus personnel's attention on the given issue (cf. Table 9.4). However, monitor indicators should reach organisational dynamics without optimisation or impression management on the part of personnel. Thus, the response rate of a safety culture questionnaire (drive indicator) is a good bonus indicator, but a mean score on employee safety commitment (monitor indicator) on the same questionnaire is a bad one. As indicators can never cover everything the organisation does, they can be considered samples of all things affecting patient safety. These samples should represent the state of affairs in the organisation as accurately as possible. By emphasising which indicators the management considers important to the personnel, these samples may become 'contaminated' and cease to represent the organisation.

Many issues dealing with subjective risk perceptions and 'gut feelings' are difficult to monitor with quantitative indicators. People may feel that 'something is wrong' or 'missing' but not be able to communicate clearly what it is (Reiman and Rollenhagen 2011). If the healthcare organisation consequently dismisses such reports as representing little more than general complaints rather than something that actually could be a vague perception of an existing risk, then potential warning signals are ignored.

Indicators as Part of Decision-making Concerning Safety Activities

The importance of taking a proactive stance on indicators and safety management cannot be stressed enough. Organisations need to take action and improve their activities continuously and without any visible signs of (increased) problems in taking care of patient safety. Monitor indicators can provide signals of improvement needs before these weaknesses actualise as adverse events. However, since adverse events are causally complex and their occurrence can never be predicted exactly, organisations should also actively seek ways to develop their capability to deliver good quality care. Furthermore, organisations should facilitate this continuous development of the healthcare system by setting appropriate drive indicators and monitoring the effects of the measured development actions on organisational activities. It is good for organisations to rely always on multiple indicators, never basing decisions on only one indicator.

Dyreborg (2009: 475) points out an important distinction between the necessary measures for dealing with lead and lag indicators: 'Decreasing lead indicator performance levels calls for improvement of existing risk control parameters, whereas decreasing lag indicator performance levels without such a lead indicator decrease, calls for a revision of the risk control, i.e., reconsidering the causal relation between lead and lag indicators.' In our model, this means that if the outcome indicators show a decrease in safety that is not explained by the monitor indicators, the underlying safety model may need revising. For example, if the number of adverse events increases even when the monitor indicators are on the desired level, one possible explanation is that the monitor indicators are not focusing on the right things. For example, they may not focus enough on organisational issues. The monitor indicators may instead focus too much on social factors, such as the attitudes and commitment of personnel, thus making the hospital subject to neglect of working practices and the organisation of work. In this situation, the organisation needs to revise its safety model to include attitudinal, cognitive and technical issues as equally relevant factors. Furthermore, if the drive indicators show an increase without a corresponding increase in the monitor indicators, the underlying safety model should be reflected on and possibly revised. This reconsideration does not necessarily mean a reconsideration of the causal relation between lead and lag, but it may mean a reconsideration of how safety is created and how an increase in safety manifests itself in the organisation. Correspondingly, a decrease in monitor indicator values requires improvements in safety management activities, steered and supported by the drive indicators.

Without an underlying safety model describing the postulated causal relations between a set of indicators, it is difficult to know why a change in some indicator values (monitor or outcome) has occurred and what the change implies for safety management or tells us about the level of patient safety in the organisation. All of our observations are theory driven, including those observations and interpretations that we make about safety in our own organisation. Of course, having an explicit safety model does not automatically mean it would be correct, or even usable, in practice. Woods (2009: 499) reminds us about the lesson from the Columbia accident investigation: 'Organizations need mechanisms to assess the risk that the organization is operating nearer to safety boundaries than it realizes – a means to monitor for risks in how the organization monitors its risks relative to a changing environment.' This monitoring of how well the organisation is monitoring its risks (second-order or meta-monitoring) is an important yet difficult endeavour. Safety indicators are one mechanism for this but they need to be complemented with an explicit safety model that emphasises the complexity of safety and its management. This means that the safety model should emphasise the importance of proactive measures to improve safety, of continuous monitoring of potential warning signals, of reporting and analysing near-misses and adverse events and also a view on the limitations of safety indicators in accomplishing all this.

The selection of countermeasures or corrective actions that match the level of threat indicated is a challenging task. In complex systems, in which events are caused by multiple causally interlinked factors, there is always a danger of either over-reacting or under-reacting to warning signals (Reiman and Rollenhagen 2011). Solutions should also always be considered at system level, not at the level of individual components – such as incompetent individuals or inadequate supervision. In any case, it is important that the indicator system has an active role in safety management. It is not uncommon for organisations to collect masses of data with little subsequent use (Reiman and Rollenhagen 2011). An indicator system is of no use if it is not integrated into a proactive safety management system.

With regard to the outcome indicators, the complexity of the system makes it impossible to deduce directly the organisational causes of the various values of the indicators. The explanation of an indicator value needs to be separated from the measurement of that value (cf. Reiman and Rollenhagen 2011; Wilkin 2010). An explanation has to build an expansive, rich, contextualised picture of the event or phenomena concerned, one that can set out the variety of causal mechanisms and processes that over time and space generated the event or phenomena and that can engage with the intentions and meanings of the actors involved (Wilkin 2010: 238). When dealing with human interaction in sociotechnical systems the quest for explanations has to be supplemented with the quest for understanding. This means seeking to identify the subjective, qualitative and more in-depth meanings of personnel concerning the topic of interest, all the while acknowledging that this act of identification is itself interpretation guided by one's own meanings and conceptions (Scherer 2003) and, in the case of safety indicators, by one's safety model. Indicators can provide clues about issues that need more in-depth investigation aimed at a deeper understanding of the dynamics of the healthcare organisation in question. The generalisability of knowledge is not the only criterion of validity, as some international indicator schemes seem to emphasise. On the contrary, each sociotechnical system is unique and this sets constraints on what can be generalised outside the context of the given system.

Acknowledging the Limitations of Indicators in Safety Management

Traditionally, safety indicators have tended to focus on the formal, quantitative and structural instead of the informal, qualitative and processual. This is not enough to guarantee safety. Proactive patient safety management also needs a focus on the qualitative aspects of organisational life, as well as on the informal practices and the continuing unfolding of meanings, values and norms in the workplace. These phenomena are difficult to grasp with indicators, especially quantitative indicators. It is important that an indicator system is understood as one imperfect tool of safety management among other imperfect tools, such as reporting systems, checklists, auditing programmes and training programmes. Together with monitor indicators, trends in outcome indicators can help envision possible wanted and unwanted futures. Drive indicators can then be used in an attempt to avoid the unwanted scenarios and steer the organisation towards the wanted future. All in all, a significant aspect of proactive safety management is a critical attitude towards the information gathered by the safety performance indicators as well as a continuous effort to improve all means of achieving information about the organisation.

There is a risk that a positive trend in safety indicators will lead to complacency and a belief that the organisation is, and will be, safe. This applies especially to outcome indicators that measure the occurrence of negative incidences. A good patient safety culture is marked by being constantly on guard, even in situations in which performance indicators point to an increased or high level of safety. Drive indicators are necessary precisely for preventing organisations from ceasing their safety work as soon as the most easily measurable and visible indications of a lack of safety disappear. In a similar vein, monitor indicators that seek to measure the presence of organisational preconditions for safety are always needed to complement outcome indicators.

Two reasons why safety management should never rely solely on indicators for information are the phenomena of normalisation of deviance (Vaughan 1996; cf. Dekker 2011) and the drift of activities (Snook 2000). Another, third reason has to do with the complex nature of the system itself.

Adverse events are non-linear and unpredictable products of the functioning of the entire system and indicators have serious challenges capturing non-linear causal relations.

Normalisation of deviance refers to a process in which small changes – new behaviours, technical/physical/social anomalies or variations that are slight deviations from the normal course of events – gradually become the norm, providing a basis for accepting additional deviance (Vaughan 1996). Normalisation of deviance produces disregard and misinterpretation – neutralisation – of potential danger signals. A signal of potential danger is information that deviates from expectations, contradicting the existing world view (Vaughan 1996: 243). This means that normalised signals are not reported since they are no longer considered danger signals but rather signs of normal work. This phenomenon is very difficult to capture by traditional outcome indicators. This is due to two issues. First, normalising practices, deviations, variations, actions and events typically focus on activity that has not yet led to adverse outcomes. Second, these deviations are typically not considered worth reporting as near misses since they are considered part of normal everyday work. Only qualitative auditing or safety culture evaluations may capture this so far successful incremental departure from previous norms. One way of striving to capture the phenomenon is to measure and improve the aspects of the organisation contributing to the normalisation (Vaughan 1996; see also Dekker 2011) – 1) scarcity of resources (that leads to the necessity of optimisation), 2) real or perceived production pressures (that lead to conflicting goals and trade-offs), 3) uncertain and unruly technology (that creates glitches such as false alarms), 4) structural secrecy (where danger signals remain local and not subjected to any credible outside scrutiny), 5) intolerance of dissenting opinions (that may challenge the new norms or raise safety concerns) and 6) distant information patterns (e.g. specialisation of knowledge so that few can judge the information generated by others and formalisation of knowledge sharing so that information instead of knowledge becomes shared).

Drift is closely associated with normalisation of deviance. While normalisation deals with normative changes in the valuation of danger signals, drift refers to local modification and adaptation of centrally designed practices. Snook (2000: 194) writes: 'Practical drift is the slow steady uncoupling of practice from written procedure ... After extended periods of time, locally practical actions within subgroups gradually drift away from originally established procedures ... Constant demands for local efficiency dictate the path of the drift.' Two of the main reasons for the danger of locally optimising working practices are the loose couplings prevalent in complex sociotechnical systems that make it possible to change one part of the social system without immediate effect on the others and the tendency of complex systems in some conditions to become tightly coupled. The risk becomes evident when different, locally adapted practices meet (Dekker 2011: 125) – when they become tightly coupled. The countermeasure to drift is an enforcement of centralised control measures for emergency situations.

For example, the deployment of centrally controlled rapid response teams at the first sign of patient decline has been considered a good practice in terms of patient safety (Dekker 2011: 125).

A complex system cannot ever be prescribed in a detailed enough way for it to be possible to devise guidelines for all possible contingencies. The introduction of guidelines can also increase the complexity of the system or introduce unexpected side effects. Rather than advocating absolute rule following, safety management should aim to develop practitioners' skills in judging when and how to adapt guidelines to local circumstances (Dekker 2011: 127). Furthermore, organisations have to monitor and understand the reasons behind the gap between guidelines and practice (ibid.). The gap is a monitor indicator that implies not a violation but rather a compliance with norms and local expectations; 'they comply with unwritten rules and operating standards that probably make good local (and clinical) sense' (Dekker 2011: 128). Again, this local understanding may not be able to capture all sources of risk.

Safety performance indicators can make explicit the issues that safety professionals would pay attention to in any case, or they can direct attention to issues that safety professionals would not otherwise focus on. They can make the underlying models of safety explicit and subject to criticism and correction. Thus, not only are safety indicators necessary, their (implicit or explicit) use is inevitable in safety management. Explicit indicators enable the critical consideration of their functioning and underlying premises. It is hoped that this will lead to the modification of either the underlying models or the chosen indicator gradually to improve the monitoring and driving of safety.

References

De Savigny, D. and Taghreed, A. (eds) 2009. *Systems Thinking for Health Systems Strengthening*. Alliance for Health Policy and Systems Research, WHO.

Dekker, S.W.A. 2005. *Ten Questions about Human Error. A New View of Human Factors and System Safety*. New Jersey: Lawrence Erlbaum.

Dekker, S. 2011. *Patient Safety. A Human Factors Approach*. Boca Raton: CRC Press.

Dyreborg, J. 2009. The causal relation between lead and lag indicators. *Safety Science*, 47, 474–5.

EPRI 2001. *Final Report on Leading Indicators of Human Performance*. 1003033. EPRI & US Department of Energy, Palo Alto, CA; Washington, DC.

EUNetPaS 2010. *Patient Safety Culture Instruments used in Member States*. European Society for Quality in Healthcare, Office for Quality Indicators.

Grabowski, M., Ayyalasomayajula, P., Merrick, J., Harrald, J.R. and Roberts, K. 2007. Leading indicators of safety in virtual organizations. *Safety Science*, 45, 1013–43.

Hale, A. 2009. Why safety performance indicators? *Safety Science*, 47, 479–80.

Herrera, I.A. 2012. *Proactive Safety Performance Indicators. Resilience Engineering Perspective on Safety Management.* Doctoral theses at NTNU, 2012:151. Norwegian University of Science and Technology, Trondheim.

Hollnagel, E. 2004. *Barriers and Accident Prevention.* Aldershot: Ashgate.

Hollnagel, E. 2008. Safety management – Looking back or looking forward. In E. Hollnagel, C.P. Nemeth and S. Dekker (eds), *Resilience Engineering Perspectives, Volume 1. Remaining Sensitive to the Possibility of Failure.* Aldershot: Ashgate, 63–7.

Hollnagel, E. and Woods, D.D. 2006. Epilogue – Resilience engineering precepts. In E. Hollnagel, D.D. Woods and N. Leveson (eds), *Resilience Engineering. Concepts and Precepts.* Aldershot: Ashgate.

Hopkins, A. 2009. Thinking about process safety indicators. *Safety Science,* 47, 460–65.

HSE 2006. *Developing Process Safety Indicators: A Step-by-step Guide for Chemical and Major Hazard Industries.* UK Health and Safety Executive.

Kjellén, U. 2009. The safety measurement problem revisited. *Safety Science,* 47, 486–9.

Kristensen, A., Mainz, J. and Bartels, P. 2007. *Establishing a Set of Patient Safety Indicators. Safety Improvement for Patients in Europe.* SImPatIE – Work Package 4. Aarhus: The ESQH-office for Quality Indicators.

Lichtenstein, B.B., Uhl-Bien, M., Marion, R., Seers, A., Orton, J.D. and Schreiber, C. 2006. Complexity leadership theory: An interactive perspective on leading in complex adaptive systems. *Emergence: Complexity and Organization,* 8, 2–12.

McDaniel, R.R. and Driebe, D.J. 2001. Complexity science and health care management. *Advances in Health Care Management,* 2, 11–36.

Mearns, K. 2009. From reactive to proactive – Can LPIs deliver? *Safety Science,* 47, 491–2.

Millar, J. and Mattke, S. 2004. Selecting indicators for patient safety at the health systems level in OECD countries. *OECD Technical Papers,* No. 18, OECD Publishing.

Pietikäinen, E., Heikkilä, J. and Reiman, T. (eds) 2012. Adaptiivinen potilasturvallisuuden johtaminen. VTT Technology 58. [In Finnish]. Available at: www.vtt.fi/inf/pdf/technology/2012/T58.pdf (last accessed on 16 April 2014).

Plsek, P.E. and Greenhalgh, T. 2001. Complexity science. The challenge of complexity in health care. *BMJ,* 323, 625–8.

Rasmussen, J. 1997. Risk management in a dynamic society: A modelling problem. *Safety Science,* 27, 183–213.

Reiman, T. and Oedewald, P. 2009. *Evaluating Safety Critical Organizations. Focus on the Nuclear Industry.* Swedish Radiation Safety Authority, Research Report 2009: 12.

Reiman, T. and Pietikäinen, E. 2012. Leading indicators of system safety – Monitoring and driving the organizational safety potential. *Safety Science,* 50, 1993–2000.

Reiman, T., Pietikäinen, E. and Oedewald, P. 2010. Multilayered approach to patient safety culture. *Quality and Safety in Health Care*, 19, 1–5, doi:10.1136/qshc.2008.029793.

Reiman, T., Pietikäinen, E., Oedewald, P. and Gotcheva, N. 2012. System modeling with the DISC framework: Evidence from safety-critical domains. *Work*, 41, 3018–25.

Reiman, T. and Rollenhagen, C. 2011. Human and organizational biases affecting the management of safety. *Reliability Engineering & System Safety*, 96, 1263–74, doi:10.1016/j.ress.2011.05.010.

Reiman, T., Silla, I. and Pietikäinen, E. 2013. The validity of the Nordic patient safety culture questionnaire TUKU. *International Journal of Risk and Safety in Medicine*, 25(3), 169–84.

Rollenhagen, C. 2010. Can focus on safety culture become an excuse for not rethinking design of technology? *Safety Science*, 48, 268–78.

Rouse, W.B. 2007. Health care as a complex adaptive system: Implications for design and management. *The Bridge*, Spring 2008, 17–25.

Scherer, A.G. 2003. Modes of explanation in organization theory. In H. Tsoukas and C. Knudsen (eds), *The Oxford Handbook of Organization Theory. Metatheoretical Perspectives*. Oxford: Oxford University Press.

Snook, S.A. 2000. *Friendly Fire. The Accidental Shootdown of U.S. Black Hawks over Northern Iraq.* New Jersey: Princeton University Press.

Vaughan, D. 1996. *The Challenger Launch Decision.* Chicago: University of Chicago Press.

Weick, K.E. 1987. Organizational culture as a source of high reliability. *California Management Review*, 29, 112–27.

Wilkin, P. 2010. The ideology of ergonomics. *Theoretical Issues in Ergonomics Science*, 11, 230–44.

Woods, D.D. 2009. Escaping failure of foresight. *Safety Science*, 47, 498–501.

Wreathall, J. 2009. Leading? Lagging? Whatever! *Safety Science*, 47, 493–4.

Zwetsloot, G.I.J.M. 2009. Prospects and limitations of process safety performance indicators. *Safety Science*, 47, 495–7.

PART III
Application and Practice

Chapter 10

Safety Culture in Practice: Assessment, Evaluation, and Feedback

Çakil Agnew and Rhona Flin

Introduction

The recognition of a high prevalence of adverse events for patients and the associated costs (Vincent 2010), resulted in healthcare organisations being advised to adopt safety management techniques used in high risk industries to improve patient safety (Kohn et al. 2000). Investigations into failures in healthcare delivery have identified weak safety culture as a contributing factor (e.g. Francis 2010, 2013) and revealed the need to measure this aspect of organisational culture. In 2007, the Health Department in Scotland launched the Scottish Patient Safety Alliance, a national initiative to improve patient safety, with the key objective, 'to drive a change in the safety culture in NHS organisations' (http://patientsafety.etellect.co.uk/programme/about/aims).

Examination of an organisation's safety culture first requires a baseline assessment of the current level of relevant cultural factors in the workplace so that interventions can be targeted and any subsequent improvements can be measured (Flin 2007). In this chapter, we describe three stages in a study from Scotland on hospital safety culture, designed to provide an initial measurement and to transfer the findings back to the risk managers of the hospitals involved. We identify the main challenges encountered and key lessons learnt for each stage.

First, as no comprehensive safety culture survey had been run before in Scottish NHS (National Health Service) hospitals, we had to identify a questionnaire for this study and to test the usability and the psychometric properties of the selected instrument in this setting (Sarac et al. 2011). Then in stage one (assessment), we measured the safety culture and related safety outcome variables in a sample of hospitals and examined the relationship between culture factors and outcomes (Agnew et al. 2013). In the second stage (evaluation), we explored the survey findings in more depth by adopting a qualitative approach. In order to achieve this aim, focus group discussions were held with healthcare professionals to investigate the extent to which survey findings reflected the reality within two of the hospitals. Finally, the third stage (feedback) used an interactive workshop with representatives from the participating hospitals to stimulate discussions on the overall findings and provide an enhanced understanding of safety culture and its measurement. Before describing the process (see Figure 10.1), we outline some of the relevant research that informed the design of the study.

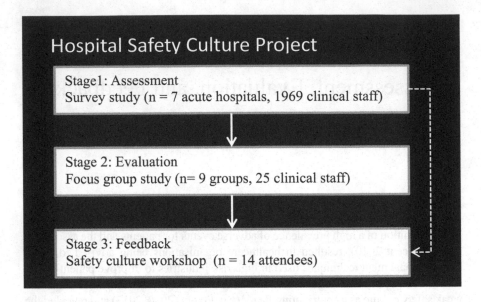

Figure 10.1   Stages of the hospital safety culture project

Several instruments have been developed to assess hospital staff's perceptions of aspects of workplace safety culture and a number of studies have reported associations between hospital safety culture and safety outcome measures (Jackson et al. 2010). Profiling the hospital safety culture scores is relatively straightforward but finding safety outcome measures for patients or workers is more challenging (Flin 2007). Different types of safety outcome data can be collected, e.g. (i) hospital incident records for staff or patients or clinical data for patients, (ii) self-reports of incidents and injuries by workers or patients and (iii) workers' safety behaviours (self-reported or observed).

Hospital Safety Data

Several studies have linked safety culture to safety data gathered by a hospital. For example, in one Israeli acute hospital (Naveh et al. 2005), lower treatment error rates were shown when staff perceived that the procedures were suitable and that managers practiced safely. Similarly, an association was found across 30 ICU units between lower safety culture scores and increased length of stay for patients, as well as a link between less favourable perceptions of management and higher mortality rates (Huang et al. 2010). Moreover, safety culture dimensions (e.g. fear of blame) were also related to higher rates of patient safety indicators (e.g. hospital discharge data) across 30 USA hospitals (Rosen et al. 2010).

Focussing on worker outcomes, Mark et al. (2007) reported a direct influence of safety culture on workers' back injuries, but not on needle-stick injuries, in a sample of 143 acute hospitals in the USA. In a study with 1,127 nurses from general medical-surgical units, negative associations were found between unit level safety culture scores and workers' back injuries, as well as with patient urinary tract infections and medication errors (Hofmann and Mark 2006). For several reasons we were not able to collect this type of safety outcome data in our study.

Self-reported Incident Data

A sample of 475 staff from 10 hospitals in Costa Rica, Gimeno et al. (2005) found that safety culture was related to self-reported work-related injuries. Another study of 789 hospital-based healthcare workers in the USA, found that experienced blood and body fluid exposure incidents for workers were lower when senior management support, safety feedback and training were perceived favourably (Gershon et al. 2000). In Japan, reduced needle-stick and sharp injuries to hospital workers were associated with safety culture factors, such as being involved in health and safety matters (Smith et al. 2010). In our Scottish hospital sample, we collected information on self-reports of worker injuries, as well as observed errors affecting patients.

Safety Behaviours

When healthcare workers' safety behaviours were assessed through observational techniques, Zohar et al. (2007) showed both group and hospital level culture as predictors of the future safety behaviours (e.g. medication and emergency safety), observed in a sample of 955 nurses in Israel. Since observations can be difficult to gather in hospitals, self-report measures are also used to assess safety behaviours such as workers' safety compliance and safety participation behaviours). Positive associations between safety culture and their self-reported measure of safety behaviours were shown in a sample of 525 employees in an Australian hospital (Neal et al. 2000). Later, improvements in these behaviours at the group level were linked to a reduction in future accident rates (Neal and Griffin 2006). A similar measure was included as a measure of workers' safety behaviour in this Scottish study.

Stage 1: Assessment: Measuring Patient Safety Culture in Hospitals

In order to identify areas of organisational culture that could be targeted to improve patient safety in Scottish hospitals, a safety culture survey was conducted. The aims of the study were first, to obtain a measure of safety culture from a sample of NHS acute hospitals in Scotland and then to test whether these culture scores were associated with clinical workers' safety behaviours and patient and worker injuries. This would also provide a measure of safety culture within the Scottish acute hospital sector and contribute an organisational cultural perspective to the current national patient safety initiative (www.scottishpatientsafetyprogramme. scot.nhs.uk/programme).

Hospital Survey on Patient Safety Culture

Several instruments were available to measure safety culture in healthcare. For this reason, rather than designing a new questionnaire for the study, it was decided to adopt one of the existing hospital safety culture measures for the survey as that would allow comparison with other UK or international data sets. The HSOPSC was selected for this project as, at the time of selection, this questionnaire had been subjected to more rigorous psychometric testing than the alternative tools (Flin et al. 2006). Since it had been used extensively in the US with hundreds of hospitals (Sorra and Battles, this volume), it provided North American benchmark data. In addition, there were reports from studies that had employed the HSOPSC in several European countries, e.g. the Netherlands (Smits et al. 2008); Norway (Olsen 2010) and Switzerland (Pfeiffer and Manser 2010) (see also Hammer and Manser, this volume). These offered data for cross-national comparisons. Furthermore, the instrument covers a wide range of safety culture dimensions (e.g. hospital management's commitment to safety, supervisory practices, teamwork) which had been shown to relate to safety outcomes in other settings, as well as in healthcare (Alahmadi 2010; Mardon et al. 2010; Van Noord et al. 2010). Finally, the questionnaire had been extensively tested with healthcare staff and therefore was not likely to present usability problems. However as the HSOPSC was originally designed for US healthcare staff, it required customisation for a Scottish NHS sample, as had been necessary in the other European studies, and a series of interviews with different staff roles was conducted for this purpose, resulting in minor amendments.

Access and Administration of Survey

The data were collected from seven acute NHS hospitals in Scotland (one per Board), between March and October 2009. Hospital size ranged between large and small – large hospitals: more than 500 beds, medium-size hospitals: between 250 and 499 and small hospitals: between 50 and 249 beds. Given the results of previous studies conducted with the HSOPSC (indicating that some items might not be suitable for non-clinical staff), it was decided to include only clinical staff in the study. The procedure for administration of the questionnaires was discussed with NHS Healthcare Improvement Scotland (HIS) and each participating Board or hospital. Both paper based and web-based surveys (SNAP 9, www.snapsurveys.com) were made available. Paper questionnaires were provided with a covering letter and a sealable envelope. No names or specific unit identifiers (e.g. ward number) were requested to enhance anonymity.

A total of 1,998 clinical staff completed the questionnaire with an estimated 21 percent response rate. Although the numbers of questionnaires sent to each participating hospital were known, it was not clear how many questionnaires were actually distributed within each unit. The estimated response rate for each hospital ranged from 4 percent to 31 percent, and these were likely to be underestimates of the response rates, as some hospital representatives later stated that not all the delivered questionnaires had been distributed. Nurses constituted the majority of the sample (52 percent) followed by Allied Health Professionals (21 percent), doctors (14 percent) and nursing or healthcare assistants (15 percent). In relation to work unit, 21 percent were from surgical units, followed by medicine (16 percent). The majority of the participants (37.5 percent) were found to have worked more than 10 years within their current hospital, and 31 percent had worked there between one and five years. Regarding their current profession, 30 percent were found to be highly experienced with more than 21 years practice. A total of 92 percent of the respondents reported having direct contact with patients and 7 percent reported not having any.

Scottish Hospital Survey on Patient Safety Culture (SHSPSC)

The overall SHSPSC questionnaire (63 questions) comprised four main components: the 44 items of the HSOPSC (described above) plus 10 safety behaviour items, two items measuring self-reported worker and patient injuries, and seven demographic questions. While the intention was to keep the questionnaire as short as possible, the HSOPSC only measures self-reports of incident reporting behaviour, so safety behaviour items needed to be incorporated. These have been shown to be key outcome variables both in industrial (Clarke 2006; Neal et al. 2000; Neal and Griffin 2006) and healthcare safety research (Zohar et al. 2007). Workers' safety participation and compliance behaviours have been linked to both safety culture and accident data (Neal and Griffin 2006). Therefore a 10-item scale was included to measure self-reports of workers' behaviours. Safety participation was assessed by four items (Neal et al. 2000) and for safety compliance, six items were incorporated from research on offshore oil installations (Mearns et al. 2003) and reworded for healthcare workers. In order to measure injuries experienced by workers and patients separately, two self-report items were used. The first item (based on the Offshore Safety Questionnaire; Mearns et al. 2001) asked how often the individual had experienced a work-related injury in this hospital in the last 12 months and the second question asked about the number of witnessed incidents in which a patient had been harmed in the last 12 months. Initial psychometric analysis (exploratory and confirmatory factor analysis) of the data set revealed evidence that the 12-factor structure of the HSOPSC performed adequately in the Scottish sample (Sarac et al. 2011). Accordingly, it was decided that no modifications were required, since this structure would allow the possibility of cross-national comparisons.

'Hospital handovers' were found to receive the lowest mean score (2.9) whereas 'Teamwork within units' scored the highest (3.7). Most of the mean scores were above 3.0, suggesting favourable perceptions of safety. For a detailed report of all the scores, see Sarac, 2011). Regarding the safety outcome data, mean scores ranged from 2.2 to 3.6. Most scores were found to be in the same range as the US and other European samples based on 12 factor structures. No major differences between four main occupational groups (AHPs, medical doctors and dentists, nurses and nursing and healthcare assistants) were found, so subsequent analyses were carried out on the total sample. It is however worth mentioning that nurses reported significantly less positive perceptions of teamwork within units than doctors. Overall, doctors were found to be more critical about most of the aspects of patient safety compared to nurses. Similar results were reported for the Norwegian sample in relation to nurses' perceptions of supervisory expectations. But contrary to the Scottish findings, Norwegian nurses were found to score lower on organisational learning and continuous improvement and higher on teamwork within units factors when compared to the rest of the occupational groups (Olsen 2010). Additionally, two of the Scottish groups (registered nurses and doctors) did not significantly differ on their perceptions about feedback and communication about error, non-punitive response to error, staffing levels, teamwork across units and overall perceptions of safety. Allied health professionals on the other hand had more favourable perceptions about staffing levels, hospital management's commitment to safety and about overall perceptions of safety than either doctors or nurses.

For various reasons (see Sarac 2011), several subsets of data did not have outcome scores and so the overall sample was reduced to six hospitals and 1,866 participants (registered nurses; Allied Health Professionals, 22 percent). Only 6 percent of the respondents were doctors. We first examined the associations between the 12 scales and the two single item outcome measures of the HSOPSC (patient safety grade and number of incidents reported). Moderate correlations were observed between the variables. The correlations between Overall Perceptions of Safety and Staffing were higher compared to rest of the scales, pointing to a possible tendency by healthcare employees to judge overall patient safety through the adequacy of staffing levels, as was also suggested by the English data using the HSOPSC (Waterson et al. 2009). On the other hand, similar to the Norwegian findings, a weak influence of safety culture dimensions on the single item 'number of incidents reported in the last 12 months' variable was observed. For this reason, it was suggested that this single item be adopted as a change measure to assess the level of incident reporting over time, rather than as a criterion variable to assess the predictive validity of the HSOPSC dimensions (Sorra and Nieva 2004), since it might not allow assessment of the actual risk levels (Olsen 2008). We also examined the influence of HSOPSC dimensions on additional safety outcome variables (see Agnew et al. 2013 for further details). Staffing levels and hospital management's support consistently predicted all the outcome variables. Every criterion measure was predicted by both the unit and hospital level dimensions. For the behavioural measures, the most significant predictor of safety compliance was the adequacy of staffing levels and, for safety participation, organisational learning accounted for most of the variance. For the self-reported single item injury measures, out of 10 predictors, only three met the criteria and were found to be negatively related to self-reports of worker and patient injury rates. Although significant, the effect sizes were smaller when examined in relation to self-reported worker and patient injuries. The difficulty of identifying accurate outcome measures in patient safety research has been previously noted, especially the challenges associated with incident reporting data (Ross et al. 2010).

Challenges

Regarding the limitations of this survey, although seven acute hospitals across Scotland were involved, the low overall response rate is of concern given the risk of selection bias in that the perceptions of safety culture reported here might not represent the non-respondents. While the response rates were comparable to an earlier Scottish NHS survey on reporting culture (Cross et al. 2007), the timing of our survey had first to be delayed for several months to avoid overlapping with the distribution of a national NHS staff survey. Then, when permission was granted, its administration coincided with the introduction of a major Scottish government initiative on patient safety (SPSP) which required various types of data collection including, in some centres, questionnaires. For future NHS hospital safety culture surveys of this type, additional effort would be required to ensure a higher level of local support and managerial involvement before data collection commenced.

More importantly, as specific work unit identifiers were not collected to enhance anonymity, it was not possible to link the responses of the individuals to their specific work location (e.g. an individual ward or clinic) beyond the general work areas (e.g. surgery department). For this reason, although safety culture has been demonstrated as a group level phenomenon (Sirriyeh et al. 2012; Zohar 2000) and the HSOPSC instrument can distinguish between hospital units (Blegen et al. 2009; Smits et al. 2008) for the current study, consequent clustering analysis at the unit level could not be conducted.

Finally, despite the limitations of adopting self-reported data as a criterion measure (i.e. biased recall, common method bias, inhibiting certain severe incidents, social desirability), objective data (i.e. via company records and incident reporting systems) has also been found to be problematic (Cooper and Phillips 1994). It has been argued (Pransky et al. 1999) that the incentives, although designed to decrease organisational injury rates by rewarding the employees, might suppress actual reporting behaviour. Additionally, technological barriers, time constraints, underestimated the consequences of minor injuries, and lack of appraisal undermine the validity of reported work-related injuries (Pransky et al. 1999). Although the alignment between self-reported accidents and national accident rate figures has been previously established on a sample of offshore platform personnel in Norway (Rundmo 1994), the same was not true for hospital workers' reporting of patient injury data (Williams et al. 2008). While investigating the adverse event rates in a large Scottish hospital, researchers reported a lack of agreement between the hospital reporting system records and the number of adverse events (AEs) identified through a review of a sample of case notes. Only 10 percent of the AEs were found to have been reported to the hospital's voluntary reporting system (Williams et al. 2008), with a similar finding reported in an English study (Sari et al. 2007).

Box 10.1   Lessons learnt: assessment process

- Paper based surveys are more useful than web based surveys for hospital staff.
- Significant local support is required to increase participation in the survey.
- Safety culture factors explained more variance in safety behaviour measures than they did variance of worker and patient injury rates.
- The difficulty of identifying accurate outcome measures in patient safety research is acknowledged.

Stage 2: Evaluation: Creating an Enhanced Understanding of Patient Safety Culture

Building on the survey results, the next phase utilised focus groups with members of the clinical staff (i.e. nurses, AHPs and doctors) of one acute hospital from each of two Scottish Health Boards. Although safety culture research tends to rely more on quantitative methods (Guldenmund 2000; Mearns et al. 2001), it is often not possible to understand why people think the way they do from survey results. In other words, questionnaires provide us with an answer to 'what' but, in order to explain 'why', qualitative methods are recommended (Guldenmund 2007) to understand further the responses to questionnaire items. Due to the limitations associated with employing safety culture surveys (i.e. the threat of non-random response bias), a recent study warned against extensive reliance on surveys and recommended triangulation of quantitative and qualitative methods to obtain a more accurate diagnosis of culture (O'Connor et al. 2011). Following the evidence from industrial safety culture research (e.g. Safety Culture Enhancement Toolbox and Safety Climate Assessment Toolkit, Cox and Cheyne 2000), as mentioned earlier, healthcare organisations were advised to adopt the safety management techniques used by high-risk industries (Kohn et al. 2000). Such techniques include qualitative methods (e.g. interviews, workshops and focus groups) as sources of supplementary information which can enhance understanding of an organisation's safety culture. As Cox and Cheyne (2000: 111) described, a mixed methods approach reveals different aspects of the organisation's culture and produces 'complementary data rather than alternatives'. Since the questionnaire data provided an overall level of shared perceptions of the staff with regards to safety culture, the principal aim was to obtain a more detailed understanding of the aspects related to 'safety culture in practice'.

Why Focus Groups in Safety Culture Research?

People develop their attitudes and perceptions about any specific construct in relation to their environment, in interaction with other people, not in isolation. In this sense, safety culture is also a concept that develops through the process of making sense of our environment, rather than a passive observation of the organisational world around us; a social-cognitive construct (Zohar and Luria 2004). The procedures, policies, behaviours of others become the source of our perceptions. In this respect, focus groups allow the researcher to examine people's different perspectives as they engage in a social network, exploring the factors contributing to participants' articulation of opinions (Kitzinger and Barbour 1999). An effective example of using focus groups as part of safety culture intervention is addressed by EUROCONTROL (the European Organization for the Safety of Air Navigation) (Eurocontrol White Paper on Safety Culture, www. eurocontrol.int/; Mearns et al. 2013 or see Kirwan and Shorrock, Chapter 17). In order to assess and improve safety culture, they organised workshops with small groups of organisational personnel, based on the findings of their safety culture survey. The programme for the workshops involved summarising questionnaire results, discussing the causes of the issues identified by survey data and generating solutions from participants. The overall data were then used to pinpoint strengths and weaknesses in respect to safety culture and to provide recommendations for improvement as part of the feedback process.

For the current project, we followed a similar approach. Firstly, taking into account the heavy workload of healthcare professionals, time limitations were acknowledged. Multiple methods were used to recruit participants for focus groups by contacting hospital risk managers and the leads of each professional group at each hospital. Participation was voluntary. The Clinical Governance and Risk Management Unit in hospitals A and D aided the recruitment via posters. Furthermore, the researcher attended two managerial meetings in hospital D in order to encourage participation in the focus groups. Participants were required to have been a member of staff within their hospital for at least 12 months.

While constructing the focus group agenda, it was important to cover both positive and negative perceptions of the culture. Consequently, based on the results of the two hospitals in this study, the most favourable and most unfavourable items were derived from their own survey data (a total of four items). The safety culture profiles of both hospitals were very similar and therefore the same items were presented to the staff members of each hospital in order to allow comparisons (participants were shown the response pattern from their own hospital data). The initial aim was to derive items from both the hospital and unit level dimensions for each category (favourable and unfavourable responses). However, the results revealed that none of the hospital level dimensions received highly positive responses (> 60 percent average positive percent score). Therefore, only unit level scale items were included for the favourable responses category (*Teamwork within units* and *Communication openness*). For the unfavourable responses category, items from both levels (unit and hospital) were represented in the focus group schedule. Following the item selection and coding, representative graphs were prepared to provide visual data (see Figure 10.2).

In order to investigate the findings of the questionnaire data (reality check), participants were first asked whether responses on the specific item reflected the reality within their hospital and to what extent did the findings fit their perceptions. Later, to determine any other additional aspects and consequences of the specific issue being discussed, participants were required to comment on how this (the theme reflected by the presented item) can have an effect on patient safety. Finally, their suggestions were requested to improve that aspect of patient safety within their hospital. In order to compare the survey findings with focus group data, at the end of each focus group participants were given the written definitions of all 12 dimensions of the HSOPSC survey. They were asked to rank (individually) the top five safety culture scales according to their relative importance in relation to patient safety.

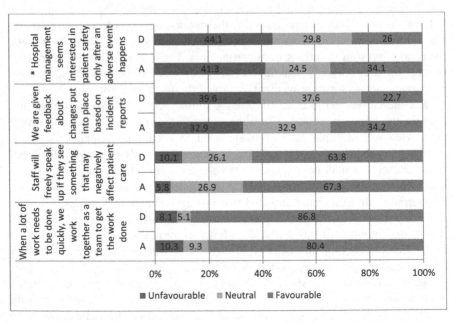

Figure 10.2   Overall results of selected four items per hospital (A and D)

Table 10.1   Questionnaire items presented and the related focus group questions

Questionnaire item	Focus group question
Favourable items	
Communication openness (Unit level)	
Staff will freely speak up if they see something that may negatively affect patient care	
Teamwork within units (Unit level)	
When a lot of work needs to be done quickly, we work together as a team to get the work done	• *Do you think this reflects the reality within your hospital? / To what extent does this fit with your perceptions?*
Unfavourable items	
Hospital management's support (Hospital level)	• *Can this affect patient safety? In what ways?*
Hospital management seems interested in patient safety only after an adverse event happens	• *Do you have any suggestions to improve this? / What actions could make a difference?*
Feedback and communication about error (Unit level)	
We are given feedback about changes put into place based on the incident reports	

In the first hospital, five discussions were held (Group 1: 4 AHPs, Group 2: 2 nurses, Group 3: 2 nurses, Group 4: 1 nurse, Group 5: 1 consultant) with a total of 10 participants. Although two of the sessions had only one participant, their responses were included given the difficulty of recruiting hospital clinical staff as participants and the time constraints. Similarly, despite the aim of recruiting clinical staff for six focus groups in the second hospital, four group discussions were conducted (Group 1: 4 AHPs, Group 2: 4 nurses, Group 3: 3 nurses, Group 4: 4 consultants) with a total of 15 clinical staff. Therefore, the overall sample constituted nine focus groups with 25 participants from three different professional groups who were in direct contact with patients at two separate hospital settings. The majority of the participants were nurses (n = 12), followed by AHPs (n = 8), and medical consultants (n = 5).

Confidentiality

Prior to group discussions, informed consent was obtained. Participants were told that the tape recordings would be destroyed after transcription and the data collected during the focus groups would be de-identified.

In order to facilitate the subsequent group discussions, participants were given 10 minutes to write down the strengths of and threats to patient safety within their hospitals. Staff at both hospital settings commented on similar aspects of patient safety within their organisations. The main strengths of their work place were judged to be the Datix system (www.datix.co.uk) which provided an opportunity to report incidents electronically, support from the management, ongoing initiatives, leadership and continuing risk assessments. Staff shortages, governmental targets (e.g. bed pressures, waiting times), financial climate, patient flow and lack of feedback on the other hand were the common weaknesses stated at both hospitals. Although, SPSP was perceived favourably and addressed as a strength in one of the hospitals, resources directed towards the programme and the expanded documentation were described as weaknesses. Finally, staff in both hospital settings acknowledged that the size of the organisation might have negative effects on implementation of the new initiatives and challenge both culture change and the communication between the staff members at different levels.

The data obtained from the group discussions illustrated the role of hospital and unit level management on safety-related outcomes perceived by the participants. Data clearly showed that the staff did not receive regular feedback based on the incident reports and also questioned the efficiency of the incident reporting system. Participants' suggestions were focused on efforts to improve the Datix (incident report) system (www.datix.co.uk) to provide feedback and a frequent comment on the consequences of a lack of feedback was loss of motivation to report incidents experienced by the staff. This might have a number of implications for safety management and for safety research. First, an organisation that cannot capture incidents, near-misses or adverse events and the reasons behind them runs the risk of failure to detect future incidents and, in turn, will not be able to establish a learning culture (Healthcare Commission 2009). Next, such perceptions illustrating the ineffectiveness of the reporting system where no feedback is given following a reported incident (Probst and Estrada 2010) might decrease the reporting behaviour of the staff. An increased discrepancy can occur between actual incidents and the overall under-reporting rates when employees do not benefit from reporting an incident (Glendon 1991; Francis 2013). In the long run, these perceptions and the behavioural patterns undermine the validity of the objective data that can be adopted for future research investigating safety-related outcomes (Williams et al. 2008).

Furthermore, in the focus groups, the communication gap between the senior management and front line staff was also acknowledged and hospital management was perceived to be reactive.

Table 10.2   Summary of the themes that emerged in the group discussions

Item	Reality check/ themes	Effects on patient safety	Improvement efforts
Unfavourable			
We are given feedback about the changes put into place based on incident reports	Lack of feedback, incident reporting system, lack of time, severity of the incident, communication, individual factors (ignorance, workload, lack of effort)	Loss of motivation, well being (demoralisation), lack of learning, sharing, failure to detect future incident, less reporting	Feedback reports, improving Datix system, more visible management (walk rounds)
Hospital management seems to be interested in patient safety only after an adverse event happens	Communication gap, visibility of the management, reactive management, lack of prioritisation of safety, pressures from the government, individual factors (lack of understanding)	Loss of motivation, lack of prioritisation of safety by the staff, well being (stress), patients at risk	More visible management (walk rounds, face to face meetings), approachable management, efficient communication, positive feedback, setting examples (role models)
Favourable			
Staff will freely speak up if they see something that may negatively affect patient care	Discrepancy between the actual and the desired behaviour, experience in the profession, confidence, hierarchy, consequences, unit manager	Risk of reoccurrence of the mistakes and incidents, poor teamwork, lack of learning, poor clinical decision making, patients at risk	Open culture, non-punitive response, setting examples (role models), approachable management, tolerance, valuing the staff
When a lot of work needs to be done quickly, we work together as a team to get the work done	Increase in teamwork, less conflict and hierarchy, better communication, differences across units, unit manager, staff shortages	Poor quality of care, incomplete tasks, lack of knowledge, possibility of mistakes, communication breakdown, less documentation	Positive feedback, importance of the unit management, morale, efficient communication, better leadership, multi-disciplinary meetings

In considering senior management, a more visible and approachable management, coupled with efficient communication, was said to be necessary in order to improve the safety-related perceptions of the staff. Patient safety literature also drew attention to these issues. For example, leadership style in an organisation was identified as a key factor in determining the organisational readiness for patient safety initiatives (Burnett et al. 2010). Transparency (sharing information about incidents and near-misses) is an important aspect of culture of safety (Leape et al. 2009) where timely feedback is a motivator for an effective incident reporting system (Benn et al. 2009). In a study of safety culture in technical operations in Air Traffic Controllers using triangulation techniques (i.e. focus groups, interviews, observations and survey), similar perceptions were revealed. Participants reported less trust in senior management compared to their immediate supervisors. Following the implementation of a non-punitive error reporting system initiative, significant improvements in safety-related employee perceptions and attitudes were demonstrated (Patankar et al. 2012).

In the current focus group study, unit level management was also discussed, being one of the favourably perceived items. More specifically, the actions of supervisors were mentioned in relation to speaking up behaviour and teamwork within units. (These findings led to a subsequent study of senior charge nurses' safety leadership, see Agnew and Flin 2014.) Participants questioned the highly favourable findings (i.e. suggesting there could be a social desirability effect) obtained from the survey study on the speaking up behaviour, and no consensus was reached on this particular item. While the amount of experience within one's profession and confidence in one's abilities were mentioned as aspects that might facilitate speaking up behaviour, the hierarchy between the occupational groups could restrain the amount and the quality of communication openness within the organisation. In this respect, reflecting on the overall progress of patient safety improvements in the last decade, the failure to achieve an open culture has been associated with reoccurrence of mistakes and poor teamwork and decision making, in addition to putting patients at risk (Leape et al. 2009).

Although the focus group discussions were limited to a few positive and negative topics derived from the survey responses, they demonstrated the multi-level conceptualisation of safety culture (Zohar and Luria 2005). For example, staff perceptions at an individual level revealed concerns about the effects of their motivation, knowledge, responsibilities and wellbeing on patient safety issues and improvement efforts. At the unit level, comments highlighted the positive role of unit/ward managers in determining the leadership and teamwork within the work area, providing feedback and ensuring the quality of communication between team members. On the other hand, senior management was frequently viewed unfavourably and dissatisfaction was reported regarding the visibility of the management team and their ineffective use of communication channels. However, participants also acknowledged the effects of institutional context (i.e. time pressures and budget costs) on management activities such as governmental targets.

These pressures (forcing prioritisation of production over safety) on the organisational decision making process might have further implications. For instance, the implicit acceptance of deviations from safety standards in order to achieve certain governmental targets could result in performance drifting towards the boundaries of safety (Dekker 2006; Francis 2013). The human factors literature also emphasises the role of external factors in the safety of care (Carayon 2007) at different levels, recognising the influence of regulatory and governmental forces on organisational activities while discussing ways to minimise human error. In this respect, data obtained from the focus groups extended our understanding of the survey findings by demonstrating the contributing factors. A key factor appears to be the effect of managerial behaviours (at each level of the organisational hierarchy) on safety related outcomes. (See Fruhen et al. 2013 for a novel method of examining senior managers' interpretations of safety culture.)

Box 10.2   Lessons learnt: focus groups

- Using examples from survey data assists in interpreting findings.
- Careful consideration of the composition of the group: may increase discussion by combining participants from similar occupational groups.
- Minimise disruption to work activities: organise close to worksites, provide refreshments/ lunch.
- If there is only one facilitator, recording is recommended rather than note taking.

Stage 3: Feedback:
Helping Managers Understand and Benchmark their Data

Following the completion of data analysis, all participating hospitals were sent a confidential feedback report explaining the background and the aims of the study, in addition to a summary of their own results. These reports provided baseline measures of each organisation's safety culture, where staff perceptions regarding different aspects of patient safety were demonstrated at the item level for each questionnaire scale. Subsequent anecdotal feedback received from the contact managers at the hospitals was that, in some cases, these confidential reports had been discussed at Board level.

As the next stage, a one day workshop ('Safety Culture in Healthcare Workshop') was organised at the University of Aberdeen to discuss the overall findings, with the involvement of the participating hospitals and different NHS Board members who had not taken part in the study but were responsible for the patient safety programme within their own organisations. The workshop was designed to provide feedback and generate discussions based on the results of the study. The aim was to enhance understanding of safety culture, present the project findings, review the methods and tools used to assess hospital safety culture, stimulate discussions around factors influencing safety culture, formulate recommendations for improving safety culture and provide an opportunity to learn from a safety culture programme in another high reliability industry (i.e. aviation).

The workshop opened with an articulation of the aims and structure of the day. The 14 participants (from different hospitals) introduced themselves and stated their expectations from the event. The first part of the workshop focused on what safety culture is and why it is important to measure it. Next, an overview of the overall project was presented. For the interactive group exercise, a case scenario was developed in order to facilitate the exchange of ideas between participants. The case was related to the results of a safety culture assessment carried out by a fictitious hospital and presented through graphs and quotes from hospital staff. Although not stated in the discussion materials, the data were derived from the overall results of the project. The participants, in small groups, were then asked to identify challenges and provide recommendations to hospital management. Participants exchanged experiences and produced suggestions, not only on how to improve areas of concern but also on how to maintain the hospital's strengths. A general focus on incident reporting and senior management was observed. For example, while each group proposed initiatives to encourage incident reporting, they also acknowledged the necessity of providing rapid feedback. In this respect, the efficiency of current communication channels in dissemination of the proposed changes and the rationale behind them were highlighted. Finally, a number of people emphasised the importance of sharing the results of the safety culture assessment at management level within their organisations.

Following the discussion session, current techniques used in the assessment of safety culture were presented. This was followed by suggestions on how to improve safety culture in participants' own organisations. Finally, the workshop closed with an invited speaker describing the safety culture programme in Air Traffic Management run by EUROCONTROL (Mearns et al. 2013).

The feedback received from attendees showed that they were highly satisfied with the information provided and the resulting discussions. The workshop was perceived as an effective way of disseminating findings by providing an opportunity to involve different Health Board members in the review of findings and their implications. On the other hand, participants suggested the involvement of more senior hospital managers in future workshops.

Box 10.3   Lessons learnt: safety culture workshops

- Workshops provide a valuable source of information for identifying solutions to improve safety culture.
- They can be used as a feedback tool to discuss the strengths and areas of concern in relation to safety.
- Action plans can be developed following workshop discussions.
- Facilitators should ensure each participant is involved in discussions.
- The case study exercise using generic results for a fictitious hospital worked well.

Conclusions

The hospital safety culture project presented in this chapter demonstrated an overview of a safety culture assessment process in a healthcare setting, using different methods. Achieving a cultural change within an organisation requires continuous investment in resources and allocating responsibilities. A one-time measurement with survey instruments is clearly not sufficient to identify the causal factors contributing to a culture of safety (Patankar et al. 2012). Longitudinal assessment of safety culture, combined with tracking and monitoring measurable performance goals will contribute to continuous learning and organisational improvement. Application of a validated safety culture instrument can be made part of the regular hospital assessment process and can also be employed as an evaluation tool pre- and post-safety related interventions (Jackson et al. 2010). In this respect, the timing of the assessment, the length of the instrument employed, the feedback given to staff and the development of action plans based on the findings are some of the factors that need special consideration in order to increase the effectiveness of the safety culture measurement process. (See Fleming 2005 or Patankar et al. 2012 for more advice on conducting safety culture measurement and feedback in healthcare.) In order to achieve a cultural change, it is crucial to be realistic and deliver clear expectations about what needs to be changed and to communicate the benefits of the desired change (Cox and Cheyne 2000). Encouraging active participation by employees in cultural transformation would also increase the ownership of staff in the implementation of a safety initiative.

Acknowledgements

This work was funded by a Scottish Funding Council Strategic Research Development Grant to the Scottish Patient Safety Research Network.

References

Agnew, C. and Flin, R. 2014. Senior charge nurses' leadership behaviours in relation to hospital ward safety: A mixed method study. *International Journal of Nursing Studies*, 51, 768–80.

Agnew, C., Flin, R. and Mearns, K. 2013. Patient safety climate and worker safety behaviours in acute hospitals in Scotland. *Journal of Safety Research*, 45, 95–101.

Alahmadi, H.A. 2010. Assessment of patient safety culture in Saudi Arabian hospitals. *Quality & Safety in Health Care*, 19, 1–5.

Benn, J., Koutantji, M., Wallace, L., Spurgeon, P., Rejman, M., Healey, A. and Vincent, C. 2009. Feedback from incident reporting: Information and action to improve patient safety. *Quality and Safety in Healthcare*, 18, 11–20.

Blegen, M.A., Gearhart, S., O'Brien, R., Sehgal, N.L. and Alldredge, B.K. 2009. AHRQ's hospital survey on patient safety culture: Psychometric analyses. *Journal of Patient Safety*, 5, 139–44.

Burnett, S., Benn, J., Pinto, A., Parand, A., Iskander, S. and Vincent, C. 2010. Organisational readiness: Exploring the preconditions for success in organisation wide patient safety improvement programmes. *Quality and Safety in Health Care*, 19, 313–17.

Carayon, P. 2007. *Handbook of Human Factors and Ergonomics in Health Care and Patient Safety*. Boca Raton, FL: CRC Press.

Clarke, S. 2006. The relationship between safety climate and safety performance: A meta-analytic review. *Journal of Occupational Health Psychology*, 11, 315–27.

Cooper, D. and Phillips, R. 2004. Exploratory analysis of the safety climate and safety behavior relationship. *Journal of Safety Research*, 35, 497–512.

Cox, S.J. and Cheyne, A. 2000. Assessing safety culture in offshore environments. *Safety Science*, 34, 111–29.

Cross, S., Whittington, C. and Miller, Z. 2007. *NHS Scotland Incident Reporting Culture. Extended Study – National Summary Report*. Edinburgh: NHS/QIS. Available at: www.nhshealthquality.org/nhsqis/3870.html (last accessed on 24 April 2014).

Dekker, S. 2006. Resilience engineering: Chronicling the emergence of confused consensus. In E. Hollnagel, D. Woods and N. Leveson (eds), *Resilience Engineering: Concepts and Precepts*. Aldershot, UK: Ashgate, 77–92.

EUROCONTROL (The European Organization for the Safety of Air Navigation) *Eurocontrol White Paper on Safety Culture*. Available at: www.eurocontrol. int/articles/safety-culture (last accessed on 24 April 2014).

Fleming, M. 2005. Patient safety culture measurement and improvement: A "How To" guide. *Healthcare Quarterly*, 8, 14–19.

Flin, R. 2007. Measuring safety culture in healthcare: A case for accurate diagnosis. *Safety Science*, 45, 653–67.

Flin, R., Burns, C., Mearns, K., Yule, S. and Robertson, E.M. 2006. Measuring safety climate in health care. *Quality and Safety in Health Care*, 15, 109–15.

Francis, R. 2010. *Independent Inquiry into Care provided by Mid Staffordshire NHS Foundation Trust January 2005 – March 2009*. London: The Stationery Office.

Francis, R. 2013. *Final Report of the Mid Staffordshire NHS Foundation Trust Public Inquiry*. London: The Stationery Office.

Fruhen, L., Mearns, K., Flin, R. and Kirwan, B. 2013. From the surface to the underlying meaning: An analysis of senior managers' safety culture perceptions. *Safety Science*, 57, 326–34.

Gershon, R.M., Karkashian, C.D., Grosch, J.W., Murphy, L.R., Escamilla-Cejudo A., Flanagan, P.A., Bernacki, E., Kasting, C. and Martin, L. 2000. Hospital safety climate and its relationship with safe work practices and workplace exposure incidents. *American Journal of Infection Control*, 28, 211–21.

Gimeno, D., Felknor, S., Burau, K.D. and Delclos, G.L. 2005. Organisational and occupational risk factors associated with work related injuries among public hospital employees in Costa Rica. *Occupational and Environmental Medicine*, 62, 337–43.

Glendon, A.I. 1991. Accident data analysis. *Journal of Health and Safety*, 7, 5–24.

Guldenmund, F.W. 2000. The nature of safety culture: A review of theory and research. *Safety Science*, 34, 215–57.

Guldenmund, F.W. 2007. The use of questionnaires in safety culture research – An evaluation. *Safety Science*, 45, 723–43.

Healthcare Commission 2009. *Investigation into Mid-Staffordshire NHS Foundation Trust*. London: Commission for Healthcare Audit and Inspection.

Hofmann, D.A. and Mark, A. 2006. An investigation of the relationship between safety climate and medication errors as well as other nurse and patient outcomes. *Personnel Psychology*, 59, 847–69.

Huang, D.T., Clermont, G., Kong, L., Weissfeld, L.A., Sexton, J.B., Rowan, K., Angus, D. and Romn, K.M. 2010. Intensive care unit safety culture and outcomes: A US multicenter study. *International Journal of Quality in Health Care*, 22, 151–61.

Jackson, J., Sarac, C. and Flin, R. 2010. Hospital safety climate surveys: Measurement issues. *Current Opinion in Critical Care*, 16, 632–8.

Kitzinger, J. and Barbour, R.S. 1999. Introduction: The challenge and promise of focus groups. In R.S. Barbour and J. Kitzinger (eds), *Developing Focus Group Research: Politics, Theory and Practice*. London: Sage, 1–20.

Kohn, L.T., Corrigan, J.M. and Donaldson, M.S. 2000. Institute of Medicine (US) Committee on Quality of Health Care in America. *To Err Is Human: Building a Safer Health System*. Washington, DC: National Academy.

Leape, L., Berwick, D., Clancy, C., Conway, J., Gluck, P., Guest, J., Lawrence, D., Morath, J., O'Leary, D., O'Neill, P., Pinkiewicz, D. and Isaac, T. 2009. Transforming healthcare: A safety imperative. *Quality and Safety in Health Care*, 18, 424–28.

Mardon, R., Khanna, K., Sorra J., Dyer, N. and Famolaro, T. 2010. Exploring relationships between hospital patient safety culture and adverse events. *Journal of Patient Safety*, 6, 226–32.

Mark, B., Hughes, L., Belyea, M., Chang, Y., Hofmann, D., Jones, C. and Bacon, C. 2007. Does safety climate moderate the influence of staffing adequacy and work conditions on nurse injuries? *Journal of Safety Research*, 38, 431–46.

Mearns, K., Kirwan, B., Reader, T., Jackson, J., Kennedy, R. and Gordon, R. 2013. Development of a methodology for understanding and enhancing safety culture in Air Traffic Management. *Safety Science*, 53, 123–33.

Mearns, K., Whitaker, S. and Flin, R. 2001. Benchmarking safety climate in hazardous environments: A longitudinal, inter-organisational approach. *Risk Analysis*, 21, 771–86.

Mearns, K., Whitaker, S.M. and Flin, R. 2003. Safety climate, safety management practice and safety performance in offshore environments. *Safety Science*, 41, 641–80.

Naveh, E., Katz-Navon, T. and Stern, Z. 2005. Treatment errors in healthcare: A safety climate approach. *Management Science*, 51, 948–60.

Neal, A. and Griffin, M.A. 2006. A study of the lagged relationships among safety climate, safety motivation, safety behavior, and accidents at the individual and group levels. *Journal of Applied Psychology*, 91, 946–53.

Neal, A., Griffin, M.A. and Hart, P.M. 2000. The impact of organizational climate on safety climate and individual behaviour. *Safety Science*, 34, 99–109.

O'Connor, P., Buttrey, S., O'Dea, A. and Kennedy, Q. 2011. Identifying and addressing the limitations of safety climate surveys. *Journal of Safety Research*, 42, 259–65.

Olsen, E. 2008. Reliability and validity of the Hospital Survey on Patient Safety Culture at a Norwegian hospital. In J. Øvretveit and P. Sousa (eds), *Quality and Safety Improvement Research: Methods and Research Practice from the International Quality Improvement Research Network (QIRN)*. Lisbon: Escola Nacional de Saúde Pública, 173–86.

Olsen, E. 2010. Exploring the possibility of a common structural model measuring associations between safety climate factors and safety behaviour in healthcare and the petroleum sectors. *Accident Analysis & Prevention*, 42, 1507–16.

Patankar, M., Brown, J., Sabin, E. and Bigda-Peyton, T. 2012. *Safety Culture: Building and Sustaining a Cultural Change in Aviation and Healthcare*. Farnham: Ashgate.

Pfeiffer, Y. and Manser, T. 2010. Development of the German version of the Hospital Survey on Patient Safety Culture: Dimensionality and psychometric properties. *Safety Science*, 48, 1452–62.

Pransky, G., Snyder, T., Dembe, A. and Himmelstein, J. 1999. Under reporting of work related disorders in the workplace: A case study and review of the literature. *Ergonomics*, 42, 171–82.

Probst, T.M. and Estrada, R.X. 2010. Accident under-reporting among employees: Testing the moderating influence of psychological safety climate and supervisor enforcement of safety practices. *Accident Analysis & Prevention*, 42, 1438–44.

Rosen, A.K., Singer, S., Zhao, S., Shokeen, P., Meterko, M. and Gaba, D. 2010. Hospital safety climate and safety outcomes: Is there a relationship in the VA? *Medical Care Research and Review*, 67, 590–608.

Ross, A., Plunkett, M. and Walsh, K. 2010. Adverse event categorisation across NHS Scotland. *Quality and Safety in Healthcare*, 19, 1–4.

Rundmo, T. 1994. Associations between safety and contingency measures, and occupational accidents on offshore petroleum platforms. *Scandinavian Journal of Work, Environment & Health*, 20, 128–31.

Sarac, C. 2011. *Safety Climate in Acute Hospitals.* PhD thesis, University of Aberdeen.

Sarac, C., Flin, R., Mearns, K. and Jackson J. 2011. Hospital survey on patient safety culture: Psychometric analysis on a Scottish sample. *BMJ Quality & Safety*, 20, 842–8.

Sari, A.B.A., Sheldon, T.A., Cracknell, A. and Turnbull, A. 2007. Sensitivity of routine system for reporting patient safety incidents in an NHS hospital: Retrospective patient case note review. *British Medical Journal*, 334, 79–81.

Scottish Patient Safety Alliance 2007. Available at: http://patientsafety.etellect. co.uk/programme/about/aims (last accessed on 24 April 2014).

Sirriyeh, R., Lawton, R., Armitage, G., Gardener, P. and Ferguson, S. 2012. Safety subcultures in healthcare organizations and managing medical error. *Health Services Management Research*, 25, 16–23.

Sorra, J.S. and Nieva, V.F. 2004. *Hospital Survey on Patient Safety Culture.* (Prepared by Westat, under Contract No. 290-96-0004). AHRQ Publication No. 04-0041. Rockville, MD.

Smith, D.R., Muto, T., Sairenchi, T., Ishikawa, Y., Sayama, S., Yoshida, A. and Townley-Jones, M. 2010. Hospital safety climate, psychosocial risk factors and needlestick injuries in Japan. *Industrial Health*, 48, 85–95.

Smits, M., Dingelhoff, I.C., Wagner, C., Wal, G. and Groenewegen, R. 2008. The psychometric properties of the 'Hospital Survey on Patient Safety Culture' in Dutch hospitals. *BMC Health Services Research*, 8, 230.

Van Noord, I., De Bruijne, M. and Twisk, J. 2010. The relationship between patient safety culture and the implementation of organizational patient safety defences at emergency departments. *International Journal of Quality in Health Care*, 22, 162–9.

Vincent, C. 2010. *Patient Safety* (second edition). Chichester: Wiley-Blackwell.

Waterson, P.E., Griffiths, P., Stride, C., Murphy, J. and Hignett, S. 2009. Psychometric properties of the hospital survey on patient safety: Findings from the UK. *Quality and Safety in Health Care*, 19, 1–5.

Williams, D., Olsen, S., Crichton, W., Witte, K., Flin, R., Ingram, J. Campbell, M., Watson, M., Hopf, Y. and Cuthbertson, B. 2008. Detection of adverse events in a Scottish hospital using a consensus-based methodology. *Scottish Medical Journal*, 53, 29–33.

Zohar, D. 2000. A group-level model of safety climate: Testing the effect of group climate on micro-accidents in manufacturing jobs. *Journal of Applied Psychology*, 85, 587–96.

Zohar, D., Livne, Y., Tenne-Gazit, O., Admi, H. and Donchin, Y. 2007. Healthcare climate: A framework for measuring and improving patient safety. *Critical Care Medicine*, 35, 1312–17.

Zohar, D. and Luria, G. 2004. Climate as a social-cognitive construction of supervisory safety practices: Scripts as proxy of behavior patterns. *Journal of Applied Psychology*, 89, 322–33.

Zohar, D. and Luria, G. 2005. A multilevel model of safety climate: Cross-level relationships between organization and group-level climates. *Journal of Applied Psychology*, 90, 616–28.

Chapter 11

The Use of the Hospital Survey on Patient Safety Culture in Europe

Antje Hammer and Tanja Manser

Introduction

The relevance of safety culture to safe operation in high-risk industries is not disputed (Cox and Flin 1998). However, the last three decades have seen extensive debates on the concepts behind, and the distinction between, the terms safety culture and safety climate (Alhemood et al. 2004; Choudhry et al. 2007; Flin et al. 2006; Guldenmund 2000; Hale 2000; Stricoff 2005).

Safety culture is an aspect of organisational culture that refers to how safety is viewed and treated in organisations (Hofinger 2008). Guldenmund characterised organisational culture as a relatively stable, multi-dimensional and hypothetical construct, which depends on shared values and norms in the work environment (Guldenmund 2000) (see also Chapter 2 by Guldenmund for details). These values and norms affect the attitudes, perceptions and behaviour of all members of an organisation (Pfaff et al. 2009). Organisational climate, however, reflects the perception of employees with regard to the organisation culture (Gershon et al. 2004). Thus, *safety climate* can be defined as the shared perceptions of employees about safety relevant aspects of their work environment (Seo et al. 2004; Zohar 1980). These perceptions provide a frame of reference, guiding employees in fulfilling their tasks and in dealing with safety issues (Schneider 1975).

Summarising the various concepts and definitions proposed, safety culture appears to be the broader, manifest concept behind the framework of safety climate. Safety culture is the source for patterns of behaviour which can be observed, described and changed (Goodmann 2004), whereas safety climate is the sum of behaviours and attitudes based on common assumptions and beliefs toward patient safety. Cox and Flin (1998) describe culture as an organisation's 'personality' while climate is seen as the organisation's 'mood'.

Measuring Safety Climate

From a measurement perspective, '… safety climate can be conceived of as a 'snapshot', or manifestation of culture' (Cox and Flin 1998; Naevestad 2009: 127–8) that can be assessed using quantitative measures, while safety culture may rather be assessed qualitatively. Nevertheless, a huge number of studies on safety culture actually measure safety climate using questionnaires (Cooper 2000). In doing so, safety climate serves as a quantifiable surrogate parameter of safety culture (Cooper 2000; Gershon et al. 2004, 2007; Guldenmund 2000).

The first measures developed to analyse safety culture in healthcare were adapted from those used in other industrial sectors in the late 1990s (Flin 2007). Following the 'functionalist paradigm' (Glendon and Stanton 2000) of organisational climate research, a number of survey instruments assessing safety climate at different levels and in different healthcare settings have been developed and published (e.g. Frankfurt Patient Safety Climate Questionnaire for General Practices (FraSiK) (Hoffmann et al. 2009); Pharmacy Safety Climate Questionnaire (PSCQ) (Ashcroft and Parker 2009). For the hospital setting the two most frequently used instruments are the Safety Attitudes Questionnaire (SAQ) (Sexton et al. 2006) (see Chapter 13 by Etchegaray and Thomas for details) and the Hospital Survey on Patient Safety Culture (HSPSC[1]) which focus on team and hospitals level respectively.

The HSPSC – developed by Sorra and Nieva (Nieva and Sorra 2003; Sorra and Nieva 2004) under an Agency for Healthcare Research and Quality grant – was designed to assess front line staff's perceptions of patient safety culture in hospital settings. The questionnaire is based on two healthcare safety culture surveys (the first was developed and administered for a medical event reporting system in Transfusion Medicine (MERS-TM) and the second was developed and administered by the Veterans Health Administration (VHA)). The final version of the original HSPSC consisted of 44 items: 42 items measuring 10 safety culture dimensions and two outcome dimensions, as well as two single item outcome measures (i.e. patient safety grade and number of events reported) (see Chapter 12 by Sorra and Battles for details). Consequently, the HSPSC covers a broad range of safety culture dimensions (Hammer 2012). Since its development in 2003, the HSPSC has been used in hospitals around the world and has also been adapted for other healthcare settings such as medical offices (Sorra et al. 2008), nursing homes (Sorra et al. 2008) and pharmacies (Franklin and Sorra 2012). The survey has been translated and adapted for use in many countries and the psychometric properties of the various versions have been evaluated in many cases.

This chapter aims to provide an overview of the use of HSPSC in hospitals throughout Europe and of the published evidence on its evaluation with regard to psychometric properties. We briefly describe HSPSC studies and discuss them with regard to similarities and differences in methodological approaches and findings. This overview will support practitioners and researchers planning to use the HSPSC in their country and identify areas for future research.

---

1   In literature several abbreviations are used for the Hospital Survey on Patient Safety Culture (e.g. HSPSC, HSOPSC, HSOPS or SOPS). In the following we will use HSPSC as abbreviation.

Although we have done our best to establish a sound basis for this overview, the selection presented is neither comprehensive nor exhaustive because it was limited to studies published before October 2013 and responses from national contact points provided to us by December 2013. Thus, we apologise to any researchers or projects that have used the HSPSC outwith those dates in a European country and are not mentioned here. Nevertheless, we hope that this overview will encourage more intense international exchange and collaboration in the future.

Our Approach to Establishing the 'State of the Art' in Europe

Identification of HSPSC Users in Europe

In order to provide a comprehensive overview of the use of the HSPSC questionnaire in European countries we have combined a formal, web-based literature search with an informal survey among individuals who had been in contact with the developers of the HSPSC regarding its potential use in a European country.

Through the initial web-based literature search on Medline and Web of Science, performed in March 2012, we identified 26 publications on the use of the HSPSC in a European country, which were published between January 2003 (when the HSPSC had originally been developed) and February 2012. We reviewed these publications focusing on instrument development (i.e. the translation and/or adaptation process), data collection process (e.g. in a pilot or field study) and psychometric evaluation (i.e. tests performed and key results of these tests) in different countries.

Through contacting individual clinicians and researchers based on a list of national contacts provided to us by Westat (http://www.westat.com), we received information on additional usage of the HSPSC, mainly for local patient safety improvement projects. In many cases the results of the survey have not been published in English, nor had any psychometric testing taken place. However, we included this information to provide a complete and balanced overview of the countries where the HSPSC has been used so far and to complement our results on the use of the HSPSC for scientific purposes.

Criteria for Describing the Use of HSPSC in Europe

Development process: translations, adaptations and innovations   Before starting in-depth analyses of psychometric properties, we reviewed papers for descriptions regarding the translation and adaptation process. We especially searched for a) information on forward-backward translations from the original American English version into the country's native language, b) adaptations such as adding or deleting items or dimensions and c) changes or revisions with regard to local context issues, such as specific meanings and terminology described in the publications. Moreover, we reviewed the papers for innovations in the methodological approach.

Evaluation of Psychometric Properties

We aimed to evaluate whether the theoretically and empirically developed factor structure of the original version of the HSPSC is comparable to results from the various European countries. In a first step, we reviewed papers for a) the conduct of a confirmatory factor analysis (CFA) (Kline 2005) and b) model refinement based on exploratory factor analyses (EFA) or principal component analyses (PCA). For this analysis we included only studies in English which used the HSPSC in a hospital setting and have already published results on psychometric properties. Studies using the HSPSC in different ambulant settings and non-English publications or manuscripts which had not been published by the end of October 2013, were excluded from further analyses.

In a second step, publications were reviewed for results on discriminant construct validity by searching for calculations of inter-correlations among the safety climate dimensions. In addition, internal consistency was reviewed by assessing Cronbach's alpha for each of the 12 dimensions. The results on Cronbach's alpha were compared to the original HSPSC version by Sorra and Nieva (2004) as well as across European countries.

The appropriateness of the psychometric properties for evaluating the psychometric properties of the HSPSC versions was assessed by common measures as described in Table 11.5 in the annex to this chapter.

Overview of the Use of the HSPSC in Europe

Overall we identified 44 different European studies that have used the HSPSC questionnaire by Sorra and Nieva (Nieva and Sorra 2003; Sorra and Nieva 2004) as a basis for their survey. These studies were conducted in 20 different European countries. Table 11.1 provides an overview of all reviewed studies using the HSPSC questionnaire in a European country.

In five studies the HSPSC that was originally designed for use in hospital settings has been adapted for surveys in ambulant care settings (e.g. general practice, medical centres, or nursing homes). Within another 20 studies, the adapted version of the questionnaire has not – or only to a limited extend – been analysed for psychometric properties or the results are not available in English. In five studies a psychometric evaluation has been conducted but the results of psychometric testing were not published by the time of writing this chapter. Finally, we identified 14 out of the 44 studies (two each in Belgium, Norway and Switzerland and one each in England, France, Germany, Scotland, Slovenia, Sweden, the Netherlands and Turkey) which have analysed and published the psychometric properties of the translated and adapted safety climate instruments.

Table 11.1   Overview of the use of the HSPSC in Europe

Country (Language)	Psychometric properties	Setting	Affiliation/organisation	Correspondence	Publications
Hospital setting, psychometric testing published					
Belgium (Dutch/ Flemish)	Yes	Hospital	Ziekenhuis Oost Limburg / University of Hasselt	Johan Hellings, Ward Schrooten	(Wenqi 2005) Master Thesis; (Hellings et al. 2007, 2010) Journal articles as part of a PhD thesis
Belgium (French)	Yes	Hospital	Faculty of Medicine and Life Sciences Hasselt University; Federal Service Public Health	Annemie Vlayen	(Pryseley 2008; Vlayen et al. 2011)
England	Yes	Hospital	Human Factors and Complex Systems (HFCS) Group, Loughborough Design School, Loughborough University	Patrick Waterson	(Murphy et al. 2007; Waterson et al. 2010)
France	Yes	Hospital	Comité de Coordination de l'Evaluation et de la Qualité en Aquitaine (CCECQA)	Phillipe Michel	(Occelli et al. 2011, 2013)
Germany	Yes	Hospital	IMVR – University of Cologne	Antje Hammer, Holger Pfaff	(Hammer et al. 2011) Journal article as part of a PhD Thesis
Norway	Yes	Hospital	Department of Media, Culture and Social Sciences, University of Stavanger	Espen Olsen	(Olsen 2007, 2008, 2010; Olsen and Aase 2010) Journal articles as part of a PhD Thesis
Norway	Yes	Hospital	Department of Anaesthesia and Intensive Care, Haukeland University Hospital and Department of Medicine, University of Bergen	Arvid Steinar Haugen	(Haugen et al. 2010, 2013) Journal articles as part of a PhD thesis
Scotland	Yes	Hospital	School of Psychology, Industrial Research Centre, University of Aberdeen	Cakil Sarac	(Agnew et al. 2013; Sarac et al. 2011)
Slovenia	Yes	Hospital	Centre for Quality and Patient Safety improvement in Health Care	Andrej Robida	(Robida 2013)

Table 11.1   Overview of the use of the HSPSC in Europe (*continued*)

Country (Language)	Psychometric Properties	Setting	Affiliation/organisation	Correspondence	Publications
Sweden	Yes	Hospital and primary care settings	Medical Management Centre; Karolinska Institutet, Stockholm	Mats Hedsköld, Marion Lindh	(Hedsköld et al. 2013)
Switzerland (German)	Yes	Hospital	ETH Zurich, in collaboration with University Hospital Zurich	Yvonne Pfeiffer, Tanja Manser	(Pfeiffer and Manser 2010) Journal article as part of a PhD thesis; (van Vegten et al. 2011)
Switzerland (French)	Yes	Hospital	Hôpital neuchâtelois, La Chaux-de-Fonds, Switzerland; Division of Clinical Epidemiology, University Hospitals of Geneva and University of Geneva, Geneva, Switzerland	François Kundig	(Kundig et al. 2011; Perneger et al. 2013)
The Netherlands	Yes	Hospital	Netherlands Institute for Health Services Research (NIVEL); EMGO Institute for Health and Care research	Cordula Wagner, Marleen Smits	(Smits et al. 2008, 2009) Journal articles as part of a PhD thesis; (Smits et al. 2012)
Turkey	Yes	Hospital (a second survey was developed but not tested for use in primary the care setting)	Balkesir University School of Medicine, Department of Public Health	Said Bodur	(Bodur and Filiz 2009, 2010)

Hospital setting, psychometric testing unpublished

Greece	Yes	Hospital	Department of Business Administration, University of Macedonia, Thessaloniki	Anastasia A. Mallidou, Vasilis Aletras	In preparation
Italy	Yes	Hospital	Azienda Ospedaliero-Universitaria	Vincenzo Parrinello	Not published
Ireland	Yes	Hospital	HSE Dublin North East	Cornelia Stuart	In preparation

Table 11.1 Overview of the use of the HSPSC in Europe (*continued*)

Country (Language)	Psychometric Properties	Setting	Affiliation/organisation	Correspondence	Publications
Finland	Yes	Hospital	Department of Nursing Science; University of Eastern Finland	Hannele Turunen	(Kuosmanen et al. 2013; Turunen et al. 2011, 2013)
Spain	Yes	Hospital	Fundacion Hospital de Calahorra (La Rioja)	Adrian Martinez, Victoria Musitu	In preparation
Hospital setting, no psychometric testing					
Croatia	No	Hospital	Croatian Society for Quality Improvement in Healthcare	Jasna Mesaric	Not published
England	No	Hospital	Faculty of Health Sciences, University of Southampton	Helen Wharam	In preparation (part of a PhD Thesis)
Finland	No	Hospital	Patient safety coordinator, South Karelia Social and Health Care District	Mari Liukka	In preparation (part of a PhD Thesis)
Germany	No	Hospital	German Agency for Quality in Medicine; Institute for Patient Safety, University of Bonn	Liat Fishman	(Kolbe et al. 2011)
Iceland	No	Hospital	Landspitali University Hospital in Iceland	Áslaug S. Svavarsdóttir	Not published
Ireland	No	Hospital	St Patrick's University Hospital	Ailish Young	Not published
Italy	No	Hospital	–	Annamaria Bagnasco	(Bagnasco et al. 2011)
Italy	No	Hospital	University of Bocconi; European Institute of Oncology (IEO)	Manuela Brusoni, Pietro Luigi Deriu, Silvia Basso	(Brusoni et al. 2009)
Italy	No	Hospital	European Master in Sustainable Regional Health System, Universita degli studi di Verona, Ospedale di San Bonifacio	Alyn Palma, Diana Pascu	Not published (Master Thesis Project)
Malta	No	Hospital	Institute of Healthcare, The University of Malta	Anthony Baldacchino	Not published (Master Thesis Project)

Table 11.1 Overview of the use of the HSPSC in Europe (*continued*)

Country (Language)	Psychometric Properties	Setting	Affiliation/organisation	Correspondence	Publications
Norway	No	Hospitals	Department of Nursing, Karlstad University, Karlstad, Sweden; Department of Nursing, Gjøvik University College, Gjøvik, Norway	Randi Ballangrud	(Ballangrud et al. 2012)
Scotland	No	Hospital	NHS Lothian	Nicola Maran, Alan Fisher	Not published
Spain	Partially, but published in Spanish	Hospital	Universidad de Murcia	Pedro Saturno	(Saturno et al. 2008; Saturno 2009)
Spain	No	Hospital	Sociedad Española de Medicina de Urgencias y Emergencias (SEMES), (Spanish Society of Emergency Medicine)	Fermin Roqueta Egea	(Roqueta Egea et al. 2011b) English; (Roqueta Egea et al. 2011a) Spanish
Spain	No	Hospital	Servicio de Medicina Preventiva y Salud Pública, Hospital Universitario Virgen de las Nieves, Granada	Manuela Skodová	(Skodová et al. 2011)
Spain	No	Hospital	Unidad de Calidad y Gestión del Riesgo Clínico, Hospital Monte Naranco, Oviedo, España	Fernando Vázquez	(Menéndez et al. 2010)
Spain	No	Hospitals	Médico especialista en Medicina Familiar y Comunitaria, Hospital Regional Universitario Carlos Haya de Málaga	Francisco Pozo Muñoz	(Muñoz and Marin 2013)
The Netherlands	Partially, but very brief	Hospital	Princess Amalia Department of Pediatrics, Isala Clinics, location Sophia	Cathelijne Snijders	(Snijders et al. 2009)
The Netherlands	No	Hospital	Department of Public and Occupational Health, VUmc-EMGO Institute for Health and Care Research, Amsterdam	Inge van Noord	(van Noord et al. 2010)

Table 11.1 Overview of the use of the HSPSC in Europe (concluded)

Country (Language)	Psychometric Properties	Setting	Affiliation/organisation	Correspondence	Publications
Denmark	No	Hospital (in addition to the HSOPS a pilot versions of the NHSOPS was used in nursing homes)	European Society for Quality in Healthcare, Office for Quality Indicators, The Danish National Clinical Quality Improvement Programme, Central Denmark Region	Solvejg Kristensen, Jan Mainz	(Øllgaardm et al. 2011) Master Thesis
Other healthcare setting					
Ireland	No	Medical centres – the questionnaire is based on both, the Hospital and Medical Office SOPS	Vhi Healthcare	Joanna Gibson	Not published
Germany	No	Nursing homes	St. Marien-Krankenhaus Siegen gem. GmbH	Karl-Hermann Menn	Not published
Spain	Yes, but not published	Primary care	Department of Health of Galicia	Clara González-Formoso	(González-Formoso et al. 2011)
The Netherlands	Yes	Primary care physicians (GP's), General practice	University Medical Center Utrecht; Netherlands Institute for Health Services Research (NIVEL)	Dorien Zwart, Cordula Wagner	(Zwart et al. 2011)
Ireland	Yes	Primary care, community and continuing care setting	HSE West – Roscommon PCCC	Fiona Garvey	(Garvey 2008) unpublished Master Thesis

Adaptations of the HSPSC to Various Languages and
National Healthcare Systems

Out of the 14 published papers on psychometric properties only a few provided information on the translation and adaptation processes. Of those who did, most described a forward-backward-translation process of the HSPSC from the original American English version into their native language (Bodur and Filiz 2010; Olsen 2007, 2008; Pfeiffer and Manser 2010). It also seems that the majority of changes or revisions in items were due to different interpretations of terminology (e.g. Bodur and Filiz 2010; Pfeiffer and Manser 2010), the addition of further items or measures of new dimensions (Hedsköld et al. 2013; Occelli et al. 2013; Pfeiffer and Manser 2010; Sarac et al. 2011), or the removal of items from the measure (e.g. Hammer et al. 2011; Waterson et al. 2010).

As part of its international use, the HSPSC has also been administered in countries such as England and Scotland where English is the native language. Even in those countries adaptations were not only necessary with regard to American versus British English but also with regard to differences in the healthcare systems and the uses of terminology. For example, in England the terms 'area' and 'unit' had to be changed to 'ward' and 'department' respectively. In Scotland the term 'event' was changed to 'incident' (Sarac et al. 2011; Waterson et al. 2010).

Innovations

Despite the HSPSC questionnaire originally being developed to measure hospital staff's perceptions of patient safety culture, we found adaptations for 1) hospital managements and 2) general practitioners in ambulant healthcare settings. Both adaptations were based on HSPSC versions already translated into the local language. In Germany, the Swiss-German version of the HSPSC (Pfeiffer and Manser 2010) which has been used in the German speaking part of Switzerland for hospital staff was adapted to form a hospital management version in order to survey medical directors of German hospitals (Hammer et al. 2011). In the Netherlands, the Dutch version of the HSPSC (COMPaZ) was adapted for use in general practice (Zwart et al. 2011). In this adaptation the researchers changed the term 'hospital unit' to 'GP practice', deleted three items and developed seven new ones. Similarly adaptations were made in Sweden, where the questionnaire has been adapted for simultaneous use in hospitals and primary care settings (Hedsköld et al. 2013).

Interestingly, one application of the HSPSC involved an adaptation for the petroleum industry to conduct a comparative study of safety climate differences between healthcare and petroleum industries (Olsen 2010; Olsen and Aase 2010). To our knowledge this is the first time that a cross-industry study has used a tool originally developed for the hospital setting.

Psychometric Testing

The following results are limited to the data available from 14 studies that have published their psychometric results. Table 11.2 gives an overview about these studies and psychometric properties used.

The first European HSPSC studies analysing psychometric properties were conducted in 2005 (Smits et al. 2008; Wenqi 2005). The number of participating hospitals ranges from one to 551 facilities between the studies. Responses in the studies range from 309 to 38,812 participants and from 22 to 92 percent response rates.

Analyses of each instrument's psychometric properties has shown that exploratory factor analysis (EFA) was conducted for two measures and confirmatory factor analysis (CFA) for two measures. In the case of 10 measures both methods were used.

In one of the 12 studies conducting a CFA, 12 separate factor analyses – one for each dimension – were performed. Within the remaining 11 studies the researchers included all questionnaire items in the confirmatory factor model. Depending on prior adaptations of the HSPSC the number of items included in the CFA models varied from 39 to 48 (42 original items and six new items). To evaluate absolute fit measures, seven studies assessed the goodness-of-fit with the Chi-squared values. Three of these studies (Pryseley 2008; Hammer et al. 2011; Robida 2013) described p values < .001. To reduce the sensitivity of the Chi-squared value to the sample size, four studies calculated the normed Chi-squared value (Chi^2/df). Three of them reached acceptable values of Chi^2/df ≤ 2.5 (Hammer et al. 2011; Pfeiffer and Manser 2010; Wenqi 2005). Alternatively, or in addition to the Chi-squared values, six studies calculated Goodness-of-fit Index (GFI), which reached an acceptable value in four of the studies (Hedsköld et al. 2013; Olsen 2008; Perneger et al. 2013; Wenqi 2005). For all 11 CFA models the Root Mean Square Error of Approximation (RMSEA) was calculated and reached acceptable values in 10 of the studies (Hammer et al. 2011; Hedsköld et al. 2013; Occelli et al. 2013; Olsen 2008; Perneger et al. 2013; Pfeiffer and Manser 2010; Robida 2013; Sarac et al. 2011; Waterson et al. 2010; Wenqi 2005). Regarding incremental model fit, eight out of 11 studies reached very good thresholds, but three studies (Occelli et al. 2013; Perneger et al. 2013; Pryseley 2008) had unacceptable values in all presented incremental model fit indices. An overview on the results of the CFAs is shown in Table 11.3.

Considering the exploratory factor analysis (EFA) we found that the dimensions 'Feedback and communication about error' and 'Communication openness' were combined into one factor in six of the studies (Bodur and Filiz 2010; Perneger et al. 2013; Pfeiffer and Manser 2010; Pryseley 2008; Robida 2013; Wenqi 2005). In five other studies (Haugen et al. 2010; Hedsköld et al. 2013; Occelli et al. 2013; Sarac et al. 2011; Smits et al. 2008) the factor 'Feedback and communication about error' merged together with the factor 'Organisational learning'. The same happened for the dimensions 'Teamwork across hospital units' and 'Hospital handoffs and transitions', which were combined into one factor in six of the studies (Occelli et al. 2013; Perneger et al. 2013; Pfeiffer and Manser 2010; Pryseley 2008; Robida 2013; Wenqi 2005). 'Overall perceptions of safety' and 'Staffing' were combined

Table 11.2   Overview of the studies that included psychometric analyses

Country (Author)	Data collection period	Population studied	Sample					Participants				Response rate in %	Psychometric analyses
			Hospital	Unit	Staff			Hospital	Unit	Staff			
Belgium *(Wenqi 2005)*	Apr–May 2005	Healthcare workers (physicians and nurses) within 1 Belgian hospital	1	–	1,757			1	–	1,323		٧٥	(1) model refinement in EFA; (2) verification of the originally hypothesised model in CFA; (3) internal consistency (Cronbach's Alpha) reported on original factor model; (4) construct validity by Pearson's correlation coefficient between sub-scale scores
Belgium *(Pryseley 2008)*	In 2006	Healthcare workers within 2 French speaking hospitals	–	–	–			–	–	–		–	(1) model refinement in EFA; (2) verification of the originally hypothesised model in CFA; (3) internal consistency (Cronbach's Alpha) reported on original factor model; (4) construct validity by Pearson's correlation coefficient between sub-scale scores
England *(Waterson et al. 2010)*	May–Jun 2006	Nursing staff (trained and untrained), allied healthcare professionals, medical, management and administrative staff in 3 large NHS acute hospitals, Trust in East Midlands	3	–	about 4,000			3	–	1,461		37	(1) model refinement in EFA; (2) verification of the originally hypothesised and the new hypothesised model in CFA; (3) internal consistency (Cronbach's Alpha) reported on original factor model
France *(Occelli et al. 2013)*	Jan 2009	Full- and part-time healthcare professionals in medical and surgical units in 7 South-western France Hospitals	7	18	–			–	–	401		77	(1) model refinement in PCA; (2) verification of the original and the new hypothesised model in CFA; (3) internal consistency (Cronbach's Alpha) reported on original factor model

Table 11.2   Overview of the studies that included psychometric analyses (*continued*)

Country (Author)	Data collection period	Population studied	Sample			Participants			Response rate in %	Psychometric analyses
			Hospital	Unit	Staff	Hospital	Unit	Staff		
Germany (*Hammer et al. 2011*)	Apr–Oct 2008	Hospital managers (medical directors) from German hospitals	1,224	–	1,224	551	–	551	45	(1) verification of the originally hypothesised model in CFA; (2) internal consistency (Cronbach's Alpha) reported on original factor model; (3) construct validity by Pearson's correlation coefficient between sub-scale scores
Norway (*Olsen 2007; Olsen 2008*)	–	Healthcare workers and other healthcare personnel within 1 Norwegian hospital	1	–	–	1	–	1,919	55	(1) verification of the originally hypothesised model in CFA. (2) internal consistency (Cronbach's Alpha) reported on original factor model; (3) construct validity by Pearson's correlation coefficient between sub-scale scores
Norway (*Haugen et al. 2010*)	Oct–Nov 2009	Operating theatre personnel (surgeons, anaesthesiologists, operating theatre nurses, nurse anaesthetists and ancillary personnel) from 1 Norwegian hospital	1	–	575	1	–	358	62	(1) model refinement in EFA; (2) internal consistency (Cronbach's Alpha) reported on original factor model; (3) construct validity by Pearson's correlation coefficient between sub-scale scores
Scotland (*Sarac et al. 2011*)	Feb–Sep 2009	Employees (nurses, allied health professionals, doctors and nursing or healthcare assistants) from 7 acute NHS hospitals in Scotland	14	–	–	7	–	1,969	22	(1) model refinement in EFA; (2) verification of the original and the new hypothesised model in CFA; (3) internal consistency (Cronbach's Alpha) reported on original factor model; (4) construct validity by Pearson's correlation coefficient between sub-scale scores

Table 11.2   Overview of the studies that included psychometric analyses (*continued*)

Country (Author)	Data collection period	Population studied	Sample Hospital	Sample Unit	Sample Staff	Participants Hospital	Participants Unit	Participants Staff	Response rate in %	Psychometric analyses
Slovenia (Robida 2013)	In 2010	Employees (clinical and non-clinical staff) from 3 acute general hospitals	3	–	1,745	3	–	1,048	60	(1) model refinement in PCA; (2) verification of the original and the new hypothesised model in CFA; (3) internal consistency (Cronbach's Alpha) reported on original factor model; (4) construct validity by Spearman's-Rho between sub-scale scores
Sweden (Hedsköld et al. 2013)	2009–2011	National data base with first time respondents; 46% of respondents represent different types of hospitals (including university, larger regional and smaller rural hospitals) and 11% of responders represent primary care centres (Total sample: 84,215; Hospital: 38,812; Primary care: 9,113)	–	–	–	–	–	38,812	–	(1) model refinement in EFA; (2) verification of the new hypothesised model in CFA; (3) internal consistency (Cronbach's Alpha) reported on original factor model; (4) construct validity by Pearson's correlation coefficient between sub-scale scores
Switzerland (Pfeiffer and Manser 2010)	Oct–Dec 2006	Employees (physicians, registered nurses, nurse assistants, medical and technical staff, management and administrative staff) in 1 Swiss University Hospital	1	61	about 6,420	1	61	3,005	47	(1) model refinement in EFA; (2) verification of the originally hypothesised model in CFA; (3) internal consistency (Cronbach's Alpha) reported on original factor model; (4) construct validity by Pearson's correlation coefficient between sub-scale scores

Table 11.2 Overview of the studies that included psychometric analyses (concluded)

Country (Author)	Data collection period	Population studied	Sample			Participants			Response rate in %	Psychometric analyses
			Hospital	Unit	Staff	Hospital	Unit	Staff		
Switzerland (Perneger et al. 2013)	Feb 2009	All employees with at least 6 months of employment or members of management from 1 public hospital in a French-speaking area of Switzerland	1	–	1,583	1	–	1,221	55–92 depending on the profession	(1) model refinement in EFA; (2) verification of the originally hypothesised model in CFA; (3) internal consistency (Cronbach's Alpha) reported on original factor model
The Netherlands (Smits et al. 2008)	Jun 2005	–	8	–	–	8	23	583	–	(1) model refinement in EFA; (2) verification of the originally hypothesised model in CFA; (3) internal consistency (Crontach's Alpha) reported on original factor model; (4) construct validity by Pearson's correlation coefficient between sub-scale scores
Turkey (Bodur and Filiz 2010)	In 2008	Employees (physicians, nurses, temporary nurses) from 3 public hospitals (1 general hospital, 1 teaching hospital, and 1 university hospital) located in the metropolitan centre of Konya Province	3	–	–	3	–	309	51–91 depending on the profession	(1) model refinement in EFA; (2) internal consistency (Cronbach's Alpha) reported on new factor model; (3) construct validity by Pearson's correlation coefficient between sub-scale scores

into one factor in four studies (Hedsköld et al. 2013; Robida 2013; Sarac et al. 2011; Waterson et al. 2010). In one study (Smits et al. 2008) the items F3 and F7 loaded slightly more on the dimension 'Teamwork across the units' instead of 'Hospital handoffs and transition'. In one more study the dimension 'Teamwork across the units' merged with 'Teamwork within units' (Waterson et al. 2010) and in two other studies 'Teamwork across the units' merged with 'Management support for patient safety' (Bodur and Filiz 2010; Hedsköld et al. 2013). Between eight and 12 dimensions out of 12 dimensions emerged in the EFAs based on 39 up to 48 (42 original items and six new items) included items. An overview of all results within the EFAs is given in Table 11.3.

Construct Validity

Eleven studies tested the construct validity of their HSPSC-based measure (for an overview see Table 11.3). The lowest inter-correlation with p = -.01 was found for 'Staffing' and 'Frequency of event reporting' within the Belgium study (Hellings et al. 2007; Wenqi 2005). However, this low inter-correlation between 'Staffing' and 'Frequency of event reporting' was found to be the lowest in seven out of 11 studies. The highest intercorrelation with p = .65 was found for 'Teamwork across hospital units' and 'Handoffs and transitions' in the Swiss study (Pfeiffer and Manser 2010). However, this high inter-correlation between 'Teamwork across hospital units' and 'Handoffs and transitions' emerged as the highest in three out of the 11 studies.

Internal Consistency

Internal consistency was reported in all 14 studies, including results from the original HSPSC version (see Table 11.4). In 13 of the studies, the alphas are based on the original HSPSC structure and one study reported Cronbach's alpha, based on scale refinement in EFA.

Only one dimension ('Frequency of event reporting') showed appropriate values of Cronbach's alpha ≥ .70 in all 14 studies. In 12 of the 13 studies 'Feedback and communication about error' showed Cronbach's alpha ≥ .70. 'Hospital management support for patient safety' and 'Supervisor/manager expectations/actions', both reached reasonably high Cronbach's alpha (≥ .68) and Cronbach's alpha ≥ .70. in 11 of 13 and 12 of 14 reported study results. These four dimensions appeared to have the strongest internal consistency. 'Teamwork within units', 'Communication openness', 'Hospital handoffs and transitions' are also strong dimensions since those three dimensions reached acceptable Cronbach's alpha values (≥ .60). However, 'Nonpunitive response to Error' showed Cronbach's alpha values < .60 in one study (Smits et al. 2008). 'Teamwork across hospital units' and 'Overall perceptions of safety' showed Cronbach's alpha values < .60 in two studies. 'Staffing' and 'Organisational learning' showed Cronbach's alpha values < .60 in several studies and seemed to be the dimensions with the worst internal consistency.

Table 11.3   Factor analyses

Country (Author)	N of questionnaires used for factor analyses	CFA	EFA	Intercorrelations of dimensions (as numbered below)*
Belgium (Wenqi 2005)	1,300	Based on 42 original items: Chi² = 1,503.8; df = 729; Chi²/df = 2.06; GFI = .931; AGFI = .915; CFI = .939, NNFI = .928; RMSR = .029; RMSEA = .033	12 factors emerged in the EFA based on 42 original items: 'Teamwork across hospital units' and 'Hospital handoffs and transitions' were combined into 1 factor; the same emerged for 'Feedback and communication about error' and 'communication openness'; questions F6 and F11 loaded high on 2 factors; the 12 factors jointly explained 60% of the variance in the responses	Intercorrelations between -.01 (9 and 12) and .57 (5 and 8)
Belgium (Pryseley 2008)	–	Based on 42 original items: Chi² = 6,441.8; p < .0001; df = 819; Chi²/df = 7.87; CFI = .639; NNFI = .621; GFI = .704; AGFI = .673; RMSEA = .077; RMR = .178	12 factors emerged in the EFA based on 42 original items: The dimensions 'Teamwork across hospital units' and 'Hospital handoffs and transitions' were combined into 1 factor ; the same emerged for the dimensions 'Feedback and communication about error' and 'Communication openness'; instead 'Overall perception of safety' split into 2 dimensions; question B2 loaded high on 2 factors; Question A9 did not load high on any of the 12 dimensions; the dimensions 'Teamwork across hospital units' and 'Hospital handoffs and transitions' did not have any item with a high load; the proposed 12 factors solution jointly explained 46% of the variance in the responses	Intercorrelations between .04 (9 and 12) and .64 (3 and 6)
England (Waterson et al. 2010)	Data were splitted for EFA and CFA	Based on 40 of the original items: Chi² = 1,907; df = 674; CFI = .91; NNFI = .89; RMSEA = .04; SRMR = .05	9 out of 12 factors emerged in the EFA based on 40 original items: 'Overall perceptions of safety' and 'Staffing' were combined into 1 factor; the factors 'Teamwork across units' and 'Teamwork within units' both dropped a single item; the factor 'Supervisor/manager expectations and actions promoting patient safety' dropped two items; Organisational learning/continuous improvement' and 'Hospital management support' were absent; the 9 (27 items remained) factors jointly explained 66.8% of the variance in the responses; Chi² = 588; df = 288; TLI = .93; CFI = .95; RMSEA = .04; SRMR = .04	–

Table 11.3 Factor analyses (*continued*)

Country (Author)	N of questionnaires used for factor analyses	CFA	EFA	Intercorrelations of dimensions (as numbered below)*
France (Occelli et al. 2013)	401	Based on 42 original items: Chi² = 1,308.4; df = 741; CFI = .848 RMSEA = .050	10 out of 12 factors emerged in the PCA based on 42 original items and 3 new items: items A7, A11 and F11 and 2 out of the 3 additional items were excluded; 'Organisational learning' and 'Feedback and communication about error' were combined into one factor; the same emerged for 'Teamwork across hospital units' and 'Hospital handoffs and transitions'; F10 became an item of 'Management support for patient safety'; based on 39 original and 1 new item: Chi² = 1,199.4; df = 685; CFI = .855; RMSEA = .049	–
Germany (Hammer et al. 2011)	547	Based on 42 original items: Chi² = 1,632.7; p < .0001; df = 753; Chi²/df = 2.17, CFI = .916; TLI = .904; RMSEA = .046; SRMR = .048	–	Intercorrelations between .13 (9 and 12) and .64 (1 and 11)
Norway (Olsen 2007, 2008)	–	Based on 42 original items: GFI = .91; AGFI = .90; CFI = .97; RMSEA = .044	–	Intercorrelations between .17 (nn) and .59 (5 and 8) (only 10 safety culture dimensions)
Norway (Haugen et al. 2010)	–	–	11 out of 12 factors emerged in the EFA based on 42 items: 'Organisational learning' and 'Feedback and communication about error' were combined into one factor	Intercorrelations between .20 (nn) and .62 (nn) (Correlations were analysed separately for unit- and hospital level factors)
Scotland (Sarac et al. 2011)	1966; data were split for EFA (n = 965) and CFA (n = 1,001)	Based on 42 original items: Chi² = 1,708.9; df = 753; CFI = .91; RMSEA = .04	10 out of 12 factors emerged in the EFA based on 42 items: items A18, F3 and F6 were excluded; 'Organisational learning' and 'Feedback and communication about error' were combined into one factor, the same emerged for 'Staffing' and 'Overall perceptions of safety'; the 10 factors (39 items remained) jointly explained 61.7% of the variance in the responses: Chi² = 1,554.5; df = 657; CFI = .91; RMSEA = .04	Intercorrelations between .13 (9 and 12) and .61 (5 and 8)

Table 11.3 Factor analyses (*continued*)

Country (Author)	N of questionnaires used for factor analyses	CFA	EFA	Intercorrelations of dimensions (as numbered below)*
Slovenia (Robida 2013)	976; data were splitted for PCA (n = 501) and CFA (n = 475)	Based on 42 original items: Chi² = 3,892.3; df = 820; p < .001; CFI = .9; RMSEA = .048	9 out of 12 dimensions emerged in the PCA based on 42 items: items A13, C6r und F11r were excluded; 'Communication openness' and 'Feedback and communication about error' were combined into one factor; the same emerged for 'Teamwork across hospital units' and 'Hospital handoffs and transitions' as well as for 'Staffing' and 'Overall perceptions of safety'; the 9 factors (39 items remained) jointly explained 57.5% of the variance in the responses: Chi² = 3,515.0; df = 703; p < .001; CFI = .9; RMSEA = .049	Intercorrelations ranged between .08 (9 and 12) and .63 (5 and 8)
Sweden (Hedsköld et al. 2013)	–	Based on 42 original and 6 new items: CFI = .91; GFI = .92; AGFI = .91; NFI = .91; NNFI = .90; RMSEA = .042	9 out of 14 factors emerged in the EFA based on 42 original and 6 new items and 14 presumed dimensions; items C4, C6, A7, A15 and A18 showed factor loading below .4; C6, A18 and F6 loaded on different factors; 'Organisational learning' and 'Feedback and communication about error' as well as 2 items from 'Communication openness' and 1 item from 'Overall perceptions of safety' loaded on 1 factor; 2 items from 'Overall perceptions of safety' and 'Staffing' were combined into one factor as well as 'Teamwork across units' and 'Management support for patient safety'; the 9 factors jointly explained 56.4% of the variance in the responses	Intercorrelations (measured by Spearman's-Rho) for the 12 original dimensions ranged between .16 (9 and 12) and .61 (3 and 6)
Switzerland (Pfeiffer and Manser 2010)	Subset of n = 568	Based on 39 of the original items: CMIN/df = 2.271; GFI = .878; NFI = .859; TLI = .901; RMSEA = .047	8 out of 12 factors emerged in the EFA based on 39 items: 'Communication openness' and 'Feedback and communication about error' were combined into one factor; the same emerged for 'Teamwork across hospital units' and 'Hospital handoffs and transitions'; 'Organisational learning' and 'Overall perceptions of safety' showed multiple cross-loadings and did not establish one factor each; the 8 factors jointly explained 59.8% of the variance in the responses	Intercorrelations between .17 (9 and 11) and .65 (3 and 6)

Table 11.3 Factor analyses (concluded)

Country (Author)	N of questionnaires used for factor analyses	CFA	EFA	Intercorrelations of dimensions (as numbered below)*
Switzerland (Perneger et al. 2013)	1,171	Based on 42 of the original items: CFI = .89; GFI => .99; NNFI = .88 SRMR = .046, RMSEA = .043	10 out of 12 factors emerged in the EFA based on 42 items: 'Communication openness' and 'Feedback and communication about error' were combined into one factor; the same emerged for 'Teamwork across hospital units' and 'Hospital handoffs and transitions'; 'Organisational learning' and 'Overall perceptions of safety' showed multiple cross-loadings and did not establish one factor each; the 10 factors (42 items) jointly explained 58% of the variance in the responses	–
The Netherlands (Smits et al. 2008)	578	12 separate factor analysis performed; analyses showed, that the items within the single factors did not consist of more than one factor	11 out of 12 factors emerged in the EFA based on 42 items; 'Organisational learning' and 'Feedback and 'Communication about error' were combined into one factor; instead of A15, A14 became an item of 'Overall perception of safety'; F3 and F7 loaded slightly more on 'Teamwork across the units' instead of 'Hospital handoffs and transition'; A15 and F6 were removed; The 11 factors (40 items remained) jointly explained 57.1% of the variance in the responses	Intercorrelations between .01 (9 and 12/ 8 and 9) and .47 (1 and 8/2 and 8) (only 11 safety culture dimensions)
Turkey (Bodur and Filiz 2010)	–	–	10 out of 12 factors emerged in the EFA based on 42 items: 'Feedback and communication about error' and 'Communication openness' were combined into 1 factor; the same emerged for 'Teamwork across units' and 'Management support for patient safety'; the 10 factors jointly explained 62.1% of the variance in the responses	Intercorrelations between .04 and .53 for the 10 factor model

Notes: * Safety Culture Dimension: 1 Hospital management support for patient safety; 2 Supervisor/manager expectations/actions; 3 Teamwork across hospital units; 4 Teamwork within units; 5 Communication openness; 6 Hospital handoffs and transitions; 7 Nonpunitive response to error; 8 Feedback and communication about error; 9 Staffing; 10 Organisational learning; Outcome dimensions: 11 Overall perceptions of safety; 12 Frequency of event reporting.

In comparison with the original version the robustness of factors was worse in Belgium (Wenqi 2005), France (Occelli et al. 2013) and the Netherlands (Smits et al. 2009) where, in all three studies, three or more of the 12 dimensions reached Cronbach's alpha values < .60. Consequently, in all three of those studies, the authors opted to make scale refinements in further EFA. In contrast, in four other studies, all 12 dimensions reached Cronbach's alpha values of at least ≥ .60 (Hammer et al. 2011; Hedsköld et al. 2013; Pfeiffer and Manser 2010; Sarac et al. 2011).

Comparing the factor structures of the various applications of the HSPSC in Europe to the original pilot tested US version, most of the 12 dimensions showed similar patterns in the Cronbach's alpha. For example, dimensions with a high Cronbach's alpha, such as 'Frequency of event reporting' received relatively high Cronbach's alpha in the European studies as well and dimensions with a low Cronbach's alpha, such as 'Staffing', also received a relatively low Cronbach's alpha. Surprisingly, the three dimensions 'Organisational learning', 'Teamwork across hospital units' and 'Overall perceptions of safety' resulted in distinctly lower Cronbach's alpha in the European HSPSC version, compared to the results of Sorra and Nieva (Sorra and Nieva 2004).

Table 11.4 Reliability (Cronbach's alpha)

Safety Culture Dimension* Country (Author)	1	2	3	4	5	6	7	8	9	10	11	12
United States *(Sorra and Nieva 2004)*	.83	.75	.80	.83	.72	.80	.79	.78	.63	.76	.74	.84
Belgium *(Wenqi 2005)*	.72	.77	.66	.66	.65	.71	.68	.78	.57	.59	.58	.85
Belgium *(Pryseley 2008)*	.77	.75	.68	.82	.72	.72	.64	.76	.52	.59	.63	.87
England *(Waterson et al. 2010)*	.69	.68	.70	.73	.67	.77	.65	.80	.58	.66	.67	.83
France *(Occelli et al. 2013)*	.73	.83	.59	.63	.62	.66	.57	.64	.46	.59	.67	.84
Germany *(Hammer et al. 2011)*	.87	.69	.78	.77	.64	.83	.73	.79	.73	.61	.73	.86
Norway *(Olsen 2007, 2008)*	.79	.77	.65	.77	.68	.65	.64	.70	.65	.51	.76	.82
Norway *(Haugen et al. 2010)*	.80	.85	.73	.75	.67	.68	.68	.73	.59	.75	.78	.82
Scotland *(Sarac et al. 2011)*	.79	.79	.70	.80	.73	.74	.77	.78	.60	.64	.71	.84
Slovenia *(Robida 2013)*	.82	.74	.74	.74	.74	.66	.61	.72	.65	.36	.65	.88
Sweden *(Hedsköld et al. 2013)*	.81	.79	.71	.75	.67	.75	.74	.76	.67	.66	.72	.87
Switzerland *(Pfeiffer and Manser 2010)*	.83	.78	.76	.73	.64	.71	.71	.79	.61	.68	.75	.88

Table 11.4 Reliability (Cronbach's alpha) (concluded)

Safety Culture Dimension*												
Country (Author)	1	2	3	4	5	6	7	8	9	10	11	12
Switzerland (Perneger et al. 2013)	.79	.75	.71	.80	.67	.78	.60	.77	.61	.57	.68	.86
The Netherlands (Smits et al. 2008)	.68	.70	.59	.66	.72	.68	.69	.75	.49	.57	.62	.79
*Turkey*** (Bodur and Filiz 2010)	–	.72	–	.83	–	.72	.71	–	.63	.68	.57	.86

Notes: References values by Sorra and Nieva (2004) were highlighted in grey; **bold values** = good values of α between .70 and .90; normal values = acceptable levels of α ≥ .60; *italic values* = unacceptable values of α < .60. *Safety Culture Dimension: 1 Hospital management support for patient safety; 2 Supervisor/manager expectations/actions; 3 Teamwork across hospital units; 4 Teamwork within units; 5 Communication openness; 6 Hospital handoffs and transitions; 7 Nonpunitive response to error; 8 Feedback and communication about error; 9 Staffing; 10 Organizational learning; Outcome dimensions: 11 Overall perceptions of safety; 12 Frequency of event reporting. ** Results after scale refinement in EFA.

Discussion and Outlook

This chapter aimed to summarise the 'state of the art' concerning the use of the HSPSC in Europe. In doing so we tried to be as inclusive and as rigorous as possible to give a complete picture of widespread and varied use and to evaluate the level of scientific evidence concerning the psychometric characteristics of the national / language versions that are currently available. Since there is no gold standard regarding which criteria to apply when evaluating patient safety climate instruments, we have systematically compared the descriptions available to us focusing on a) the development process (i.e. translations, adaptations and innovations) and b) psychometric testing (i.e. internal consistency and construct validity). There are a few things that we would like to highlight before suggesting areas for future research and collaboration across Europe.

Balancing Adoption of an Existing Safety Climate Measure and the Need for Adaptation to National Characteristics

The available evidence from studies conducted in Europe suggests that the HSPSC instruments developed on the basis of the original US version have to be adapted carefully to other national and/or healthcare contexts regarding terminology but also for more systems related issues. For example, in some European countries (e.g. Switzerland) the use of agency staff in nursing is currently relatively uncommon. Whether teams are comprised of flexible or stable members of staff may affect the safety issues encountered. Also, the role of hospital management and the way it is organised differs slightly between hospital types but also as a consequence of national or regional regulations. These issues need to be considered carefully when adapting, conducting and interpreting the results of a patient safety climate survey. We also believe that these decisions should be backed up by careful analyses of the psychometric characteristics of the respective safety climate instruments.

Standards for Psychometric Testing of Safety Climate Measures

The procedure of psychometric testing varies between the different studies included in this chapter. While many conducted no psychometric tests at all, two studies conducted a confirmatory factor analysis (CFA), two performed an exploratory factor analysis (EFA), and 10 studies did both. Despite this difficulty, we found only one study (Pryseley 2008) out of 10 which performed a CFA in order to test the original factor structure of the HSPSC questionnaire, with several unacceptable thresholds for both absolute and incremental fit indices. With regards to the EFAs, the dimensions 'Staffing', 'Communication openness', 'Organisational learning' and 'Teamwork across hospital units', appeared to be less stable.

Future Directions for the Use of HSPSC in Europe

Considering a Cross-national Perspective

While the focus of almost all studies included in this chapter was the adaptation of a patient safety climate measure from the USA to their own national and healthcare contexts, many of the questions we will have to answer in future might be better addressed at a cross-national level. At the level of the survey instrument this would require a certain amount of consistency regarding the use of terminology and the addition or removal of items – a minimal shared item set. Taking cross-national similarities and differences into account (especially with regard to internal consistencies of the various safety climate dimensions) might help to improve further the overall quality of the HSPSC and to further our understanding of factors at the level of healthcare systems that may have a considerable impact on safety climate.

Establishing a European Network

Cross-national studies require effective coordination within and between countries. We believe that the first step is a European network of HSPSC users and, to a certain degree, the process of writing this chapter has contributed to such a network by identifying researchers and practitioners with an interest in safety climate assessment. Such a network would also help to connect groups within countries that have developed parallel versions (e.g. Spain and Italy).

Research Needs

In general, research needs regarding patient safety climate can be grouped into those related to instrument development and those aiming to establish empirical links between safety climate and outcomes or between interventions and changes in safety climate. Many papers have discussed these in great detail, such as the need to combine 'leading and lagging indicators' (i.e. those providing information about potential risks and those identified *after* an accident has occurred) (Payne 2009).

We would thus like to focus on research needs related to the use of the HSPSC in Europe. Some are clearly related to the cross-national perspectives that have been discussed previously. Another characteristic of some European countries is that there are several language regions within the same country. Often these language regions or groups are described as sub-cultures but at the same time national regulations frequently do not account for potential differences. Based on the review of studies, we have also identified innovations that go beyond a mere language adaptation of the HSPSC and open up new research perspectives. For example, the hospital management version by Hammer et al. 2011 could be used to study (mis-)matches between top management and staff views and how these have an impact on safety performance.

In summary, this chapter highlights that the uptake of the HSPSC in Europe has been rather impressive and that psychometric testing has provided us with the knowledge needed to evaluate the measurement qualities of the various national / language versions. We believe that this is an excellent basis from which we can move forward at both national and cross-national levels.

Acknowledgements

We would like to acknowledge support from the European researchers contacted during our research for their feedback on our requests and the provision of more detailed information regarding their work.

Annex

Table 11.5   Appropriateness of psychometric properties

Model Fit Index	Description	Criterion (N > 250 and m ≥ 30)
Absolute fit indices		
Chi²	Indicates the difference between the observed and the expected covariance matrices	
df	Degree of Freedom	
p		< .05
Chi²/df or CMIN/df	Normed Chi-squared value	< 2.5
RMSEA	Root Mean Square Error of Approximation	< .07 (with CFI ≥ .90)
SRMR	Standardised Root Mean Residual	< .08 (with CFI ≥ .92)
RMSR/ RMR	Root Mean Square Residual	The range of RMSR is expressed in the terms of the range of the answer scales and therefore difficult to compare
GFI	Goodness of Fit Index	> .90
AGFI	Goodness of Fit Index Adjusted for degrees of Freedom	> .90

Table 11.5   Appropriateness of psychometric properties (*continued*)

Model Fit Index	Description	Criterion (N > 250 and m ≥ 30)
Incremental fit indices		
NFI	Normed Fit Index	≥ .95
CFI	Comparative Fit Index	≥ .90
TLI or NNFI	Tucker–Lewis Index or Non-normed Fit Index	≥ .90
Inter-correlation (Campbell and Fiske 1959)		
Criterion	Description	
r	Pearson's correlation coefficient	≥ .70
*r*ₛ	Spearman's-Rho	≥ .70
Internal Consistency (Campbell and Fiske 1959)		
Criterion	Description	
α	Cronbach's alpha	A good value of α range between .70 and .90 (Campbell and Thompson 2007;Kline 2005; Klingner et al. 2009); based on Sorra and Nieva (Sorra and Nieva 2004) levels of α ≥ .60 were considered acceptable

Source: Confirmatory Factor Analyses (Kline 2005). For thresholds of acceptable fit see Hair et al. (Hair et al. 2006), Bollen (Bollen 1989), Kline (Kline 2005) and Hu, L.T. and Bentler (Hu and Bentler 1999).

References

Agnew, C., Flin, R. and Mearns, K. 2013. Patient safety climate and worker safety behaviours in acute hospitals in Scotland. *Journal of Safety Research*, 45, 95–101.

Alhemood, A.M., Genaidy, A.M., Shell, R., Gunn, M. and Shoaf, C. 2004. Towards a model of safety climate measurement. *International Journal of Occupational Safety and Ergonomics*,10(4), 303–18.

Ashcroft, D.M. and Parker, D. 2009. Development of the pharmacy safety climate questionnaire: A principal components analysis. *Quality and Safety in Health Care*, 18(1), 28–31.

Bagnasco, A., Tibaldi, L., Chirone, P., Chiaranda, C., Panzone, M.S., Tangolo, D., Aleo, G., Lazzarino, L. and Sasso, L. 2011. Patient safety culture: An Italian experience. *Journal of Clinical Nursing*, 20(7–8), 1188–95.

Ballangrud, R., Hedelin, B. and Hall-Lord, M.L. 2012. Nurses' perceptions of patient safety climate in intensive care units: A cross-sectional study. *Intensive and Critical Care Nursing*, 28(6), 344–54.

Bodur, S. and Filiz, E. 2009. A survey on patient safety culture in primary healthcare services in Turkey. *International Journal for Quality in Health Care*, 21(5), 348–55.

Bodur, S. and Filiz, E. 2010. Validity and reliability of Turkish version of 'hospital survey on patient safety culture' and perception of patient safety in public hospitals in Turkey. *BMC Health Services Research*, 10, 28.

Bollen, K.A. 1989. *Structural Equations with Latent Variables.* New York: Wiley.

Brusoni, M., Deriu, P.L., Panzeri, C. and Trinchero, E. 2009. Un metodo di indagine sulla safety culture per la sicurezza dei servizi sanitari in Italia'. *Mecosan*, 69, 63–85.

Campbell, D.A. and Thompson, M. 2007. Patient safety rounds: Description of an inexpensive but important strategy to improve the safety culture. *American Journal of Medical Quality*, 22(1), 26–33.

Campbell, D.T. and Fiske, D.W. 1959. Convergent and discriminant validation by the multitrait-multimethod matrix. *Psychological Bulletin*, 56(2), 81–105.

Choudhry, R.M., Fang, D. and Mohamed, S. 2007. Developing a model of construction safety culture. *Journal of Management in Engineering*, 23(4), 207–12.

Cooper, M.D. 2000. Towards a model of safety culture. *Safety Science*, 36, 111–36.

Cox, S. and Flin, R. 1998. Safety culture: Philosopher's stone or man of straw?. *Work and Stress*, 12(3), 189–201.

Flin, R. 2007. Measuring safety culture in healthcare: A case for accurate diagnosis. *Safety Science*, 45(6), 653–67.

Flin, R., Burns, C., Mearns, K., Yule, S. and Robertson, E.M. 2006. Measuring safety climate in health care. *Quality and Safety in Health Care*, 15(2), 109–15.

Franklin, M. and Sorra, J. 2012. *Pharmacy Survey on Patient Safety Culture: User's Guide.* Westat: Rockville.

Garvey, F. 2008. *Exploring Patient Safety Culture in a Primary Community and Continuing Care Setting.* Dublin: International School of Healthcare Management, Royal College of Surgeons in Ireland.

Gershon, R.R.M., Stone, P.W., Bakken, S. and Larson, E. 2004. Measurement of organizational culture and climate in healthcare. *Journal of Nursing Administration*, 34(1), 33–40.

Gershon, R.R.M., Stone, P.W., Zeltseri, M., Faucett, J., Macdavitti, K. and Chou, S.S. 2007. Organizational climate and nurse health outcomes in the United States: A systematic review. *Industrial Health*, 45, 622–36.

Glendon, A.I. and Stanton, N.A. 2000. Perspectives on safety culture. *Safety Science*, 34, 193–214.

González-Formoso, C., Martin-Miguel, M.V., Fernández-Dominguez, M.J., Rial, A., Lago-Deibe, F.I., Ramil-Hermida, L., Pérez-Garcia, M. and Claveria, A. 2011. Adverse events analysis as an educational tool to improve patient safety culture in primary care: A randomized trial. *BMC Family Practice*, 12, 50.

Goodmann, G.R. 2004. A fragmented patient safety concept: The structure and culture of safety management in health care. *Nursing Economics*, Jan-Feb, 44–6.

Guldenmund, F.W. 2000. The nature of safety culture: A review of theory and research. *Safety Science*, 34(1–3), 215–57.

Hair, J., Black, W., Babin, B., Anderson, R. and Tatham, R. 2006. *Multivariate Data Analysis* (sixth edition). Upper Saddle River, New Jersey: Prentice Hall.

Hale, A. 2000. Culture's confusions. *Safety Science*, 34, 1–14.

Hammer, A. 2012. *Zur Messung von Sicherheitskultur in deutschen Krankenhäusern*. Köln: Universität zu Köln.

Hammer, A., Ernstmann, N., Ommen, O., Wirtz, M., Manser, T., Pfeiffer, Y. and Pfaff, H. 2011. Psychometric properties of the Hospital Survey of Patient Safety Culture for hospital management (HSOPS_M). *BMC Health Services Research*, 11, 165.

Haugen, A.S., Softeland, E., Eide, G.E., Nortvedt, M.W., Aase, K. and Harthug, S. 2010. Patient safety in surgical environments: Cross-countries comparison of psychometric properties and results of the Norwegian version of the Hospital Survey on Patient Safety. *BMC Health Services Research*, 10, 279.

Haugen, A.S., Softeland, E., Eide, G.E., Sevdalis, N., Vincent, C.A., Nortvedt, M.W. and Harthug, S. 2013. Impact of the World Health Organization's Surgical Safety Checklist on safety culture in the operating theatre: A controlled intervention study in Norway'. *British Journal of Anaesthesia*, 110(10), 807–15.

Hedsköld, M., Pukk-Härenstam, K., Berg, E., Lindh, M., Soop, M., Øvretveit, J. and Sachs, M.A. 2013. Psychometric properties of the hospital survey on patient safety culture, HSOPSC, applied on a large Swedish health care sample. *BMC Health Services Research*, 13(1), 332.

Hellings, J., Schrooten, W., Klazinga, N. and Vleugels, A. 2007. Challenging patient safety culture: Survey results. *International Journal of Health Care Quality Assurance*, 20(7), 620–32.

Hellings, J., Schrooten, W., Klazinga, N.S. and Vleugels, A. 2010. Improving patient safety culture. *International Journal of Health Care*, 23(5), 489–506.

Hoffmann, B., Domanska, O.M., Müller, V. and Gerlach, F.M. 2009 Developing a questionnaire to assess the safety climate in general practices (FraSiK): Transcultural adaption – A method report. *Zeitschrift für Evidenz, Fortbildung und Qualität im Gesundheitswesen*, 103(8), 521–9.

Hofinger, G. 2008. Sicherheitskultur. In M. Schrappe and U. Hölscher (eds), *Proceedings Dachkongress Medizintechnik und Patientensicherheit 6.-7.3.2008*. Universität Münster: Münster.

Hu, L. and Bentler, P.M. 1999. Cutoff criteria for fit indexes in covariance structure analysis: Conventional criteria versus new alternatives. *Structural Equation Modeling: A Multidisciplinary Journal*, 6(1), 1–55.

Kline, R.B. 2005. *Principles and Practice of Structural Equation Modeling*. New York: Guilford.

Klingner, J., Moscovice, I., Tupper, J., Coburn, A. and Wakefield, M. 2009. Implementing patient safety initiatives in rural hospitals. *Journal of Rural Health*, 25(4), 352–7.

Kolbe, M., Diedenhofen, H., Fishman, L., Renner, D., Dzyck, A., Günther, R., Hermes, R., Lessing, C. and Thomeczek, C. 2011. High 5s project: Hospital survey on patient safety culture: Results from Germany, *Proceeding: 2. Kongress Patientensicherheit 2011, Stiftung für Patientensicherheit Schweiz,* Basel.

Kundig, F., Staines, A., Kinge, T. and Perneger, T.V. 2011. Numbering questionnaires had no impact on the response rate and only a slight influence on the response content of a patient safety culture survey: A randomized trial. *Journal of Clinical Epidemiology,* 64(11), 1262–5.

Kuosmanen, A., Tiihonen, J., Repo-Tiihonen, E., Eronen, M. and Turunen, H. 2013. Patient safety culture in two Finnish state-run forensic psychiatric hospitals. *Journal of Forensic Nursing,* 9(4), 207–16.

Menéndez, M.D., Martínez, A.B., Fernandez, M., Ortega, N., Díaz, J.M. and Vazquez, F. 2010. Walkrounds y Briefings en la mejora de la seguridad de los pacientes. *Revista de Calidad Asistencial,* 25(3), 153–60.

Munoz, F.P. and Marin, V.P. 2013. Evaluación de la cultura de seguridad del paciente en el ámbitode un área sanitaria. *Revista de Calidad Asistencial,* 28(6), 329–36.

Murphy, J., Hignett, S., Griffiths, P. and Durbridge, M. 2007. Measuring patient safety culture in a UKNHS hospital with an American assessment tool. *Contemporary Ergonomics 2007,* 445–50.

Naevestad, T.-O. 2009. Mapping research on culture and safety in high-risk organizations: Arguments for a sociotechnical understanding of safety culture. *Journal of Contingencies and Crisis Management,* 7(2), 126–36.

Nieva, V.F. and Sorra, J. 2003. Safety culture assessment: A tool for improving patient safety in healthcare organizations. *Quality and Safety in Health Care,* 12, 1117–23.

Occelli, P., Quenon, J.-L., Hubert, B., Kosciolek, T., Haarau, H., Pouchadon, M.-L., Amalberti, R., Auroy, Y., Salmi, L.-R., Matthieu Sibé, P., Parneix, P. and Michel, P. 2011. Development of a safety culture: Initial measurement at six hospitals in France. *Journal of Healthcare Risk Management,* 30(4), 42–7.

Occelli, P., Quenon, J.-L., Kret, M., Domecq, S., Delaperche, F., Claverie, O., Castets-Fontaine, B., Amalberti, R., Auroy, Y., Parneix, P. and Michel, P. 2013. Validation of the French version of the Hospital Survey on Patient Safety Culture questionnaire. *International Journal for Quality in Health Care,* 25(4), 459–68.

Øllgaardm, B., Lønfeldt Jakobsen, C. and Gyldenhegn Ejlsborg, T. 2011. unpublished Masterprojekt, Syddansk Universitet.

Olsen, E. 2007. [Workers' perceptions of safety culture at a hospital]. *Tidsskr Nor Laegeforen,* 127(20), 2656–60.

Olsen, E. 2008. Reliability and validity of the hospital survey on patient safety culture at a Norwegian hospital. In J. Øvretveit and P.J. Sousa (eds), *Quality and Safety Improvement Research: Methods and Research Practice from the International Quality Improvement Research Network (QIRN).* Lisbon: National School of Public Health, 173–86.

Olsen, E. 2010. Exploring the possibility of a common structural model measuring associations between safety climate factors and safety behaviour in health care and the petroleum sectors. *Accident Analysis and Prevention*, 42(5), 1507–16.

Olsen, E. and Aase, K. 2010. A comparative study of safety climate differences in healthcare and the petroleum industry. *Quality and Safety in Health Care*, 19(Suppl. 3), i75–i79.

Payne, S. 2009. *How Can Gender Equity be Addressed through Health Systems?* Geneva: WHO.

Perneger, T.V. 2004. Adjustment for patient characteristics in satisfaction surveys. *International Journal for Quality in Health Care*, 16(6), 433–5.

Pfaff, H., Hammer, A., Ernstmann, N., Kowalski, C. and Ommen, O. 2009. [Safety culture: Definition, models and design]. *Zeitschrift für Evidenz, Fortbildung und Qualität im Gesundheitswesen*, 103(8), 493–7.

Pfeiffer, Y. and Manser, T. 2010. Development of the German version of the hospital survey on patient safety culture: Dimensionality and psychometric properties. *Safety Science*, 48(2010), 1452–62.

Pryseley, A.N. 2008. *The Validation of the French Translation of the Hospital Survey on Patient Safety Culture Questionnaire*. Unpublished Report, CenStat, Interuniversity Institute for Biostatistics and Statistical Bioinformatics, Hasselt, Universiteit Hasselt.

Robida, A. 2013. Hospital survey on patient safety culture in Slovenia: A psychometric evaluation. *International Journal for Quality in Health Care*, 25(4), 469–75.

Roqueta Egea, F., Tomás Vecina, S. and Chanovas Borras, M.R. 2011a. Cultura de seguridad del paciente en los servicios de urgencias: resultados de su evaluación en 30 hospitales del Sistema Nacional de Salud español. *Emergencias*, 23, 356–64.

Roqueta Egea, F., Tomás Vecina, S. and Chanovas Borras, M.R. 2011b. Patient safety culture in 30 Spanish hospital emergency departments: Results of the agency for healthcare research and quality's hospital survey on patient safety culture. *Emergencias*, 23, 356–64.

Sarac, C., Flin, R., Mearns, K. and Jackson, J. 2011. Hospital survey on patient safety culture: Psychometric analysis on a Scottish sample. *BMJ Quality and Safety*, 20, 842–8.

Saturno, P.J. 2009. *Análisis de la cultura sobre seguridad del paciente en el ámbito hospitalario del Sistema Nacional de Salud Español*. Madrid: Ministerio de Sanidad y Política Social.

Saturno, P.J., Da Silva Gama, Z.A., de Oliveira-Sousa, S.L., Fonseca, Y.A., de Souza-Oliveira, A.C., Castillo, C., Lopez, M.J., Ramon, T., Carrillo, A., Iranzo, M.D., Soria, V., Saturno, P.J., Parra, P., Gomis, R., Gascon, J.J., Martinez, J., Arellano, C., Gama, Z.A.D., de Oliveira-Sousa, S.L., de Souza-Oliveira, A.C., Fonseca, Y.A. and Ferreira, M.S. 2008. [Analysis of the patient safety culture in hospitals of the Spanish National Health System]. *Med Clin Monogr (Barc)*, 131(Suppl. 3), 18–25.

Schneider, B. 1975. Organizational climates: An essay. *Personnel Psychology*, 28, 447–79.

Seo, D.-C., Torabi, M.R., Blair, E.H. and Ellis, N.T. 2004. A cross-validation of safety climate scale using confirmatory factor analytic approach. *Journal of Safety Research*, 35(4), 427–45.

Sexton, J., Helmreich, R., Neilands, T., Rowan, K., Vella, K., Boyden, J., Roberts, P. and Thomas, E. 2006. The Safety Attitudes Questionnaire: Psychometric properties, benchmarking data, and emerging research. *BMC Health Services Research*, 6, 44.

Skodová, M., Velasco Rodriguez, M.J. and Fernández Sierra, M.A. 2011. Opinión de los profesionales sanitarios sobre seguridad del paciente en un hospital de primer nivel. *Review de Calidad Asistencial*, 26(1), 33–8.

Smits, M., Christiaans-Dingelhoff, I., Wagner, C., van der Wal, G. and Groenewegen, P.P. 2008. The psychometric properties of the 'Hospital Survey on Patient Safety Culture' in Dutch hospitals. *BMC Health Services Research*, 8, 230.

Smits, M., Wagner, C., Spreeuwenberg, P., Timmermans, D.R.M., van der Wal, G. and Groenewegen, P.P. 2012. The role of patient safety culture in the causation of unintended events in hospitals. *Journal of Clinical Nursing*, 21(23–24), 3392–401.

Smits, M., Wagner, C., Spreeuwenberg, P., van der Wal, G. and Groenewegen, P.P. 2009. Measuring patient safety culture: An assessment of the clustering of responses at unit level and hospital level. *Quality and Safety in Health Care*, 18(4), 292–6.

Snijders, C., Kollen, B.J., van Lingen, R.A., Fetter, W.P.F., Molendijk, H. and Neosafe, S.G. 2009. Which aspects of safety culture predict incident reporting behavior in neonatal intensive care units? A multilevel analysis. *Critical Care Medicine*, 37(1), 61–7.

Sorra, J., Franklin, M. and Streagle, S. 2008. *Survey User's Guide: Nursing Home Survey on Patient Safety Culture*. Rockville: Agency for Healthcare Research and Quality U.S. Department of Health and Human Services.

Sorra, J. and Nieva, V. 2004. *Hospital Survey on Patient Safety Culture*. Rockville, MD: U.S. Department of Health and Human Services.

Stricoff, R.S. 2005. Understanding safety's role in culture and climate. *Occupational Hazards*, 67(12), 25–7.

Turunen, H., Partanen, P., Kvist, T., Miettinen, M. and Vehviläinen-Julkunen, K. 2013. Patient safety culture in acute care – a web-based survey of nurse managers' and registered nurses' views in four Finnish hospitals. *International Journal of Nursing Practice*, 19(6), 609–17.

Turunen, H., Partanen, P., Mäntynen, R., Kvist, T., Miettinen, M. and Vehviläinen-Julkunen, K. 2011. *Management Support for Patient Safety: A Three Year Follow Up Study for Hospital Staff in Four Finnish Acute Care Hospitals*. 10th ENDA Congress, 5–8 October 2011, Rome.

van Noord, I., de Bruijne, M.C. and Twisk, J.W.R. 2010. The relationship between patient safety culture and the implementation of organizational patient safety defences at emergency departments. *International Journal for Quality in Health Care*, 22(3), 162–9.

van Vegten, A., Pfeiffer, Y., Giuliani, F. and Manser, T. 2011. Patientensicherheitsklima im Spital: Erfahrungen mit der Planung, Organisation und Durchführung einer Mitarbeitervollbefragung. *Zeitschrift für Evidenz, Fortbildung und Qualität im Gesundheitswesen*, 105, 734–42.

Vlayen, A., Hellings, J., Claes, N., Peleman, H. and Schrooten, W. 2011. A nationwide Hospital Survey on Patient Safety Culture in Belgian hospitals: Setting priorities at the launch of a 5-year patient safety plan. *BMJ Quality & Safety*, 21(9), 760–67.

Waterson, P., Griffiths, P., Stride, C., Murphy, J. and Hignett, S. 2010. Psychometric properties of the Hospital Survey on Patient Safety Culture: Findings from the UK. *Quality and Safety in Health Care*, 19(5), e2.

Wenqi, L. 2005. *Validation of the Questionnaire on Patient Safety Culture*. Hasselt: Universiteit Hasselt.

Zohar, D. 1980. Safety climate in industrial organizations: Theoretical and applied implications. *Journal of Applied Psychology*, 65(1), 96–102.

Zwart, D., Langelaan, M., van de Vooren, R.C., Kuyvenhoven, M.M., Kalkman, C.J., Verheij, T.J.M. and Wagner, C. 2011. Patient safety culture measurement in general practice: Clinimetric properties of SCOPE. *BMC Family Practice*, 12(1), 117.

Chapter 12

Lessons from the AHRQ Hospital Survey on Patient Safety Culture

Joann Speer Sorra and James Battles

Background

Safety culture has been an essential element of the patient safety movement from its beginning. Lessons learned about safety culture from other high-hazard or high-risk industries such as aviation, nuclear power and the petrochemical industry have influenced our approach to dealing with risks and hazards in healthcare (Weick 1987; Hofstede et al. 1990; Roberts 1990). One of the perplexing issues in examining organisational safety culture is that there is no common understanding about what culture is (Westrum and Adamski 1999). Westrum (2004) has noted that culture is defined as an organisation's pattern of response to the problems and opportunities it encounters. He provides three dominant types of organisational response that shape an organisation's culture – pathological, bureaucratic, and generative. Organisations with a pathological orientation are characterised as controlling, with low cooperation. Such cultures conceal problems, assuming everything is fine, and information to the contrary is not welcomed. In a bureaucratic organisation there is modest cooperation with a narrow focus on following rules and regulations. In a generative organisation there is high cooperation and innovation and risks and hazards are identified. Westrum's typology has proven to be a useful framework for understanding high hazard organisations.

The US Institute of Medicine (IOM) in its landmark report *To Error is Human* recommended that healthcare organisations develop a culture of safety in order to improve the quality and safety of care provided (Kohn et al. 2000). The US Agency for Healthcare Research and Quality (AHRQ) was designated as the lead federal agency to address issues related to patient safety and medical error through its reauthorisation legislation in 1999. In 2001, the US Senate Committee on Appropriations requested that AHRQ describe how it was responding to applicable recommendations in the IOM's report, *To Err is Human* (Senate Report 2000). In addition to AHRQ's specific role in patient safety, the President of the United States directed the establishment of a Quality Interagency Coordination Task Force (QuIC) to recommend actions that various federal agencies could take, either together or separately, to address patient safety. The QuIC report, *Doing What Counts for Patient Safety: Federal Actions to Reduce Medical Errors and Their Impact* (Quality Interagency Coordination Task Force 2000) described more than 100 actions that the QuIC and its participating agencies would undertake either alone, or together with the private sector and state governments. The QuIC successfully brought together federal agencies to ensure that they worked in a coordinated way toward the common goal of improving the quality and safety of healthcare in the US.

One of the challenges that AHRQ and the QuIC faced was to address the issue of safety culture, as identified by the IOM, and translate that directive into practical and workable programmes and projects. One of the first and foremost questions was how an organisation would measure safety culture. There was a critical need to create a shared understanding of just what safety culture was in healthcare and to develop a shared mental model of it. Team training researchers have most clearly articulated theories involving shared cognition in general and definitions of shared mental models in particular. Initial theorising on shared mental models suggests that, for teams to coordinate their actions, they must possess commonly held knowledge structures that allow them to predict behaviour based on shared performance expectations (Cannon-Bowers et al. 1995). The key issues for AHRQ were to identify the critical elements or dimensions of patient safety culture and determine whether those dimensions could be articulated in such a way as to be measured effectively.

Another key issue was whether culture pertained to the organisation as a whole or also to its constituent parts (Westrum and Adamski 1999). In healthcare this issue of the organisation as a whole versus its parts is critically important. Healthcare organisations are composed of multiple, differentiated, autonomous microsystems. Clinical care is most often delivered at the clinical microsystem level (Batalden et al. 1997; Mohr et al. 2004). Edmondson (2004) points out the importance of the clinical unit manager as a shaper of unit-level culture and organisational learning. Thus each clinical microsystem or unit may well have a distinct culture that is different from another clinical microsystem. A critical challenge in assessing hospital patient safety culture is the ability to measure it at the clinical microsystem level, as well as being able to examine the overall patient safety culture of the entire organisation, or macrosystem level, in a consistent and effective manner.

The medical errors work group of the QuIC, with AHRQ's lead, issued a contract to develop and deploy a valid and reliable instrument that would be available in the public domain and could be used by all components of the federal government, as well as by healthcare institutions and researchers. In order to carry out the necessary development, a contract was awarded to Westat, a private research organisation. A definition of safety culture was adopted for instrument development:

The safety culture of an organization is the product of individual and group values, attitudes, perceptions, competencies, and patterns of behavior that determine the commitment to, and the style and proficiency of, an organization's health and safety management. Organizations with a positive safety culture are characterized by communications founded on mutual trust, by shared perceptions of the importance of safety, and by confidence in the efficacy of preventive measures (as cited in Nieva and Sorra 2003: ii18).

Survey Development

Westat conducted a number of activities to inform the development of the Hospital Survey on Patient Safety Culture. One of the first tasks was a literature review to examine existing culture surveys that had been developed by the Veteran's Administration (Burr et al. 2002) and federally-funded grant programmes from the National Institutes of Health (NIH) (Sorra et al. 2008), as well as other research efforts on safety climate and culture across various industries (Guldenmund 2000; Lee and Harrison 2000; Sexton et al. 2006; Zohar 2000). The purpose of the literature review was to identify dimensions to be included in the safety culture construct. Empirical and theoretical articles were reviewed in areas such as safety management and accidents in nuclear, aviation and manufacturing industries, employee health and safety, organisational climate and culture, safety climate and culture and medical error and event reporting.

Based on the literature review, key dimensions of hospital patient safety culture were identified and items drafted to measure those dimensions. Items were written to obtain a staff-level perspective of the extent to which a hospital organisation's culture supports patient safety and event reporting. In addition, most of the items were focused on the respondent's own work area or unit because unit-level culture is more salient and relevant and has the most immediate influence on staff attitudes and behaviors. Since culture varies across units, it was important to focus respondents on their own unit's culture by asking them to identify and select their unit first and then answer the questions in the survey about that unit. However, some patient safety culture issues cut across units, so the last part of the survey focused specifically on hospital-wide patient safety culture, including handoffs and transitions, perceptions of management support and teamwork across units.

As further background for the survey, Westat conducted telephone and in-person interviews with hospital nurses, staff and physicians to assess whether the survey dimensions covered all relevant aspects of patient safety culture or whether any new dimensions were needed. Based on these interviews, there was a general consensus that the draft dimensions and items appeared to measure key patient safety concepts.

The draft survey was then cognitively tested with over a dozen hospital staff. Cognitive testing involved asking individuals to complete the survey and provide comment about their answers in one-on-one, in-person or telephone interviews. The purpose of cognitive testing was to assess respondent comprehension and interpretation of items, to determine how they arrived at their answers and to find out if there were any problems with the instructions or item wording. Results from cognitive testing led to dropping or rewording some dimensions and items, resulting in a revised version of the survey that was pilot tested.

The survey was pilot tested in 2003 with 1,437 hospital staff from 21 hospitals across six states in the US. The pilot data were analysed to determine the instrument's psychometric properties and drop poorer-performing dimensions and items. The goal of the analysis was to develop a shorter, revised instrument by identifying conceptually meaningful, independent and reliable dimensions or composites, with three to four items measuring each composite. The final Hospital Survey on Patient Safety Culture, released by AHRQ in 2004 along with related toolkit materials, includes 12 composites (shown in Table 12.1) and 42 items, plus additional background questions (Sorra and Nieva 2004). Subsequent multi-level psychometric analyses with a larger data set, confirmed the psychometric properties of the instrument (Sorra and Dyer 2010).

After the release of the hospital survey and the success of its subsequent adoption and use, AHRQ later funded the development of patient safety culture surveys for use in nursing homes in 2008, medical offices in 2009, community pharmacies in 2012 and in ambulatory surgical centres (not completed as of publication date). The SOPS surveys are currently being used by thousands of healthcare organisations each year, with the AHRQ SOPS web site pages receiving an average of over 22,000 visits per month in 2012 (SOPS pages on AHRQ Web site 2012).

Table 12.1   Patient Safety Culture composites and definitions

Patient Safety Culture Composite	Definition: The extent to which ...
1. Communication openness	Staff freely speak up if they see something that may negatively affect a patient, and feel free to question those with more authority
2. Feedback and communication about error	Staff are informed about errors that happen, given feedback about changes implemented, and discuss ways to prevent errors
3. Frequency of events reported	Mistakes of the following types are reported: (1) mistakes caught and corrected before affecting the patient, (2) mistakes with no potential to harm the patient, and (3) mistakes that could harm the patient, but do not
4. Handoffs and transitions	Important patient care information is transferred across hospital units and during shift changes
5. Management support for patient safety	Hospital management provides a work climate that promotes patient safety and shows that patient safety is a top priority
6. Nonpunitive response to error	Staff feel that their mistakes and event reports are not held against them, and that mistakes are not kept in their personnel file

7.	Organizational learning– Continuous improvement	Mistakes have led to positive changes and changes are evaluated for effectiveness
8.	Overall perceptions of patient safety	Procedures and systems are good at preventing errors and there is a lack of patient safety problems
9.	Staffing	There are enough staff to handle the workload and work hours are appropriate to provide the best care for patients
10.	Supervisor/manager expectations and actions promoting safety	Supervisors/managers consider staff suggestions for improving patient safety, praise staff for following patient safety procedures, and do not overlook patient safety problems
11.	Teamwork across units	Hospital units cooperate and coordinate with one another to provide the best care for patients
12.	Teamwork within units	Staff support each other, treat each other with respect, and work together as a team

The AHRQ Surveys on Patient Safety Culture are now used locally, nationally, and internationally. Most users are healthcare organisations and systems, survey vendors that service healthcare organisations, or healthcare researchers. The surveys are used to:

1. Raise staff awareness about patient safety
2. Diagnose and assess the current status of patient safety culture
3. Identify strengths and areas for patient safety culture improvement
4. Examine trends in patient safety culture change over time
5. Evaluate the cultural impact of patient safety initiatives and interventions
6. Conduct internal and external comparisons.

US Comparative Database

After the survey's release, hospitals wanted to compare results to determine whether their own scores were above or below norms for other hospitals. In response to those requests, AHRQ established a comparative database for the survey in 2007. The goal of the database was to produce aggregate statistics to enable hospitals to compare their survey results to other hospitals overall and by bed size, teaching status, ownership and region. Results can also be compared by hospital work areas/ units, staff positions and whether staff have direct interaction with patients.

The comparative database has been a tremendous resource for information about patient safety culture in US hospitals. Hospital de-identified data from the database has been made available to researchers. In addition, hospital-identifiable data has been made available to researchers interested in linking patient safety culture to other measures of quality and safety. Research proposals requesting identifiable data have to be approved by both AHRQ and hospitals that submit data to the database.

Since the first annual comparative database report, published in 2007, which included data voluntarily submitted from 382 US hospitals, the number of hospitals and staff respondents included in the database report has grown each year. The Hospital SOPS 2012 Comparative Database Report displays results from 1,128 hospitals and 567,703 hospital staff respondents (Sorra et al. 2012). This large number of hospitals provides for a much more reliable and comprehensive set of benchmarks.

What We Have Learned from the Database

Hospital work areas and staff asked to complete the survey   When the Hospital SOPS was developed, it was not specifically designed and tested for use with non-clinical staff like those in housekeeping, facilities, or human resources. Yet once the survey was released, it became very clear that hospitals wanted to survey all staff from all units and departments, with the understanding that every staff member plays an important role in ensuring patient safety. By being attentive and aware of patient safety risks, in an environment that encourages open communication and learning, even non-clinical staff can help prevent medication errors, patient identification errors and many other types of errors. Since one of the uses of the survey is as an education and awareness tool, it makes sense for hospitals to conduct the survey in a broad way across units and staff positions.

Because the work areas listed in the survey are very general, they typically do not correspond well to the numerous departments that most hospitals have. In addition, the staff positions included in the survey are almost all clinical positions. The database therefore includes a high percentage of respondents selecting 'Other' as their work area or unit (about 30 percent) and 'Other' as their staff position (about 20 percent). In 2014, 10 years after the survey's initial release, AHRQ and Westat began working on a revision of the Hospital SOPS survey to update the areas assessed by the survey and revise item wording. In the updated survey, AHRQ will reexamine work areas and staff positions and add more categories to try to reduce the large number of respondents selecting 'Other'.

Our latest guidance to hospitals is that they administer the survey using their own, customised work areas and staff positions so they can produce survey results in ways that are most useful to them. Many hospitals want to produce and distribute results to their specific works areas and review results by staff positions. When submitting to the database, we ask hospitals simply to cross-reference their customised work areas and staff positions to the more general ones included in the survey so that comparative results can be generated.

Frequency of survey administration   Initially many hospitals administered the patient safety culture survey every year. However, more recently we are seeing hospitals wait longer to readminister the survey, with an average of 20 months between administrations. Hospitals are moving to an every-other-year schedule because changes in survey scores, and in culture, occur slowly. Often there is little change between scores from year to year. Because administering the survey after more than two years might miss important changes in culture over time, hospitals are alternating years to administer the survey. A two-year schedule is also more practical since it can take months to prepare for data collection, administer the survey, analyse the results (particularly by unit), distribute and discuss the results throughout the hospital, develop action plans and then implement improvement initiatives. Less frequent administration allows time for patient safety initiatives and improvements to take effect before reassessment of culture is done. However, there are situations in which the survey could be readministered more frequently or within the same year. For example, if the survey is used to evaluate the impact of a specific intervention, it can be administered before and after the intervention. The key is to plan ahead how frequently the survey should be administered based on the purposes and uses of the data.

Mode of survey administration   When the first comparative database report was released in 2007, most hospitals administered paper surveys (56 percent), followed by web (25 percent) and mixed mode (both paper and web – 19 percent). By 2012, with rapidly advancing technology, web surveys became the predominant mode (66 percent), followed by paper (21 percent) and mixed mode (13 percent). The one caveat in this trend is that paper surveys still seem to get the highest response rates in US hospitals – on average at least 10 percentage points higher than other modes. So if response rates are a concern and a hospital has the capability and resources to conduct a paper survey, it is still best to administer it by paper. But if costs are an issue, it can often be very cost-effective to administer a web survey. User's Guides, available for each of the SOPS surveys on the AHRQ web site (www.ahrq.gov), contain tips and guidance on how best to administer the survey and present results.

Similar patterns of results   Overall, the pattern of patient safety culture results across US hospitals tends to be similar from hospital to hospital. The highest-scoring composites – Teamwork within Units, Supervisor and Management Support for Patient Safety – tend to be the same, as do the lower-scoring composites – Nonpunitive Response to Error and Handoffs and Transitions. What seems to differ is the overall level of scores from one hospital to the next. Higher-performing hospitals have higher scores across all·of the composites compared to lower-performing hospitals but, within each hospital, the profile of areas that are strengths and that need improvement tends to be the same. Figure 12.1 shows an example comparing scores from a high, average and low-scoring hospital to show the similarity in the pattern of highs and lows. The composites that are strengths and those that need improvement tend to be the same for the high, average and low-scoring hospitals. This pattern reflects that fact that hospitals face similar challenges when it comes to patient safety, but that some hospitals have organisational cultures that enable them to handle those challenges across the board better than others.

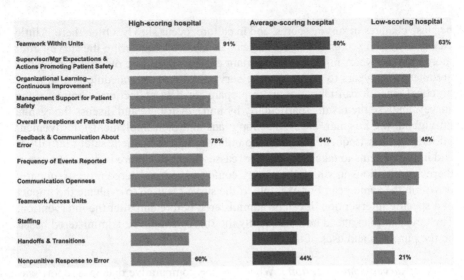

Figure 12.1   Example of the similar pattern of strengths and areas for improvement for high-, average- and low-scoring hospitals

Variation across units and hospitals   In line with theory that stipulates that culture exists at many levels, we also find that scores on the AHRQ patient safety culture survey vary across units and across hospitals. Sorra and Dyer (2010) found good multilevel psychometric properties for the survey at the unit and hospital levels, indicating the survey can appropriately be used to assess culture at these different levels. Campbell et al. (2010) administered the Hospital SOPS across 57 units in an acute care hospital and found that scores varied markedly across units within the hospital. They emphasise the importance of a unit-level examination of results, in a way that resembles the recommendations of other patient safety culture researchers (Deilkas and Hofoss 2010; Huang et al. 2010). Scores on the survey composites also tend to vary considerably across hospitals. In the Hospital SOPS 2012 Comparative Database, the average difference between the highest-scoring and lowest-scoring hospitals on the survey's 12 composites was 57 percentage points. The highest-scoring hospitals had more than 80 percent of respondents answer positively on the survey and the lowest-scoring hospitals had less than 50 percent of respondents answer positively.

Findings from the Hospital SOPS 2012 Comparative Database Report

Next, we describe highlights of results from the Hospital SOPS 2012 Comparative Database Report (Sorra et al. 2012). As mentioned earlier, the 2012 database report presents results from 1,128 hospitals and 567,703 hospital staff respondents. Database scores on the survey are calculated by computing the percent positive response which is the percentage of respondents that answered an item in a positive way. As an example, for positively-worded items (e.g. 'Our procedures and systems are good at preventing errors from happening'), the percent positive score is the percentage of respondents within a hospital who answered Strongly agree or Agree. For negatively-worded items (e.g. 'We have patient safety problems in this unit') it is the percentage of respondents who answered Strongly disagree or Disagree. Note that there are also frequency response scales used in the survey but scores are calculated in the same way. Composite-level scores are calculated by averaging the percent positive scores on the items within a composite.

Results by staff position Administrators/managers were the most positive on patient safety culture with the highest scores on 11 of 12 composites (an average of 74 percent positive). Administrators/managers were up to 10 percentage points more positive than the next highest staff position on three composites: Feedback and Communication About Error, Communication Openness and Nonpunitive Response to Error. Administrators/managers were also most positive in rating their work area/unit as Excellent or Very Good (86 percent positive). Other research by Singer et al. (2008) similarly found that senior managers perceive safety climate more positively than other staff.

Pharmacists were least positive on patient safety culture overall with an average percent positive score of 68 percent. Pharmacists also had the lowest ratings of their work area/unit as Excellent or Very Good (68 percent). However, pharmacists had the highest percentage of staff reporting one or more events in the past 12 months (71 percent).

Results by work area/unit Rehabilitation units had the highest scores on 10 of 12 composites (an average of 69 percent), and appear to have the best patient safety culture of any hospital work area. Rehabilitation units had the highest ratings of their work/area unit as Excellent or Very Good (85 percent positive). Rehabilitation also had the lowest percentage of staff reporting one or more events in the past 12 months (39 percent).

Emergency departments (EDs) scored lowest on 10 of 12 patient safety culture composites with an average of 57 percent positive. EDs scored particularly low on Staffing, with only 50 percent of ED staff responding positively. Most ED staff indicated they work in 'crisis mode trying to do too much, too quickly' (only 38 percent strongly disagreeing/disagreeing with this statement) and only 44 percent agreed that they have enough staff to handle their workload. This finding is supported by other research studies that highlight issues with handoffs and transitions in emergency care, which can threaten the safety and quality of patient care (Beach et al. 2003; Behara et al. 2005).

Hospital pharmacies had the lowest scores on Handoffs and Transitions of any department, with an average of only 30 percent of staff responding positively. Most pharmacy staff indicated that 'things "fall through the cracks" when transferring patients from one unit to another' (only 22 percent strongly disagreeing/disagreeing with this statement). Pharmacy units also indicated there were problems in the exchange of information across units (only 30 percent positive) and that shift changes were problematic for patients (only 33 percent positive). These findings from the pharmacy area support research that shows that medication errors are one of the most common types of adverse events occurring in hospitals (Institute of Medicine 2006).

Results by hospital characteristics Smaller hospitals (6–24 beds) had more positive patient safety cultures than larger hospitals (400 beds or more). Hospitals with 49 beds or fewer had the highest percentage of respondents who gave their work area/unit an Excellent or Very Good patient safety grade (80 percent) compared to larger hospitals with 400 beds or more (71 percent). Interestingly, this finding that smaller facilities have a more positive patient safety culture is also found in data from the comparative databases on the AHRQ Nursing Home and AHRQ Medical Office Surveys on Patient Safety Culture. Smaller nursing homes (49 or fewer beds) had the highest patient safety culture scores and medical offices with only one or two providers also had the highest scores. With regard to these results, some health systems with hospitals of various sizes have commented that they do not think their smaller hospitals are necessarily safer for patients than larger hospitals, so it is not clear why smaller facilities seem to score consistently higher on patient safety culture assessments. More research is needed to understand whether there is some type of bias that is related to facility size or a smaller number of staff or respondents, or whether smaller facilities really are safer for patients.

Findings on Response to Error and Event Reporting

One of the unique aspects of the Hospital SOPS has been its strong emphasis on aspects of culture related to the organisation's response to error and reporting of errors and mistakes. What we have learned is that the majority of staff do not report events, which is unfortunate but not surprising. Over time, our understanding about event reporting has evolved from blame-free culture, to nonpunitive culture, to what is now widely understood as 'just' culture. The just culture model recognises three classes of human fallibility: 1) human error – which is inadvertent, 2) at-risk behavior – taking shortcuts that lead to increased risk and 3) reckless behavior – where someone deliberately makes choices that put others in harm's way (AHRQ Web M&M 2007). Just cultures create clear boundaries for employee behavioural choices, focus on improving systems rather than blaming individuals and support a learning organisation where employees feel safe reporting errors and the organisation learns from errors to prevent them recurring.

Nonpunitive response to error Nonpunitive response to error remains the one patient safety culture composite that consistently scores at the bottom for virtually all hospitals, year after year. Staff continue to worry that the mistakes they make are kept in their personnel file, that individuals are 'written up' rather than the problem and that their mistakes are held against them. Many hospitals have commented that, although they have taken steps to address staff perceptions of a punitive culture, it has been difficult to see improvement in scores on the survey. In 2010, Westat developed new survey items focusing more on just culture principles. These items were pilot tested with a number of hospitals engaged in a collaborative programme to improve just culture. The new items showed promise but still need more revision and testing. The revision to the original AHRQ Hospital SOPS will build on this work to assess important aspects of 'just' culture.

Event reporting   Event reporting is important for hospitals to understand patient safety risks and enable them to take steps to prevent and mitigate harm to patients. However, the status of event reporting in the US still shows considerable underreporting. A 2012 report from the Office of the Inspector General found that US hospital incident reporting systems captured only an estimated 14 percent of the patient harm events experienced by Medicare beneficiaries (Department of HHS 2012).

When examining event reporting results in the Hospital SOPS 2012 database by work area, intensive care units (ICUs) had the highest percentage of staff (64 percent) that had reported one or more events in the past 12 months; rehabilitation had the lowest at 39 percent. When examining event reporting by staff position, pharmacists (71 percent) and nurses (68 percent) had the highest percentages of reporting. Only 37 percent of physicians, physician assistants and nurse practitioners indicated they had reported events. In addition, more staff with direct patient interaction reported events (50 percent) compared to staff without direct patient interaction (30 percent). Staff indicated that near misses that had potential to harm the patient, but did not, were more likely to be reported always or most of the time (74 percent) compared to mistakes that had no potential to harm (59 percent) or events that were caught and corrected before affecting the patient (57 percent).

In response to the US Patient Safety and Quality Improvement Act of 2005 (PSQA, PL-109-41) (US Congress 2005), AHRQ established Patient Safety Organizations (PSO) which are entities to which health systems, hospitals and other healthcare organisations can send event report data, knowing that is safe from legal discovery. AHRQ also established a Privacy Protection Center (PPC) to which PSOs could submit patient safety event data using common formats, also developed by AHRQ. A National Patient Safety Databases contractor was hired to analyse the de-identified event data but, as of 2012, PSOs had yet to submit data. AHRQ is also exploring how and in what manner patients and family members can report patient safety events by supporting a pilot test of a consumer reporting system that will be operated in conjunction with an established PSO. While consumer or patient reporting is not currently covered by the PSQA, it is an area that can add to our understanding of patient safety events from the patient's perspective.

One way to measure the impact of the PSQA and the PSO contribution to patient safety is to examine the effects of protected reporting on staff perceptions of error. If levels of event reporting improve within an organisation, the responses to cultural assessment should also change, particularly since event reporting benchmarks have, over the past several years, changed very little in the database.

Farley et al. (2008) surveyed risk managers from over 2,000 US hospitals asking them about event reporting and concluded that: 1) only 32 percent of hospitals have established environments that support reporting, 2) only 13 percent have broad staff involvement in reporting adverse events and 3) only 20–21 percent fully distribute and consider summary reports on identified events. Significant underreporting of events means that hospitals are unable fully to identify the incidence and prevalence of patient safety risks within their systems. Work must continue on the event reporting front, as well as with the information that is gleaned and analysed from event reports. Action is also needed on policy and system changes needed within hospitals to support staff in preventing errors.

Trends over time   The Hospital SOPS 2012 Comparative Database Report includes a trending chapter that describes patient safety culture change over time for 650 hospitals that administered the survey and submitted data to the database more than once. It compares results from their previous and most recent administrations. The trending chapter of the report is the most difficult one to prepare and draw conclusions from. We attempt to summarise changes in patient safety culture scores but we observe many types of changes in the data. Some hospitals have scores that increase or decrease significantly but most changes are less than 5 percentage points, which we classify as 'no change' because such small changes in scores could occur by chance rather than from real patient safety culture improvement. Comparing hospitals' previous and most recent scores, the overall average percent positive scores on the survey composites increased by only 1 percentage point (range: 0 to 2 percentage points). What is clear from this trending data is that culture change, in the aggregate, is small and slow to improve over time. Yet these aggregated conclusions do not adequately capture what is happening at the individual hospitals with regard to culture change. We are still struggling with how best to analyse and interpret this change data and how to do so when there are data from three points in time or more. Hospitals are also struggling with how to interpret their own change data.

One nagging concern that has been expressed by a number of survey users has been that, when they administer the Hospital SOPS for a second time, some find that their scores decrease rather than increase, despite efforts to improve patient safety. Some users have suggested that perhaps hospital staff are initially positive because their standards of what constitutes good patient safety culture change over time. The theory is that once a hospital engages staff and undertakes patient safety improvement initiatives that examine work practices and behaviours, staff think more critically about patient safety and rate their unit and hospital lower the second time around. What is not clear is whether hospitals find that their scores on all composites go down, or just on some composites, and the degree to which scores are declining. Is the decline less than 5 percentage points or more like 10? Are they are seeing improvements in some areas, but focussing more on the declining scores since they do not expect any areas to decline over time? When we have attempted systematically to examine change over multiple points in time in the database, we do not find a consistent result with regard to decreases in scores at Time 2. There seems to be a lot of variability across hospitals with regard to changes in scores, with the one consistent finding being that changes are small, regardless of their direction.

Another possible explanation for changes in scores is that the respondents may be different. If a hospital has a very low response rate initially, when the survey is readministered it is likely that many of the respondents will be different because so few answered the survey the first time. In addition, if there is high turnover, that could also explain changes in scores. Clearly more work needs to be done in this area to describe and explain changes in patient safety culture scores over time and identify the reasons for such change. It is likely that to truly understand culture change, a qualitative assessment involving focus groups and interviews is needed to understand the various and interactive factors that contribute to changes in patient safety culture.

Using the Survey to Improve Patient Safety Culture

Despite the great data and information that are now available as a result of the Hospital SOPS database, one big deficit has been in our understanding of what hospitals are doing between their patient safety culture survey assessments. What initiatives are hospitals implementing? How successful are those initiatives in improving patient safety culture and, ultimately, patient safety?

Once hospitals use the survey to identify areas for improvement, their next step is to work out what they can do, what actions they can take, to improve patient safety culture. A resource list describing dozens of patient safety initiatives is available to survey users on the AHRQ web site, but hospitals still have to identify what will be effective and which initiatives are likely to be successful in their facilities. For the Hospital SOPS 2011 Comparative Database (Sorra et al. 2008), we asked 456 trending hospitals what types of patient safety initiatives they had implemented between survey assessments. The top five initiatives are shown in Table 12.2.

Table 12.2   Percentages of Hospital SOPS trending hospitals that implemented these patient safety improvement activities

1. Improved fall prevention programme	56%
2. Conducted root cause analysis	52%
3. Implemented SBAR Communication (Situation-Background-Assessment-Recommendation)	51%
4. Improved compliance with Joint Commission National Patient Safety Goals	50%
5. Held educational/patient safety fair for staff	47%
6. Made changes to policies/procedures	46%
7. Implemented patient safety walkarounds	44%

The difficulty in assessing the effectiveness of patient safety interventions is the same as that for all programme evaluations; it is difficult to find rigorous studies that measure programme effectiveness quantitatively. There have been a number of recent reviews of patient safety initiatives and their effectiveness. Morello et al. (2012) reviewed over 2,000 articles and found only 21 studies meeting their inclusion criteria for study rigour. They concluded that there is some evidence to support the theory that leadership walk rounds and multi-faceted unit-based programmes may have a positive impact on patient safety climate.

In 2013, AHRQ released *Making Health Care Safer II: An Updated Critical Analysis of the Evidence for Patient Safety Practices* (Shekelle et al. 2013). This report is an update of an original report *Making Health Care Safe*, released in 2001. These two reports define a safe practice with the following statement: 'A Patient Safety Practice is a type of process or structure whose application reduces the probability of adverse events resulting from exposure to the healthcare system across a range of procedures.'

In the newest report, patient safety culture was included as one of the safe practices that were reviewed for its evidence, both as a practice and in terms of the context sensitivity of the practice and its adoption. In reviewing the evidence on patient safety culture as a practice, Weaver et al. (2013) suggest that developing a culture of safety is a core element of many efforts to improve patient safety and healthcare quality. Their systematic review identified and assessed interventions used to promote safety culture or climate in acute care settings. Based on this review the authors indicate that there is evidence suggesting that interventions can improve perceptions of safety culture and potentially reduce patient harm.

Relationships between Patient Safety Culture and Other Measures

An early criticism of patient safety culture survey data was that there was not sufficient evidence that culture was related to medical error, patient safety or quality or patient outcomes. There are still very few empirical studies linking the Hospital SOPS to these important outcomes and many more studies are needed. However, there have been a few key studies showing positive relationships between Hospital SOPS scores and outcomes.

Mardon et al. (2010) examined relationships between the Hospital SOPS and rates of in-hospital complications and adverse events measured by the AHRQ Patient Safety Indicators (PSIs). Higher patient safety culture scores were associated with lower adverse event rates (average correlation, r = -.36), including infections due to medical care and postoperative complications. The strongest finding was that hospitals scoring higher on Hospital Handoffs and Transitions across units, had lower adverse event rates (based on PSI composite score average correlation, r = -.50).

In another study, Sorra et al. (2012) examined relationships between Hospital SOPS and patients' assessments of hospital care, as measured by the Consumer Assessment of Healthcare Providers and Systems (CAHPS) Hospital Survey.

Hospitals with higher patient safety culture scores had better patient experience scores (correlation, r = .41). The strongest relationship was between the Hospital SOPS measure of the adequacy of staffing and patients' perceptions of the responsiveness of staff. Interestingly, the Hospital SOPS measures were not significantly correlated with patients' overall ratings of the hospital or the patients' willingness to recommend the hospital.

Other studies have examined the relationship between patient safety climate and outcomes using other culture surveys. Using the Patient Safety Climate in Healthcare Organizations (PSCHO) survey, Rosen et al. (2010) found safety climate overall in Veterans Administration hospitals was not related to the AHRQ Patient Safety Indicators (PSIs) or to the PSI composite, although a few individual dimensions of safety climate were associated with specific PSIs. They also found that the perceptions of frontline staff were more closely aligned with PSIs than those of senior managers. Another study by Singer et al. (2009), using the PSCHO survey, found that hospitals with a better safety climate overall had lower relative incidence of PSIs.

Clearly more empirical studies are needed linking patient safety culture scores to patient safety and quality outcome data. Such studies should examine linkages at the hospital level and at the unit level where relationships may be stronger. The difficulty of this research is that large numbers of hospitals with both safety culture and outcome data are needed to detect these relationships.

Extensions to Other Healthcare Settings

Given the widespread adoption and use of the Hospital SOPS, AHRQ funded the development of patient safety culture surveys for use with staff in nursing homes, medical offices and community pharmacies. AHRQ is also supporting comparative databases for these surveys. These databases rely on voluntary submission of survey data by health systems, vendors and facilities. The Nursing Home SOPS Comparative Database Report (Sorra et al. 2011) was released in 2011 based on data from 16,155 nursing home staff respondents in 226 nursing homes. The Medical Office SOPS Comparative Database Report (Sorra et al. 2012) was released in 2012 based on data from 23,679 medical office staff respondents in 934 medical offices. AHRQ is also funding the development of a survey for use in outpatient procedure/surgery facilities.

The measurement of safety culture has now become a core component of patient safety and additional patient safety culture assessment instruments will be needed to accommodate organisational settings that cover the continuum of care in multiple settings. It will also be necessary to examine differences, as well as similarities, in staff perceptions of patient safety culture across different settings of care. In addition, it is necessary to examine in these other healthcare settings the relationships between patient safety culture and patient perceptions of care, as well as clinical outcome measures, as has been done in the hospital setting.

International Users

The need for a valid and reliable instrument to assess safety culture is not restricted to the United States. The AHRQ Hospital SOPS instrument has been widely adopted in over 45 countries around the world and the instrument has been translated into more than 20 languages. It has become the de-facto international survey for measuring safety culture in healthcare. While other industries such as aviation, nuclear power and petrochemicals have been interested in assessing safety culture there is neither a standard nor agreement as to what to measure. However, in healthcare, the AHRQ instrument is a standard and it has created a shared mental model of safety culture that has become almost universal in healthcare throughout the world.

The World Health Organization (WHO) High 5s project was launched in 2006 to focus on medication reconciliation, management of concentrated injectables and correct site surgery in selected hospitals in six countries over five years (WHO 2013). Part of the measurement and evaluation of the programme involved administering the Hospital SOPS. In 2009 and 2010, the the United Kingdom, France, the Netherlands, Germany, Singapore and Australia participated in the project. Hospital SOPS data were collected from 23,520 respondents in 59 hospitals and AHRQ provided feedback reports to hospitals and country-wide comparative reports comparing hospitals within each country. While results were presented comparing survey scores across countries, those results were interpreted with caution, with more attention paid to within-country comparisons.

Other international efforts have included the EUNetPaS (European Union Network for Patient Safety) which was officially launched in 2008 in Utrecht, the Netherlands to establish an umbrella network of all European Union (EU) Member States to encourage collaboration in patient safety. EUNetPaS sought to establish common principles at the EU level, integrating knowledge, experiences and expertise from member states and offering support to countries that were less advanced in patient safety. EUNetPaS published a two-volume report in 2010 reviewing patient safety culture instruments (Kristensen and Bartels 2010). The AHRQ Hospital SOPS was one of only three patient safety culture instruments that was officially recommended after an extensive review of available tools.

In a few studies comparing results across countries (Bodur and Filiz 2009; Wagner et al. 2013), the general conclusion seems to be that the US tends to score higher than other countries on most of the survey's composites. However, one study in Taiwan found that staff results were higher than the US on many of the composites (Chen et al. 2010).

Differences in scores across countries may be due to a number of factors other than true differences in patient safety culture, making direct comparisons between countries complicated. The first factor is the quality and comparability of translations. Translations must convey the intent of the original items rather than merely directly translating the words and translations should undergo cognitive testing and validation before use. Most international administrations of the survey have shown similar factor structures to the US version and acceptable psychometric properties: Olsen (2008) in Norway; Smits et al. (2008) in the Netherlands; Itoh et al. (2011) in Japan; and Sarac et al. (2011) in Scotland. However, a few studies have shown divergent factor structures compared to the US: Pfeiffer and Manser (2010) with a German version administered in Switzerland; and Waterson et al. (2010) with a modified English version administered in the UK. There may also be significant cultural differences across countries in how respondents interpret the survey items and the way in which they use the response scales. In addition, there are differences in the way healthcare is structured and provided across countries that could affect the interpretation of items and comparability of responses. Overall, international feedback on the survey has been positive and is evidenced by numerous publications based on administrations around the world.

To assist international users, the AHRQ web site on the SOPS surveys has an international page that lists the countries and translations that exist, provides translation guidelines for those interested in developing a translation and has information specific to each SOPS survey to help translators understand the original intent of the US English items when developing translations. Users have also translated and administered the other SOPS surveys. AHRQ provides technical assistance to international users, connecting them with one another and retaining information about translations of the SOPS surveys that have been developed and administered.

Conclusions

Safety culture is a critical component of patient safety. Assessment of safety culture has now become an accepted and even required measurement of patient safety. The AHRQ Hospital Survey on Patient Safety Culture (Hospital SOPS) is widely used by health systems, hospitals and researchers to assess patient safety culture in the US and internationally. The US Hospital SOPS comparative database represents the largest, non-proprietary publicly available compilation of patient safety culture survey data in healthcare, or any industry, that we are aware of thus far. The efforts of AHRQ have led to healthcare being the only industry that regularly publishes and reports safety culture data. International adoption and translation of the survey have led to the development of a shared mental model, across nations, about what constitutes patient safety culture in healthcare around the globe. Despite some variation in the factor structure and psychometrics of the survey across countries, the survey seems to hold up well internationally with different healthcare delivery models and settings. The tool is raising awareness about what aspects of patient safety culture must be attended to and what areas in particular should be the focus for improvement and it has opened dialogue about patient safety culture and event reporting around the world.

References

AHRQ 2001. *Making Health Care Safer: A Critical Analysis of Patient Safety Practices*, AHRQ Publication No. 01-E057. AHRQ: Rockville, MD. Available at: http://archive.ahrq.gov/clinic/tp/ptsaftp.htm (last accessed on 24 April 2014).

AHRQ Web M&M. *In Conversation with ... David Marx, JD.* Oct 2007. Available at: http://webmm.ahrq.gov/perspective.aspx?perspectiveID=49 (last accessed on 30 January 2013).

Batalden, P.B., Mohr, J.J., Nelson, E.C., Plume, S.K., Baker, G.R., Wasson, J.H., Stoltz, P.K., Splaine, M.E. and Wisniewski, J.J. 1997. Continually improving the health and value of health care for a population of patients: The panel management process. *Quality Management in Health Care*, 5, 41–51.

Beach, C., Croskerry, P. and Shapiro, M. 2003. Profiles in patient safety: Emergency care transitions. *Academic Emergency Medicine*, 10(4), 364–7.

Behara, R., Wears, R.L., Perry, S.J., Eisenberg, E., Murphy, L., Vanderhoef, M., Shapiro, M., Beach, C., Croskerry, P. and Cosby, K. 2005. A conceptual framework for studying the safety of transitions in emergency care. Advances in patient safety: From research to implementation. *AHRQ Publication*, No. 05-0021-2, 2, 309–21.

Bodur, S. and Filiz, E. 2009. A survey on patient safety culture in primary healthcare services in Turkey. *International Journal for Quality in Health Care*, 21(5), 348–55. Available at: http://intqhc.oxfordjournals.org/content/21/5/348.full.pdf+html (last accessed on 28 April 2014).

Burr, M., Sorra, J., Nieva, V. and Famolaro, T. 2002. *Analysis of the Veteran's Administration (VA) National Center for Patient Safety (NCPS) Patient Safety Questionnaire.* Unpublished technical report.

Campbell, E., Singer, S., Kitch, B., Iezzoni, L. and Meyer, G. 2010. Patient safety climate in hospitals: Act locally on variation across units. *Joint Commission Journal on Quality and Patient Safety*, 36(7), 319–26.

Cannon-Bowers, J.A., Tannenbaum, S.I., Salas, E. and Volpe, C.E. 1995. Defining competencies and establishing team training requirements. In R.A. Guzzo and E. Salas (eds), *Team Effectiveness and Decision Making in Organizations*. San Francisco: Jossey-Bass, 333–80.

Chen, I.C. and Li, H.H. 2010. Measuring patient safety culture in Taiwan using the Hospital Survey on Patient Safety Culture (HSOPSC). *BMC Health Services Research*, 7(10), 152. Available at: http://www.biomedcentral.com/1472-6963/10/152/ (last accessed on 30 January 2013).

Deilkås, E. and Hofoss, D. 2010. Patient safety culture lives in departments and wards: Multilevel partitioning of variance in patient safety culture. *BMC Health Services Research*, 10(85) doi:10.1186/1472-6963-10-85.

Department of Health and Human Services, Office of Inspector General. 2012. *Hospital* incident reporting systems do not capture most patient harm. Office of Inspector General: Dallas, TX. OEI-06-09-00091 (last accessed on 30 January 2013).

Departments of Labor, Health and Human Services, Education, and Related Agencies Committee Appropriation Bill, 2001. Senate Report 106–293. May 12, 2000, 195–8.

Edmondson, A.C. 2004. Learning from failure in health care: Frequent opportunities, pervasive barriers. *Qual Saf Health Care*, 13, ii3–ii9.

Farley, D., Haviland, A., Champagne, S., Jain, A., Battles, J., Munier, W. and Loeb, J. 2008. Adverse-event-reporting practices by US hospitals: Results of a national survey. *Quality and Safety in Health Care*, 17, 416–23. doi:10.1136/qshc.2007.024638.

Guldenmund, F.W. 2000. The nature of safety culture: A review of theory and research. *Safety Science*, 34, 215–57.

Hofstede, G., Neuijen, B., Ohayv, D. and Sanders, G. 1990. Measuring organizational cultures: A qualitative and quantitative study across twenty cases. *Administrative Science Quarterly*, 35, 286–316.

Huang, D.T., Clermont, G., Kong, L., Weissfeld, L.A., Sexton, J.B., Rowan, K.M. and Angus, D.C. 2010. Intensive care unit safety culture and outcomes: A US multicenter study. *International Journal for Quality in Health Care*, 22(3), 151–61.

Institute of Medicine 2006. Preventing medication errors: Quality chasm series. P. Aspden, J. Wolcott, J. Bootman and L. Cronenwett (eds). The National Academies Press: Washington, DC.

Itoh, S., Seto, K., Kigawa, M., Fujita, S. and Hasegawa, T. 2011. Development and applicability of Hospital Survey on Patient Safety Culture (HSOPS) in Japan. *BMC Health Serv Res*, 11, 28. doi:10.1186/1472-6963-11-28. Available at: http://www.biomedcentral.com/1472-6963/11/28 (last accessed on 30 January 2013).

Kohn, L., Corrigan, J. and Donaldson, M. (eds). 2000. *To Err is Human: Building a Safer Health System.* Institute of Medicine Report. Washington, DC: National Academies Press.

Kristensen, S. and Bartels, P. 2010. *Use of Patient Safety Culture Instruments and Recommendations.* EUNetPaS (European Union Network for Patient Safety) European Society for Quality in Healthcare – Office for Quality Indicators, Denmark.

Lee, T. and Harrison, K. 2000. Assessing safety culture in nuclear power station. *Safety Science*, 34, 61–97.

Mardon, R., Khanna, K., Sorra, J., Dyer, N. and Famolaro, T. 2010. Exploring relationships between hospital patient safety culture and adverse events. *Journal of Patient Safety*, 6(4), 226–32.

Mohr, J.J., Batalden, P. and Barach, P. 2004. Integrating patient safety into the clinical microsystem. *Quality and Safety in Health Care*, 13(SupplII), ii34–ii38.

Morella, R.T., Lowthian, J.A., Barker, A.L., McGinees, R., Dunt, D. and Brand, C. 2012. Strategies for improving patient safety culture in hospitals: A systematic review. *BMJ Quality and Safety*, doi:10.1136/bmjqs-2011-000582. Online 31 July 2012 (last accessed on 30 January 2013).

Nieva, V.F. and Sorra, J.S. 2003. Safety culture assessment: A tool for improving patient safety in healthcare organizations. *Quality and Safety in Healthcare*, 12(Suppl II), ii17–ii23.

Olsen, E. 2008. Reliability and validity of the Hospital Survey on Patient Safety Culture at a Norwegian hospital. In J. Øvretveit, and P. Sousa (eds), *Quality and Safety Improvement Research: Methods and Research Practice from the International Quality Improvement Research Network*. Lisbon: Escola Nacional de Saúde Pública, 173–86.

Pfeiffer, Y. and Manser, T. 2010. Development of the German version of the Hospital Survey on Patient Safety Culture: Dimensionality and psychometric properties. *Safety Science*, 48, 1452–62.

Quality Interagency Coordination Task Force (QuIC). 2000. *Doing What Counts for Patient Safety: Federal Actions to Reduce Medical Errors and Their Impact*. Available at: http://archive.ahrq.gov/quic/report/mederr2.htm (last accessed on 30 January 2013).

Roberts, K.H. 1990. Some characteristics of one type of high reliability organization. *Organization Science*, 1(2), 160–76.

Rosen, A.K., Singer, S., Zhao, S., Shokeen, P., Meterko, M. and Gaba, D. 2010. Hospital safety climate and safety outcomes: Is there a relationship in the VA? *Medical Care Research Review*, 67(5), 590–608.

Sarac, C., Flin, R., Mearns, K. and Jackson, J. 2011. Hospital Survey on Patient Safety Culture: Psychometric analysis on a Scottish sample. *BMJ Quality and Safety*, 20(10), 842–8.

Sexton, J.B., Helmreich, R.I., Neilands, T.B., Rowan, K., Vella, K., Boyden, J., Roberts, P. and Thomas, E. 2006. The Safety Attitudes Questionnaire: Psychometric properties, benchmarking data, and emerging research. *BMC Health Services Research*, (6), 44.

Shekelle, P.G., Wachter, R.M., Pronovost, P.J., Schoelles, K., McDonald, K.M., Dy, S.M., Shojania, K., Reston, J,, Berger, Z., Johnsen, B., Larkin, J.W., Lucas, S., Martinez, K., Motala, A., Newberry, S.J., Noble, M., Pfoh, E., Ranji, S.R., Rennke, S., Schmidt, E., Shanman, R., Sullivan, N., Sun, F., Tipton, K., Treadwell, J.R., Tsou, A., Vaiana, M.E., Weaver, S.J., Wilson, R. and Winters, B.D. 2013. Making health care safer II: An updated critical analysis of the evidence for patient safety practices. *Comparative Effectiveness Review No. 211*. (Prepared by the Southern California-RAND Evidence-based Practice Center under Contract No. 290-2007-10062-I.) AHRQ Publication No. 13-E001-EF. Rockville, MD: Agency for Healthcare Research and Quality. March 2013. Available at: www.ahrq.gov/research/findings/evidence-based-reports/ptsafetyuptp.html (last accessed on 30 January 2013).

Singer, S., Falwell, A., Gaba, D. and Baker, L. 2008. Patient safety climate in US hospitals: Variation by management level. *Medical Care*, 46(11), 1149–56.

Singer, S., Lin, S., Falwell, A., Gaba, D. and Baker, L. 2009. Relationship of safety climate and safety performance in hospitals. *Health Serv Res*, 44(2 Pt 1), 399–421.

Smits, M., Christiaans-Dingelhoff, I., Wagner, C., van der Wal, G. and Groenewegen, P. 2008. The psychometric properties of the 'Hospital Survey on Patient Safety Culture' in Dutch hospitals. *BMC Health Services Research*, 8, 230. Available at: http://www.biomedcentral.com/1472-6963/8/230 (last accessed on 30 January 2013).

SOPS pages on AHRQ Website visit statistics from unpublished contract monthly report prepared by Westat, Rockville, MD, under Contract No. HHSA 290200720024C. July 2012.

Sorra, J.S. and Dyer, N. 2010. Multilevel psychometric properties of the AHRQ hospital survey on patient safety culture. *BMC Health Serv Res*, 10, 199.

Sorra, J., Famolaro, T., Dyer, N., Khanna, K. and Nelson, D. 2011. *Hospital Survey on Patient Safety Culture: 2011 User Comparative Database Report* (Prepared by Westat, Rockville, MD, under Contract No. HHSA 290200710024C). Rockville, MD: Agency for Healthcare Research and Quality; March 2008. AHRQ Publication No. 11-0030.

Sorra, J., Famolaro, T., Dyer, N., Khanna, K. and Nelson, D. 2011. *Nursing Home Survey on Patient Safety Culture: 2011 Comparative Database Report* (AHRQ Publication No. 11-0071). Rockville, MD: Agency for Healthcare Research and Quality. Available at: http://www.ahrq.gov/qual/nhsurvey11/nhsurv111.pdf (last accessed on 28 April 2014).

Sorra, J., Famolaro, T., Dyer, N., Nelson, D. and Smith, S. 2012. *Hospital Survey on Patient Safety Culture: 2012 User Comparative Database Report* (Prepared by Westat, Rockville, MD, under Contract No. HHSA 2902007110024C). Rockville, MD: Agency for Healthcare Research and Quality; February 2012. AHRQ Publication No. 12-0017.

Sorra, J., Famolaro, T., Dyer, N., Smith, S., Liu, H. and Ragan, M. 2012. *Medical Office Survey on Patient Safety Culture: 2012 User Comparative and Database Report* (AHRQ Pub. No. 12-0052). Rockville, MD: Agency for Healthcare Research and Quality. Available at: http://www.ahrq.gov/qual/mosurvey12 (last accessed on 28 April 2014).

Sorra, J., Khanna, K., Dyer, N., Mardon, R. and Famolaro, T. 2012. Exploring relationships between patient safety culture and patients' assessments of hospital care. *Journal of Patient Safety*, 8(3), 131–9.

Sorra, J.S. and Nieva, V.F. 2004. *Hospital Survey on Patient Safety Culture.* (Prepared by Westat, under Contract No. 290-96-0004). AHRQ Publication No. 04-0041. Rockville, MD: Agency for Healthcare Research and Quality.

Sorra, J., Nieva, V., Rabin-Fastman, B., Kaplan, H., Schreiber, G. and King, M. 2008. Staff attitudes about event reporting and patient safety culture in hospital transfusion services. *Transfusion*, 48, 1934–42.

US Congress 2005. *Patient Safety and Quality Improvement Act of 2005*, S. 544, enacted by the 109th Congress. Washington: US Government Printing Office.

Wagner, C., Smits, M., Sorra, J. and Huang, C.C. 2013. Assessing patient safety culture in hospitals across countries. *International Journal for Quality in Health Care Advance Access*. 1–9. Available at: http://intqhc.oxfordjournals.

org/content/early/2013/04/08/intqhc.mzt024.full.pdf?keytype=ref&ijkey=ejro 415vNX9LRvU (last accessed on 30 January 2013).

Waterson, P., Griffiths, P., Stride, C., Murphy, J. and Hignett, S. 2010. Psychometric properties of the Hospital Survey on Patient Safety Culture: Findings from the UK. *Quality and Safety in Health Care*, 19(5), e2.

Weaver, S.J., Lubomksi, L.H., Wilson, R.F., Pfoh, E.R., Martinez, K.A. and Dy, S.M. 2013. Promoting a culture of safety as a patient safety strategy. *Annals of Internal Medicine*, 158, 369–74.

Weick, K.E. 1987. Organization culture as a source of high reliability. *California Management Review*, 29, 112–27.

Westrum, R. 2004. A typology of organizational cultures. *Quality and Safety in Health Care*, 13(Suppl II), ii22–ii27.

Westrum, R. and Adamski, A. 1999. Organization factors associated with safety and mission success in aviation environments. In D.J. Garland, J.A. Wise and V.D. Hopkins (eds), *Aviation Human Factors*. Hillsdale, NJ: Lawrence Erlbaum.

World Health Organization (WHO) 2013. Action on patient safety – High 5s. Available at: http://www.who.int/patientsafety/implementation/solutions/high5s/en/index.html (last accessed on 30 January 2013).

Zohar, D. 2000. A group-level model of safety climate: Testing the effect of group climate on micro-accidents in manufacturing jobs. *Journal of Applied Psychology*, 85, 587–96.

Chapter 13
Safety Attitudes Questionnaire: Recent Findings and Future Areas of Research

Jason Etchegaray and Eric Thomas

Patient safety researchers have developed several surveys to assess patient safety climate, including the Safety Attitudes Questionnaire (SAQ), AHRQ's Hospital Survey on Patient Safety Culture (SOPS; see Chapter 12 for more detailed information), and Patient Safety Climate in Healthcare Organizations. Our chapter provides an in-depth examination of the SAQ in terms of what it measures, key findings and limitations and future areas in need of research.

What is the SAQ?

Sexton et al. (2006) explained that the SAQ was created to assess six primary dimensions of the work environment based on Donabedian's (1988) and Vincent et al.'s (1998) frameworks about safety and quality, respectively. The six dimensions are: job satisfaction, perceptions of management, safety climate, stress recognition, teamwork climate and working conditions (see Table 13.1). There are many different versions of the SAQ – ambulatory, ICU, labour and delivery, OR, etc. – and it has been translated into multiple languages and administered in numerous geographic locations. The dimensions consist of four (perceptions of management, stress recognition and working conditions), five (job satisfaction), six (teamwork climate) or seven (safety climate) items that are answered on a 5-point, Likert-type scale ranging from 1 = disagree strongly and 5 = agree strongly with an NA = not applicable option. When possible, researchers aggregate survey responses at the unit level because climate is conceptualised as a shared belief about the work environment.

Table 13.1   Dimensions of SAQ and example items

Dimension	Definition	Example survey item
Safety climate	Environment emphasises safety	I would feel safe being treated here as a patient.
Teamwork climate	Environment focuses on teamwork	The physicians and nurses here work together as a well-coordinated team.
Job satisfaction	Positive feelings about job	This is a good place to work.
Perceptions of management	Management provides needed support	Management supports my daily efforts.
Stress recognition	Role stress plays at work	I am less effective at work when fatigued.
Working conditions	Overall quality of work environment	This hospital does a good job of training new personnel.

Findings: SAQ as a Measure of Interventions

Sexton et al. reported favourable psychometric results for the SAQ, with strong evidence for reliability and validity. Subsequent studies (Etchegaray and Thomas 2012; Profit et al. 2012; Schwendimann et al. 2012; Weng et al. 2012) have yielded similar results at the individual and unit level. In addition to psychometric findings, much research exists about the SAQ as a measure of intervention success. Integrated within this area of research is examination of the SAQ as a predictor of outcomes. We review both of these areas of research.

By conducting a non-exhaustive literature review in PubMed that examined the following key words – safety culture, safety climate, culture interventions – we discovered that researchers have adopted two primary approaches to examining change: 1) implementing a specific intervention or 2) implementing a comprehensive intervention. We first discuss specific interventions – Executive Walk Rounds, checklists, structured interdisciplinary rounds, and medical team training – and then comprehensive interventions. While our main focus was on discussing interventions where the SAQ was also measured, we included some examples of interventions that assessed culture with a different survey instrument to provide a broader perspective about the association between culture and interventions.

Executive Walk Rounds (EWRs)

Thomas et al. (2005) and Frankel et al. (2005) have examined the impact of EWRs on perceptions of safety climate. While hospitals typically implement EWRs differently, EWRs share some common features – 1) they are led by hospital executives, 2) executives solicit input from frontline staff about improvement opportunities on the unit, 3) executives ask frontline staff about methods to improve unit safety and 4) executive leaders reiterate their focus on improving unit safety. The discussions that take place during each EWR lead to specific improvement actions on the unit and then feedback is shared with the unit. By involving executives and frontline staff in solving issues on units, Thomas et al. reasoned that the hospital might benefit from more awareness about safety issues through an increase in safety climate.

Providers from a large, urban teaching hospital in Texas completed the Safety Climate Survey before and after intervention, where the intervention consisted of EWRs implemented at the hospital. Six executives were selected to assist in the EWRs. Each executive visited a hospital unit for which they were responsible three times, with the visits being spaced one month apart. The hospital's Patient Safety Officer and one professional from Performance Improvement accompanied the executive during each EWR visit, which lasted for up to 60 minutes. The results revealed that nurses who received EWRs had significantly higher safety climate scores after EWRs than nurses who did not receive EWRs. Results examining differences between other providers based on whether they received EWRs were not significant within provider group. Overall, this study showed that EWRs can have a positive impact on safety climate scores for nurses.

Frankel et al. (2005) documented their experiences in implementing 'WalkRounds' in four Partners HealthCare hospitals from 2001 to 2003. Across the four hospitals, 1,433 comments were collected from 233 one-hour WalkRounds. Most of the comments were focused on safety concerns around equipment (30 percent), with communications (13 percent), pharmacy (7 percent), and workforce (6 percent) the next leading categories. All four hospitals addressed many of the comments (see Table 4 in Frankel et al.'s article for a detailed list of changes made). Based on their learning from these WalkRounds, Frankel et al. developed a four-week stepwise approach that can be used when implementing WalkRounds. In Week One, leadership, safety/quality professionals and middle management are educated about systems science, high reliability and the concept of WalkRounds. A pilot WalkRound is conducted and personnel are taught how to enter data collected from the WalkRounds into a computerised database. In Week Two, the hospital must identify a committee responsible for reviewing WalkRounds data, assigning teams to address the data and sharing learning with patient safety professionals. In Week Three, everyone in the hospital will be notified about WalkRounds; a feedback mechanism will be developed to share data with those participating in WalkRounds and reports will be developed to be shared with the operations committee and quality committee. In Week Four, leaders participating in WalkRounds will be identified, WalkRounds will be scheduled and key stakeholders (i.e. leaders, patient safety professionals and operations committee) will create and sign performance agreements about their intent to participate actively in the WalkRounds process.

In a subsequent study, Frankel et al. (2008) administered the SAQ – a survey that is currently used in hospital settings throughout the world – before and after the introduction of rigorous WalkRounds in seven hospitals and one ambulatory center in Massachusetts. Of these eight healthcare settings, only two hospitals implemented the WalkRounds in a 'rigorous' manner. The research team used Frankel et al's (2006) seven step framework to rigorously implementing WalkRounds; the seven steps included: 1) preparation, 2) scheduling, 3) conducting WalkRounds, 4) tracking, 5) reporting, 6) feedback, and 7) measurement. The research team trained senior leaders from the participating healthcare settings on this framework through a half-day session that included a presentation, discussion about managing data from the WalkRounds, explanations about how to address safety issues identified from them and instruction about how to conduct WalkRounds.

The research team focused on changes in safety climate scores from the SAQ. In the first of two hospitals that implemented WalkRounds in a rigorous way, the safety climate scale score significantly increased between pre- and post-WalkRounds. The increase in safety climate scale scores between pre- and post-WalkRounds was not significant for the second hospital. However, both hospitals showed significant pre-post changes for four of the safety climate items. Those items were focused on: 1) colleagues' encouragement to report patient safety concerns, 2) knowing which channels to use to ask patient safety questions, 3) learning from others with respect to errors and 4) being able to discuss errors that happened in their unit. In terms of data from WalkRounds, these researchers reported that most patient safety concerns were focused on equipment/facility and communication. Overall, this study shows that safety climate perceptions can be improved if WalkRounds are conducted in a rigorous manner.

Checklists

Haynes et al. (2011) decided to test whether safety and teamwork climate changed as a result of OR personnel using the WHO Safe Surgery Saves Lives checklist. OR-specific safety and teamwork climate survey items from the SAQ were administered to providers working in ORs (i.e. surgeons, nurses, anesthesiologists, etc.) before and after implementation of this checklist. Two of the survey items – 1) importance for patient safety of briefing OR personnel before surgery and 2) colleagues' encouragement to report patient safety concerns – and the overall scale these researchers used showed significant differences before and after implementing the checklist. These results provide support for the theory that safety and teamwork climate positively changing as a result of implementing surgical checklists.

Structured Interdisciplinary Rounds

O'Leary et al. (2010) noted that Thomas et al.'s (2003) interesting research findings that physicians tended to view their teamwork with nurses as high while nurses tended to view their teamwork with the same physicians as low revealed an important area that needed to be addressed in the culture literature. O'Leary et al.

tried to address this research area by examining how perceptions of teamwork culture can be improved through the use of collaborative work processes, such as structured interdisciplinary rounds (SIDRs). Previous research found that physicians perceive higher levels of collaboration when using interdisciplinary rounds. SIDRs bring the patient care team members together to discuss the plan of care for a patient, using a structured communication tool focused on 1) overall plan of care, 2) discharge plans and 3) patient safety. For this study, SIDRs occurred every morning for about 30 minutes and involved two leaders (nurse manager and unit medical director) who discussed new patients with unit nurses, physicians, pharmacists, social workers and case managers. Two teaching units in a tertiary care teaching hospital in Illinois were randomised to the control or experimental condition, with the experimental condition receiving the instruction in how to use SIDR. Physicians and nurses were surveyed several weeks after SIDR was implemented in the experimental condition. Similar to previous research, irrespective of whether physicians belonged to the control or experimental group, they perceived that nurses displayed high levels of teamwork. When nurses in the control group rated physicians on teamwork, their ratings were low, which is consistent with Thomas et al.'s earlier findings. However, when nurses in the experimental group rated the physicians with whom they worked on teamwork, their ratings were significantly higher than the control group. These findings show that SIDR is an effective process for improving nurse perceptions of the collaboration and teamwork experienced with physicians.

Teamwork Interventions

Wolf et al. (2010) documented their experiences with the use of medical team training (MTT) to improve teamwork and performance in the operating room (OR). MTT is based on the crew resource management framework and focuses on situational awareness, direct communication and human factors. These researchers administered the SAQ before and after implementing MTT. MTT was initiated with a full day interactive training session for OR staff, surgeons, nurses, anesthesiologists and other staff. The research team developed a briefing/debriefing protocol that was designed to be used before surgeries started and which allowed all team members to understand the intended goals. Surgical teams also conducted postoperative debriefings before leaving the OR to discuss problems and lessons learned and to recognise teamwork shown during the surgery. Although not significant, improvements were found between pre- and post-assessments of teamwork culture and safety culture. Additionally, two other dimensions of the SAQ – perceptions of management and working conditions – showed significant increases between pre- and post-tests.

Watts et al. (2010) administered the SAQ before and after implementation of MTT in Veterans Affairs hospitals. They found that 26 of the 63 hospitals showed improvement in some aspect of the SAQ, with most improving in the perceptions of management dimension of the SAQ. Carney et al. (2011) also studied the relationship between MTT and safety culture, using a pre-post design in medium- and high-complexity facilities. They found that MTT improved safety culture perceptions across participants and reduced pre-survey differences between medium- and high-complexity facilities.

Comprehensive Interventions

In addition to previous research efforts that focus on specific interventions, described above, some researchers have recently started looking at comprehensive interventions to improve safety with the general belief that a multidimensional perspective is needed to improve safety. To date, the most impactful comprehensive intervention has been the Comprehensive Unit-based Safety Program (CUSP). Researchers and practitioners implement CUSP through a combination of education and experiential learning that consists of five primary steps: 1) educating staff about safety science, 2) having staff identify safety risks, 3) partnering with senior leaders to address safety risks, 4) using a structured set of questions to learn from defects and 5) implementing tools that will lead to communication and teamwork improvement. Timmel et al. (2010) documented their experiences with CUSP, which is focused on unit-level interventions to improve safety culture. From 2006 to 2008, Timmel et al. found statistically significant improvements in teamwork culture and safety culture after implementing CUSP on a surgical floor in an academic medical centre in Maryland. Sexton et al. (2011) also demonstrated improvement in safety and teamwork culture scores as a result of CUSP in a pre-post design used in ICUs.

Verschoor et al. (2007) documented their comprehensive approach to creating a culture of safety. Their approach was based primarily on tools developed by the Institute for Healthcare Improvement (IHI). To identify discussion topics for the safety briefings, the research team collected staff reports of risks and hazards for six weeks. Staff reported a total of 46 risks and hazards in their environment. IHI's Safety Briefings model was piloted in one area of a children's hospital and then subsequently implemented in other areas. There are six main principles of safety briefings according to this model: 1) information about safety issues is used to improve systems, not punish people, 2) the briefings must be brief (i.e. no more than five minutes in length), 3) facilitation of safety issues based on published safety alerts might help staff learn about what types of safety issues to discuss in these briefings, 4) the briefings must be so simple that any staff can lead them, 5) the briefings can use discussion prompts (i.e. 'What environmental factors might lead to patient risk?'), which encourage staff to consider safety risks in their environment and 6) professionals from all disciplines participate in the briefings. Implementation of safety briefings led to the identification of some best practices, such as unit using notetakers to document what was discussed in the briefings and publicly displaying the issues identified during them so that the process is transparent. A pædiatric intensive care unit integrated the briefings into individual patient case reviews each day to maximise ability to identify patient risks.

Overall, the research team reported that it took about one year for these briefings to 'become entrenched in the unit culture' (Verschoor et al. 2007: 84).

In addition to safety briefings, the research team used IHI's Patient Safety Leadership Walkrounds, based on many of the same principles as the safety briefings. Additionally, units started to include staff in root cause analyses. Finally, the Comprehensive Unit-Based Safety Program (CUSP) developed by researchers at Johns Hopkins was pilot tested for two years, with an electronic version (eCUSP) also employed. eCUSP allows 24 hour access to a patient safety portal that includes educational materials, information about safety culture and a place for anonymously reporting safety problems as well as documenting changes made to address these problems. An innovative feature of eCUSP is that leaders receive real-time updates about patient safety risks entered by staff. Leaders can then create teams to address these issues and share learning by publishing completed projects for all those who access eCUSP to see. To support the effectiveness of these interventions, these researchers identified significant increases in safety culture after implementation of Safety Briefings, Walkrounds and eCUSP.

Benn et al. (2009) and Pinto et al. (2011) reported results from their work on the Safer Patients Initiative (SPI). SPI is a collaboration between the UK Health Foundation and the US Institute for Healthcare Improvement, aimed at improving care processes by implementing a comprehensive programme consisting of evidence-based practices, continuous improvement methodology and safety culture development (interested readers should view Table 1 in either of these publications to learn about project aims, timeframe, and programme tools). Benn et al. reported results from four healthcare facilities in England, Northern Ireland, Scotland and Wales that were surveyed after implementing SPI, while Pinto et al. provided results from 20 facilities in the same geographic areas. In both studies, the researchers asked members of the programme improvement teams – including senior executives, SPI coordinators, clinical operational leaders and other staff – to complete surveys asking about various organisational factors. Benn et al. asked participants to indicate which organisational factors (including safety culture) were most affected by SPI, with results revealing that safety culture was perceived as the most affected of eight quality and safety dimensions. Pinto et al. also reported descriptive statistics showing that culture was affected by SPI.

Pettker et al. (2009) aimed to create a 'comprehensive culture of safety' by improving safety through reducing patient injuries and liability losses. They created and implemented a patient safety programme in tertiary academic centre in the Eastern United States. Their patient safety programme included eight components: 1) external expert review of their services to identify safety improvement opportunities, 2) development of practice standard protocols and guidelines to address issues identified in the external expert review, 3) creation of a new nursing position – patient safety nurse – responsible for examining quality and clinical outcomes and leading root cause analyses where necessary, 4) implementation of an electronic anonymous event reporting system focused on falls, device issues and medication errors, 5) creation of an on-call hospitalist role responsible for ensuring adequate coverage of patients and resident supervision, 6) formation of an obstetric patient safety committee to understand adverse events, 7) administration of the SAQ to measure safety culture, and 8) use of team training based on crew resource management principles. Pettker et al. found that teamwork climate and safety climate improved significantly from start to end of the implementation of this comprehensive safety programme. Significant findings emerged for employees, as well as physicians and nurses. Overall, their results showed that teamwork and safety culture scores reflected a positive change that occurred as a result of the comprehensive patient safety programme they implemented.

Table 13.2   Safety Attitudes Questionnaire: summary of study findings

Author	Intervention type	Intervention length	Setting	Culture survey used	Changes in culture scores
Thomas et al. (2005)	Executive Walk Rounds	12 weeks	Inpatient clinical units in a tertiary care teaching hospital	Safety Climate Survey	
Frankel et al. (2008)	Executive Walk Rounds	Two years	Two acute care hospitals	Safety Attitudes Questionnaire	Hospital A: increased from 62% to 77%; Hospital B: increased from 46% to 56%
Haynes et al. (2011)	Checklists	Within three months	Eight hospitals in the Safe Surgery Saves Lives programme	Safety Attitudes Questionnaire	On a 5-point scale, safety culture increased from 3.91 to 4.01
O'Leary et al. (2010)	Structured Inter-disciplinary Rounds	Six months	Teaching service unit in a tertiary care teaching hospital	Safety Attitudes Questionnaire	Teamwork culture scores higher on intervention unit (82.4%) than control unit (77.3%)
Wolf et al. (2010)	Medical Team Training	One Day	Operating room	Safety Attitudes Questionnaire	No significant change

Table 13.2   Safety Attitudes Questionnaire: summary of study findings
(*continued*)

Author	Intervention type	Inter-vention length	Setting	Culture survey used	Changes in culture scores
Watts et al. (2010)	Medical Team Training	One Day	Operating room	Safety Attitudes Questionnaire	Teamwork culture increased from 66% to 72%; safety culture increased from 67% to 73%
Jones et al. (2008)	Error Reporting System	Two years	Critical access hospitals	Hospital Survey on Patient Safety Culture	Overall perception of patient safety increased from 69% to 72%; other dimensions of survey showed comparable increases
Velji et al. (2008)	Situation-Background-Assessment-Recommendation	Six to twelve months	Clinical unit in a rehabilitation hospital	Hospital Survey on Patient Safety Culture	Eight of the survey dimensions improved by at least 5%
Adams-Pizarro et al. (2008)	Targeted culture intervention	Across two years	Hospitals participating in ED, ICU and OR collaboratives	Hospital Survey on Patient Safety Culture	OR teams improved on seven of 12 culture dimensions, ICU teams on three dimensions, and ED teams on one dimension
Tupper et al. (2008)	Targeted culture intervention	Across two years	Small rural hospitals	Hospital Survey on Patient Safety Culture	Significant improvements in nine of 12 culture dimensions

Table 13.2   Safety Attitudes Questionnaire: summary of study findings
(*concluded*)

Author	Intervention type	Inter- vention length	Setting	Culture survey used	Changes in culture scores
Timmel et al. (2010)	CUSP	Across one year	Surgical floor of an academic medical centre	Safety Attitudes Questionnaire	Teamwork culture improved from 65% to 71% and safety culture improved from 61% to 69%
Sexton et al. (2011)	CUSP	Across one year	ICUs in the Keystone ICU Project	Safety Attitudes Questionnaire	Safety culture improved from 43% to 52%
Verschoor et al. (2007)	CUSP/eCUSP	Not specified	Neuroscience unit	Patient Safety Group Culture Survey	Increases (ranging from 12% to 36%) for five of six safety culture items
Benn et al. (2009)	Safer Patients Initiative	Unclear though the Initiative ran for a total of five years	Acute care organisations	Compared perceptions of safety climate to 18 performance dimensions	Safety climate improvement attributed to Safer Patients Initiative
Pinto et al. (2011)	Safer Patients Initiative	One year	NHS organisations	One item asked about safety of care improvement	Safety of care improvement attributed to Safer Patients Initiative
Pettker et al. (2009)	Patient safety programme with eight components	Two years	Tertiary-level academic medical centre	Safety Attitudes Questionnaire	Teamwork culture improved from 39% to 55% and safety culture improved from 33% to 55%

Conclusions and Summary

There were two overall findings from the literature we reviewed: 1) both specific and comprehensive interventions have been shown to improve culture and 2) safety culture surveys appear to be sensitive to changes between pre- and post-tests. The findings are particularly encouraging given that safety culture is a concept that has been measured for a relatively short time in healthcare settings. One advantage of the comprehensive interventions like CUSP and SPI is that they incorporated leadership support/involvement as part of the intervention. Aligning leadership expectations with intervention goals is important in maximising the probability of intervention success.

Readers should be cautious of two issues with respect to the findings we have summarised. First, many of the studies used pre-post designs. While pre-post designs can provide interesting information, one cannot determine causality based on them because many unmeasured variables might have affected the dependent variable between the pre- and post- timepoints. The second issue is that recent research argues for the importance of accounting for different types of context in these types of studies. Kaplan et al. argued that microsystem, unit and organisational level contexts need to be taken into account while Taylor et al. reported that, in addition to culture, researchers need to factor in the structural characteristics of organisations, external factors and management tools. While the studies we reviewed examined culture, other aspects of context were not taken into account or reported in the majority of the studies.

The purpose of this chapter was to review interventions reported by healthcare researchers to improve safety culture. Two main types of interventions were identified in our review – those focused on specific aspects of healthcare improvement and those adopting a more comprehensive approach. Research based on both types of interventions paints an equally optimistic picture – safety culture can be improved by making changes to organisational practices. The pre-post design was used by most researchers, which allowed for direct safety culture comparisons before and after implementation of interventions. Potential limitations of pre-post designs include: 1) employee attrition, when different employees respond to surveys at the 'post' time period compared to those who responded at the 'pre' time period and 2) other unmeasured variables that might have played a role in culture score changes (e.g. change in leadership). Despite these limitations, the studies reviewed have contributed to the establishment of a strong foundation for safety culture being viewed as an important outcome that is sensitive to detecting changes. Much work remains to be done, however. More research is needed that links safety culture with healthcare outcomes and the use of interventions might be one avenue for making this happen. That is, research that attempts to improve safety culture and healthcare outcomes by changing organisational practices might be one way to address this need. The most optimistic finding from this review is that safety culture surveys were able to detect changes when these interventions were implemented. As the researchers reviewed in this chapter have demonstrated, safety culture is valuable when viewed as an outcome by itself.

References

Adams-Pizarro, I., Walker, Z., Robinson, J., Kelly, S. and Toth, M. 2008. Using the AHRQ Hospital Survey on Patient Safety Culture as an intervention tool for regional clinical improvement collaboratives. *Advances in Patient Safety: New Directions and Alternative Approaches*, 1–4.

AHRQ Publication Nos. 08-0034 (1–4). 2008. Agency for Healthcare Research and Quality, Rockville, MD. Available at: http://www.ahrq.gov/qual/advances2/, 1–20 (last accessed on 4 May 2014).

Benn, J., Burnett, S., Parand, A., Pinto, A., Iskander, S. and Vincent, C. 2009. Perceptions of the impact of a large-scale collaborative improvement programme: Experience in the UK Safer Patients Initiative. *Journal of Evaluation in Clinical Practice*, 15, 524–40.

Carney, B.T., West, P., Neily, J., Mills, P.D. and Bagian, J.P. 2011. Changing perceptions of safety climate in the operating room with the Veterans Health Administration medical team training program. *American Journal of Medical Quality*, 26, 181–4.

Donabedian, A. 1988. The quality of care. How can it be assessed? *JAMA*, 260, 1743–8.

Etchegaray, J.M. and Thomas, E.J. 2012. Comparing two safety culture surveys: Safety Attitudes Questionnaire and Hospital Survey on Patient Safety. *BMJ Quality and Safety*, 21(6), 490–98.

Frankel, A., Grillo, S.P., Baker, E.G., Huber, C.N., Abookire, S., Grenham, M., Console, P., O'Quinn, M., Thibault, G. and Gandhi, T.K. 2008. Patient Safety Leadership Walkrounds™ at Partners HealthCare: Learning from implementation. *Joint Commission Journal on Quality and Patient Safety*, 31, 423–37.

Frankel, A.S., Grillo, S. and Pittman, M.A. 2006. *Patient Safety Leadership Walk-Rounds(TM) Guide*. Chicago: Health Research and Educational Trust.

Frankel, A.S. et al. 2008. Revealing and resolving patient safety defects: The impact of leadership Walkrounds on frontline caregiver assessments of patient safety. *Health Services Research*, 43, 2050–66.

Haynes, A.B., Weiser, T.G., Berry, W.R., Lipsitz, S.R., Breizat, A.H., Dellinger, E.P., Dziekan, G., Herbosa, T., Kibatala P.L., Lapitan, M.C., Merry, A.F., Reznick, R.K., Taylor, B., Vats, A. and Gawande, A.A. 2011. Changes in safety attitude and relationship to decreased postoperative morbidity and mortality following implementation of a checklist-based surgical safety intervention. *BMJ Quality & Safety*, 20, 102–7.

Jones, K.J., Skinner, A., Xu, L., Sun, J. and Mueller, K. 2008. *The AHRQ Hospital Survey on Patient Safety Culture: A Tool to Plan and Evaluate Patient Safety Programs*. In Henriksen, K., Battles, J.B., Keyes, M.A. and Grady, M. (eds), *Advances in Patient Safety: New Directions and Alternative Approaches (Vol. 2: Culture and Redesign)*. Rockville (MD): Agency for Healthcare Research and Quality (US), 1–22.

Kaplan, H.C., Provost, L.P., Froehle, C.M. and Margolis, P.A. 2012. The model for understanding success in quality (MUSIQ): Building a theory of context in healthcare quality improvement. *BMJ Quality and Safety*, 21, 13–20.

O'Leary, K.J., Wayne, D.B., Haviley, C., Slade, M.E., Lee, J. and Williams, M.V. 2010. Improving teamwork: Impact of structured interdisciplinary rounds on a medical teaching unit. *Society of General Internal Medicine*, 25, 826–32.

Paine, L.A., Rosenstein, B.J., Sexton, J.B., Kent, P., Holzmueller, C.G. and Pronovost, P.J. 2011. Assessing and improving safety culture through an academic medical centre: A prospective cohort study. *Quality and Safety in Health Care*, 19(6), 547–54.

Pettker, C.M., Thung, S.F., Norwitz, E.R., Buhimschi, C.S., Raab, C.A., Copel, J.A., Kuczynski, E., Lockwood, C.J. and Funai, E.F. 2009. Impact of a comprehensive patient safety strategy on obstetric adverse events. *American Journal of Obstetrics & Gynecology*, 200, 492.e1–92.e8.

Pinto, A., Benn, J., Burnett, S., Parand, A. and Vincent, C. 2011. Predictors of the perceived impact of a patient safety collaborative: An exploratory study. *International Journal for Quality in Health Care*, 23, 173–81.

Profit, J., Etchegaray, J.M., Peterson, L.A., Sexton, J.B., Hysong, S.J., Mei, M. and Thomas, E.J. 2012. The Safety Attitudes Questionnaire as a tool for benchmarking safety culture in the NICU. *Archives of Disease in Childhood: Fetal and Neonatal Edition*, 97, F127–32.

Schwendimann, R., Zimmermann, N., Kung, K., Ausserhofer, D. and Sexton, B. 2013. Variation in safety culture dimensions within and between US and Swiss Hospital Units: An exploratory study. *BMJ Quality and Safety*, 22(1), 32–41.

Sexton, J.B., Helmreich, R.L., Neilands, T.B., Rowan, K., Vella, K., Boyden, J., Roberts, P.R. and Thomas, E.J. 2006. The Safety Attitudes Questionnaire: Psychometric properties, benchmarking data, and emerging research. *BMC Health Services Research*, 6, 44.

Sexton, J.B., Berenholtz, S.M., Goeschel, C.A., Watson, S.R., Holzmueller, C.G., Thompson, D.A., Hyzy, R.C., Marsteller, J.A., Schumacher, K. and Pronovost, P.J. 2011. Assessing and improving safety climate in a large cohort of intensive care units. *Critical Care Medicine*, 39(5), 934–9.

Taylor, S.L., Dy, S., Foy, R., Hempel, S., McDonald, K.M., Ovretveit, J., Pronovost, P.J., Rubenstein, L.V., Wachter, R.M. and Shekelle, P.G. 2011. What context features might be important determinants of the effectiveness of patient safety practice interventions? *BMJ Quality and Safety*, 20, 611–17.

Thomas, E.J., Sexton, J.B. and Helmreich, R.L. 2003. Discrepant attitudes about teamwork among critical care nurses and physicians. *Critical Care Medicine*, 31(3), 956–9. doi: 10.1097/01.CCM.0000056183.89175.76.

Thomas, E.J., Sexton, J.B., Neilands, T.B., Frankel, A. and Helmreich, R.L. 2005. The effect of executive walk rounds on nurse safety climate attitudes: A randomized trial of clinical units. *BMC Health Services Research*, 5, 28.

Timmel, J., Kent, P.S., Holzmueller, C.G., Paine, L., Schulick, R.D. and Pronovost, P.J. 2010. Impact of the Comprehensive Unit-Based Safety Program (CUSP) on safety culture in a surgical inpatient unit. *Joint Commission Journal on Quality and Patient Safety*, 36, 252–60.

Tupper, J., Coburn, A., Loux, S., Moscovice, I., Klingner, J. and Wakefield, M. 2008. Strategies for improving patient safety in small rural hospitals. *Advances in Patient Safety: New Directions and Alternative Approaches*, 1–4, AHRQ Publication Nos. 08-0034 (1–4). Agency for Healthcare Research and Quality, Rockville, MD. Available at: http://www.ahrq.gov/qual/advances2/ (last accessed on 14 May 2014).

Velji, K., Baker, G.R., Fancott, C., Andreoli, A., Boaro, N., Tardif, G., Aimone, E. and Sinclair, L. 2008. Effectiveness of an adapted SBAR communication tool for a rehabilitation setting. *Healthcare Quarterly*, 11, 72–9.

Verschoor, K.N., Taylor, A., Northway, T.L., Hudson, D.G., Van Stolk, D.E., Shearer, K.J., McDougall, D.L. and Miller, G. 2007. Creating a safety culture at the Children's and Women's Health Centre of British Columbia. *Journal of Pediatric Nursing*, 22, 81–6.

Vincent, C.A., Taylor-Adams, S. and Stanhope, N. 1998. Framework for analyzing risk and safety in clinical medicine. *British Medical Journal*, 316, 1154–7.

Watts, B.V., Percarpio, K., West, P. and Mills, P.D. 2010. Use of the Safety Attitudes Questionnaire as a measure in patient safety improvement. *Journal of Patient Safety*, 6, 206–9.

Weng, R.H., Huang, C.Y., Huang, J.A. and Wang, M.H. 2012. The cross-level impact of patient safety climate on nursing innovation: A cross-sectional questionnaire survey. *Journal of Clinical Nursing*, 21(15–16), 2262–74.

Wolf, F.A., Way, L.W. and Stewart, L. 2010. The efficacy of medical team training: Improved team performance and decreased operating room delays: A detailed analysis of 4863 cases. *Annals of Surgery*, 252, 477–85.

Chapter 14

Experiences from a Nationwide Safety Culture Measurement using the HSPSC within Belgium

Annemie Vlayen, Ward Schrooten, Johan Hellings,
Margareta Haelterman and Hilde Peleman

Introduction

In 2004, the Agency for Healthcare Research and Quality (AHRQ) released the Hospital Survey on Patient Safety Culture (HSOPS) to help hospitals to assess their patient safety culture. Since then, hundreds of hospitals across American and European countries have implemented the survey. The HSOPS has also been validated for use within Belgian hospitals (French, Flemish and German translations). This chapter describes the findings and experiences from a nationwide baseline and follow-up safety culture measurement within Belgian hospitals and discusses opportunities for benchmarking results. The measurement of safety culture is part of the multi-annual federal programme which aims at improving the quality and safety of care in Belgian acute, psychiatric and long-term care hospitals.

Launch of a National Quality and Patient Safety Programme in Belgian Hospitals

Language Context and Organisation of Hospital Care in Belgium

Belgium is a federal state comprising three communities, three regions and four language areas: the Dutch, Bilingual (Brussels-Capital), French and German linguistic areas. Dutch is spoken by around 59 percent of the population, French by around 40 percent and German by fewer than 1 percent. The country is divided into Dutch-speaking Flanders in the north and French-speaking Wallonia in the south. Brussels is bilingual, but its dominant language is French. German is spoken in nine communities close to Germany (http://www.belgium.be/en/). Currently, Belgium has 190 hospitals, of which 104 are acute, 64 are psychiatric and 22 are long-term care hospitals, located in the regions of Flandria (53 percent), Brussels (14 percent) and Wallonia (33 percent). Acute hospitals consist of university hospitals, general hospitals 'with university character' and other non-university hospitals. Belgium has seven university hospitals, one for each medical school that offers an entire medical education. Psychiatric hospitals are exclusively designed for psychiatric care. Specialised or long-term care hospitals provide chronic treatment and/or rehabilitation of patients with, for example, cardiopulmonary diseases, locomotive diseases, neurological disorders, palliative care needs, NOS chronic diseases and psycho-geriatric conditions. They can be considered as mixed general hospitals (Gerkens et al. 2010).

Clinical Risk Management in Belgian Hospitals

At an early stage, the Belgian government focused its attention relating to clinical risk management in hospitals on specific domains with well-known risks. The government completed legislative work in these domains and created a framework for risk management, including the establishment of a Committee for Hospital Hygiene (1987), a Medico-Pharmaceutical Committee, a Committee for Medical Materials in the context of recognition of the hospital pharmacy (Order in Council. 04/03/1991) and the establishment of a Blood Transfusion Committee (Order in Council. 16/04/2002).

Quality management in hospitals was organised at the regional level and was only regulated within the region of Flanders. The Flemish Decree on Quality of Care (1997) stated that hospital management should appoint a quality coordinator to be responsible for implementing and coordinating all quality initiatives in the hospital. In 2003, the Flemish Decree was renewed, shifting the focus onto the higher accountability of hospital management through, for instance, conducting a self-evaluation of quality assurance policy within the hospital. Quality policy programmes typically included topics such as patient satisfaction and waiting times and aimed at reducing patient falls, decubitus injuries, blood transfusion risks and hospital infections. The hospital infection control programme was often run by a physician and an infection control nurse. Although several specific aspects of patient safety were covered within the quality policy or the hospital infection control programme, patient safety was not a high priority within the hospitals.

Federal Contracts on Quality and Patient Safety in Belgian Hospitals

Since July 2007, the Belgian government has provided yearly additional financing (7.81 million Euros in 2013) for the implementation of quality and patient safety initiatives in the acute, psychiatric and long-term care hospitals (www. patient-safety.be). The federal programme aims at promoting and supporting the coordination of initiatives based on Donabedian's triad: (1) development of a safety management system (structure), (2) analysis of intramural and integrated care processes (processes) and (3) development and use of quality and safety indicators (results).

A safety management system contains several interacting elements of which five are considered essential in the Belgian federal programme:

1. Implementation of a hospital wide notification and learning system for incidents and near-misses
2. Measuring safety culture (e.g. using the Hospital Survey on Patient safety Culture, HSOPS)
3. Analysing (near-) incidents (e.g. using Root Cause Analysis, RCA)
4. Classification of incidents (e.g. using the International Classification for Patient Safety of the World Health Organization, ICPS)
5. Prospective risk assessment of healthcare processes (e.g. using Healthcare Failure Mode and Effects Analysis, HFMEA).

A national guidance and support plan assists participating hospitals in the implementation of quality and patient safety initiatives. For instance, workshops were organised on the measurement of safety culture, the use of the ICPS, HFMEA, RCA, care process management and indicators.

Some Results of the First 5-Year Plan on Quality and Patient Safety (2007–12)

The enthusiastic enrolment of 179 hospitals (92 percent) in the federal programme increased general awareness of patient safety and allowed hospitals to develop, step-by-step, a quality and patient safety plan. Although the programme is labour intensive, the multi-annual approach is highly appreciated by hospitals, leading to increased compliance with quality and safety initiatives and a higher participation of hospitals in networks in which best practice is exchanged.

The federal government collects information annually on adherence to the programme and the results of initiatives in participating hospitals and provides feedback by publishing a national report of the aggregated data. Based on information obtained from the participating hospitals, several improvements were observed after five years. For instance, a positive evolution was noted in the reporting and analysis of (near-) incidents. The number of hospitals which applied a hospital-wide notification and learning system for (near-) incidents increased from 80 percent (in 2008) to 94 percent (in 2013) and 98 percent of hospitals applied the method of retrospective analysis (only 31 percent in 2008). In total, 87 percent of hospitals established a committee to coordinate all activities regarding quality and patient safety. Also, since the elaboration of the federal programme in 2007, two nationwide safety culture measurements have been introduced successfully.

Measuring Safety Culture in Belgian Hospitals

*Selection, Translation and Psychometric Validation of the
Hospital Survey on Patient Safety Culture (HSOPS)*

In 2005, Hospital Oost-Limburg was a leading hospital in the area of patient safety in Flanders. The hospital's patient safety programme aimed at creating awareness throughout the hospital by conducting an assessment of hospital safety culture. Although, several safety culture instruments were internationally available, the HSOPS was selected because of the availability of a comprehensive report of scale development and its widespread application within American hospitals (http://www.ahrq.gov/professionals/quality-patient-safety/patientsafetyculture/index.html). Also, later studies by Colla (Colla et al. 2005) and Flin (Flin et al. 2006) reviewed different safety culture and climate surveys and concluded that the HSOPS is an acceptable instrument which can be applied to both healthcare staff and non-clinicians.

The American questionnaire was translated into Dutch (Flemish) using context-specific terminology. Psychometric properties of the questionnaire were investigated by item analysis, exploratory factor analysis, confirmatory factor analysis using structural equation models, reliability analysis, analysis of composite scores and inter-correlations (Hellings et al. 2007; Wenqi 2005). The Flemish validation report showed results comparable to the original questionnaire, confirming 10 patient safety culture dimensions and two outcome measures. A manual (protocol) was written and a data collection software tool was designed, which automatically checked inclusion criteria for each record and provided a hospital report and pivot table to explore the results.

Rolling Out the HSOPS across Belgium

Flemish safety culture research was further rolled out to four other hospitals (Hellings et al. 2007). It measured whether safety culture changed after the implementation of specific improvement strategies (Hellings et al. 2010). Lessons and recommendations from this study attracted the attention of the Belgian federal government public health services. In order to measure safety culture in Belgian hospitals, the survey needed to be made ready for use within different types of settings, taking into account multiple language contexts. Besides the Dutch (Flemish) translation of the HSOPS, the survey instrument was translated and validated in French for use within hospitals located in the region of Wallonia and Brussels. The German translation of the HSOPS, which has also been used within Germany and Switzerland, was placed at the disposal of the Belgian German speaking hospitals.

With the launch of the national quality and safety programme in 2007, all Belgian hospitals were required to measure safety culture at a regular basis, using the HSOPS. The main objectives of the nationwide safety culture measurement were to raise awareness and to measure the perceptions of healthcare staff on different patient safety aspects. The HSOPS was selected for measuring safety culture in Belgian hospitals because it covers a broad range of patient safety aspects and previous research had demonstrated that the psychometric properties of the HSOPS are good. In addition, the instrument lends itself well to internal and external benchmarking. A collective approach enables hospitals to learn from each other and helps to identify possibly significant strengths and weaknesses. In addition, repeated assessments after several years can track the evolution of safety culture.

A first (baseline) measurement was organised between 2007 and 2009 for hospitals participating in the federal programme on quality and safety. A second (follow-up) nationwide measurement in 2011 aimed at assessing the evolution of safety culture (Vlayen et al. 2014). The Belgian measurement toolkit contains the validated version of the HSOPS (in Dutch, French and German) and a manual (protocol). The protocol is comparable to the original version and imposes a timescale of 13 weeks with the encouragement to send two reminders to non-responders (www.patient-safety.be).

Benchmarking Belgian Safety Culture Perceptions: Opportunities, Practical Approach and Experiences

Rationale for Comparing Safety Culture Perceptions

Currently, safety culture research is dominated by the use of survey instruments and there is an increasing trend of benchmarking safety culture perceptions in order to diagnose areas for improvement. Although there is no straightforward classification of a good or bad, a safe or unsafe, an acceptable or unacceptable position, benchmarking safety culture perceptions across healthcare organisations introduces a normative element into safety culture. This approach shifts the focus from a safety culture which is defined as 'the way things are done around here' towards safety culture as an ideal standard or model that is 'the way things should or ought to be done'. This normative character of safety culture is two-sided. On the one hand, it should help to mould the future image of the organisation through offering insight into aspects of patient safety that could be improved over time – by answering the question 'Where do we stand in the present and what do we want to be or achieve in the future?' This can also be referred to as internal benchmarking. On the other hand, assessing the external position in relation to similar organisations can help in identifying areas of strength and weakness.

The principle of benchmarking safety culture perceptions in Belgian hospitals is based on the respondents' 'positive attitude' towards patient safety. As such, the comparative report only considers explicitly positive answers of hospital staff towards different safety culture dimensions. This approach has the limitation that neutral or negative perceptions are not separately taken into account, but it provides the opportunity of comparing positive scores across organisations and creating a 'patient safety profile' which can be used as a starting point for targeted actions.

Practical Approach of Benchmarking Safety Culture Data

Hospitals participating in the federal programme on quality and patient safety were invited to join in a benchmarking initiative (2008 and 2011) on a voluntary, confidential and free of charge basis in order to provide a patient safety profile for internal learning. The comparative database is managed by a neutral academic institution and is not accessible by governmental authorities. An MS Access-based instrument, designed by the Hospital Oost-Limburg (Hellings et al. 2007), allowed hospitals to standardise data entry, automate application of exclusion criteria for respondents and analyse results. The Access tool automatically filters out questionnaires where an entire section is incomplete, fewer than half of the items throughout the survey are answered or all items are scored identically. Additionally, the tool provides the possibility of instantly creating a hospital report with an overview of respondent characteristics and hospital scores on the different items and 12 composite dimensions. Workshops were organised for participating hospitals in which objectives and tools for conducting the safety culture measurement were explained. Technical assistance was available during periods of data collection.

Belgian Comparative Database and Feedback Reports

The Belgian safety culture benchmark database includes 115,827 records drawn from 176 hospitals. Of those, 143 hospitals participated in the first benchmarking (2008) and 141 in the second (2011). Several hospitals participated once only in the comparative research. In addition, seven hospitals underwent a hospital fusion in the period between the two measurements, which reduced the number of participants for the second measurement. The survey response rates were similar to those for the AHRQ survey with 53.7 percent for the baseline and 50.4 percent for follow-up benchmarking. Reminders were an important driver in the survey in order to get a satisfactory response rate. Like other studies (Bodur and Filiz 2010; Campbell et al. 2010), lower response rates were observed for physicians (31.8 percent) in comparison with other professional groups (53.6 percent), and this might be an important indicator of lower involvement of medical staff in patient safety initiatives. Within Belgian hospitals, with the exception of a few university hospitals, physicians are not usually recruited as employees but operate as independent practitioners remunerated on a fee-for-service basis. Therefore, confidentiality and anonymity for respondents are recommended in this type of research in order to achieve the highest possible participation.

For both the baseline and follow-up benchmark initiatives, a comparative report using unique and anonymous hospital codes was generated by type of hospital (acute, psychiatric and long-term care) in order to provide a patient safety profile for internal learning (Vlayen et al. 2012). The comparative report describes the hospital and respondent characteristics, the response rates with an individual positioning of each hospital and positive dimensional scores calculated at hospital level. In addition, positive dimensional scores were provided for hospital units and professional groups according to the categories of the HSOPS. Finally, trending statistics were calculated for hospitals which participated twice in the benchmark research.

Safety culture perceptions were found to be more favorable in psychiatric and long-term care hospitals in comparison with acute hospitals. Generally, safety culture perceptions were low in Belgian hospitals, with more favorable perceptions for the dimensions of 'Teamwork within units', 'Supervisor/ manager expectations and actions promoting safety', 'Organisational learning and continuous improvement' and 'Communication openness'. In contrast, 'Handoffs and transitions', 'Staffing', 'Management support for patient safety', 'Non-punitive response to error' and 'Teamwork across units' showed a high potential for improvement. The positive dimensional scores of the Belgian acute hospitals (2011) are presented in Figure 14.1. An example of the comparative report with the positioning of hospitals is shown in Figure 14.2.

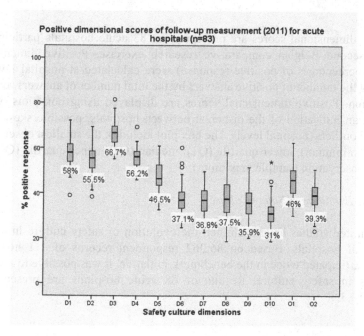

Figure 14.1   Positive dimensional scores of 83 acute hospitals (2011)

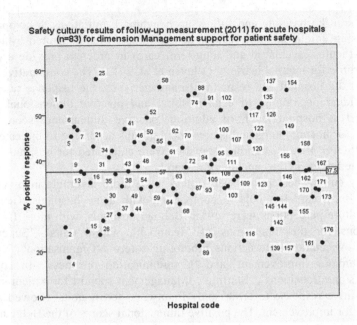

Figure 14.2   **Positive dimensional scores for dimension 'Management support for patient safety'**

Positive dimensional scores are presented for 83 acute hospitals participating in the second Belgian comparative research exercise. Positive dimensional scores (percentages of positive response) were calculated at hospital level by dividing the number of positive answers by the total number of answers for each dimension. Positive dimensional scores are displayed using box plots, which provide an indication of the dispersal between hospitals, possible skewing of data and outliers (hospital level). The box plot includes: the smallest observation (sample minimum), lower quartile (Q1), median (Q2), upper quartile (Q3) and largest observation (sample maximum).

Tracking Evolution of Safety Culture

Only limited studies have examined the evolution of safety culture in a large sample of hospitals. Based on 86,262 respondent records of 111 hospitals, which participated twice in the benchmark initiative, it was possible to measure changes in safety culture. Results of 68 acute hospitals are presented in Figure 14.3.

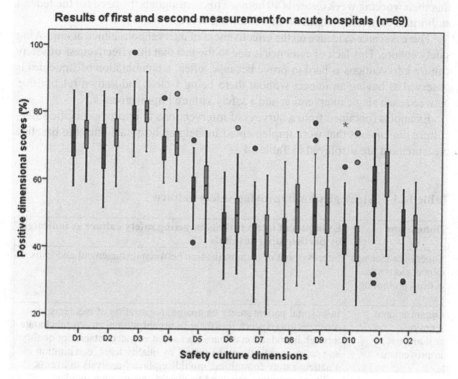

Figure 14.3 **Results of the baseline and follow-up safety culture measurement in 68 Belgian acute hospitals**

Results show a positive evolution on most safety culture dimensions, with a remarkably significant improvement for 'Management support for patient safety' (+8.5 percent for acute hospitals). The progress for this dimension demonstrates the growing involvement of hospital management in Belgian hospitals, which is an essential precondition in achieving safe care. However, despite the federal programme's focus on quality and safety, perceptions of 'Handoffs and transitions' and 'Frequency of events reported' were shown to have declined significantly (-2 percent for both dimensions within acute hospitals). The decline for these dimensions could be explained by the fact that the greater attention paid to these areas within the federal programme may have raised the awareness of hospital staff. This could explain the more critical evaluation of these dimensions. However, these areas warrant continuous attention. Also, the dimension of 'Staffing' seemed to be less susceptible to progress and was identified as a major problem within geriatrics, the operation room, internal and surgical units and, in particular, for the nursing professions. Analysis of demographic items confirms the problem of staffing, since more than a quarter of Belgian hospital staff indicate that their working week exceeds 40 hours. This area signals the need for the federal authorities to invest in higher (nurse) staffing levels.

There is scarce evidence on the effectiveness of interventions aimed at improving safety culture. This lack of evidence is due to the fact that the effectiveness of safety culture interventions is hard to prove because, often, a combination of strategies is observed as having an impact without there being a clear and known relationship between each single intervention and a safety culture improvement.

Examples (obtained from a survey) of interventions to improve specific safety culture dimensions that were implemented in Belgian hospitals after the baseline measurement are displayed in Table 14.1.

Table 14.1   Strategies for improving safety culture

Dimensions	Examples of interventions targeting safety culture as indicated by participating hospitals
Supervisor/manager expectations and actions promoting safety	Improvement of communication between management and units
Organisational learning – continuous improvement	In-hospital patient safety campaign; registration of incidents; sensitisation (posters, dashboard); organisational structural change through introduction of care teams that are accountable for quality and safety; medical record review by quality team; constitution of a patient safety committee; multidisciplinary analysis of events; audits of hospitals units and feedback; encouraging incident reporting
Teamwork within units	Designation of unit team leaders; triage on emergency care; optimalisation of hospital unit briefings; implementation of a safe surgery checklist
Communication openness	Communication plan on quality and safety issues; alignment of communication between hospital management and units; presence of hospital management during team meetings
Feedback and error communication	Feedback of incident reporting; communication of specific patient safety issues (e.g. hemovigilance); mandatory education of new staff on patient safety; discussion of feedback incident reports with units on regular basis in order to implement improvements; patient safety column in hospital magazine; patient safety dashboard via intranet; designation of incident administrator
Non-punitive response to error	Involvement of head nurses in feedback and discussion of events; patient safety committee is responsible for communication of patient safety issues to hospital management and hospital staff; education on incident reporting; stimulating a culture of openness and reporting; ending blame and shame culture; drafting a patient safety organogram to enlarge involvement of all hospital committees; sensitisation of head nurses in non-blaming job evaluations; assignment of external company as responsible for incident registration and data processing

Table 14.1   Strategies for improving safety culture (*concluded*)

Dimensions	Examples of interventions targeting safety culture as indicated by participating hospitals
Staffing	Support of mobile teams to reduce high workloads; international recruitment of nurses; enhancement of medical staff; clinical receptionists; coaching of new staff; implementation of two night shifts on geriatric, oncology and respiratory units; additional administrative support for nursing care
Management support for patient safety	Communication of safety culture data; elaboration of a hospital-wide safety plan with SMART objectives for each hospital unit; patient safety on agenda of board meetings; development of patient safety charter; establishment of a patient safety committee; reorganisation of quality and safety policy; head physician in lead of root cause analysis of incidents; discussion of patient safety indicators at board meetings; patient safety committee is accountable for incident reporting system; organisation of patient safety symposium; organisation-wide patient safety campaign
Teamwork across units	Mapping and improving transfer processes; examination for all hospital units of which information is needed; implementation and evaluation of electronic medical records; exchanging hospital staff across units if necessary
Handoffs and transitions	Mapping and improving transfer processes; implementation and evaluation of electronic medical records; implementation and evaluation of protocols for patient identification wrist bands; implementation of nursing transfer checklist
Overall perceptions of patient safety	Hospital-wide patient safety campaign; elaboration of hospital-wide procedure book; implementation of targeted actions based on incident reporting; safety walk rounds; elaboration of accreditation processes; patient safety alert weeks; assigning quality labels to hospital units
Frequency of events reported	Designation of responsible persons for analysing incidents; raising awareness on reporting specific types of adverse events; sensitisation campaigns for incident reporting on each unit

Safety Culture and Group Culture

So far, safety culture measurements in healthcare organisations have been generally limited to the diagnosis of problem areas and raising awareness of patient safety. The implementation of improvement strategies that are tailor-made for target groups is still frequently deficient. Research within the more complex healthcare organisation of the hospital demonstrated that safety culture varies across hospital units (Deilkas and Hofoss 2010; Huang et al. 2007; Kaafarani et al. 2009; Singer et al. 2009; Smits et al. 2009). This finding could imply that patient safety interventions should be aimed at the local the level of the hospital unit, rather than centralising safety programmes at the organisation-wide level. However, there is conflicting evidence as to the extent to which demographic characteristics of healthcare professionals influence safety culture perceptions (Gallego et al. 2012; Singer et al. 2009).

Clearly, there is a need to measure sources of variation in safety culture perceptions relating to individual and hospital characteristics within hospitals, in order to implement targeted interventions (Jackson et al. 2010). For instance, it is hypothesised that members with the same educational background share a common set of cultural features. In addition, hospitals comprise many different types of wards and units, with a high diversity in offered services, patient populations, organisational structures and protocols, which might explain variability in patient safety culture perceptions. So it can be assumed that safety culture is associated with specific professions and with the levels of complexity and intrinsic hazards associated with healthcare delivered in different work areas (Singer et al. 2009).

The extent to which differences in safety culture could be explained by profession, work area and work experience, was examined, based on Belgian safety culture data (Vlayen et al. 2013). A high variance in safety culture was observed between hospital units – for instance, respondents working in multiple units, the operating theatre and the emergency department have lower perceptions of safety culture compared with hospital staff working in rehabilitation, geriatrics and psychiatric units. Of course, high intrinsic hazard units usually have a high turnover of patients and are dealing with more complex tasks. This might result in a more frequent witnessing of unsafe patient care and lead to a more negative evaluation of safety culture within these units. In addition, it was found that staff working in multiple units, medical-technical services and pharmacies have lower perceptions concerning 'Handoffs and transitions', referring to items such as the loss of information when transferring patients from one unit to another or during shift changes.

Alongside variations in safety culture between hospital units, a considerable disparity in safety culture perceptions was observed between professional groups and within disciplines. For instance, results indicated an important gap in perception between clinical leaders and assistants. Head nurses were shown to have more positive perceptions of patient safety compared to nurses and nursing aids. Equally, physician heads of department were shown to have more favourable safety culture perceptions compared to physicians and physicians in training, and pharmacists and pharmacy assistants also have distinct safety culture perceptions. A possible explanation for this wide variance in safety culture might be that clinical heads tend to overestimate their units' safety performance. This is shown by more positive ratings of dimensions related to the unit level in which the role of the manager/leader is crucial (e.g. Supervisor/manager expectations and actions promoting safety, Organisational learning-continuous improvement, Teamwork within hospital units and Communication openness).

Finally, it was observed that work experience, including years working in the current hospital, hospital unit and profession, has only limited influence on safety culture perceptions. Inexperienced staff, working for less than one year in the hospital had, overall, more positive perceptions of patient safety compared to their more experienced colleagues. Inexperience or a lower awareness of patient safety could be reflected in these more positive ratings.

Safety Culture in Psychiatric Hospitals

Although patient safety requires a general approach, there are unique safety issues in psychiatric settings. Within psychiatric care, the focus of patient safety is often centred on a safe physical environment in which chances for patients to hurt themselves are reduced. For example, a safe physical environment enhances fall prevention and medication safety. But, on the other hand, a safe psychological environment for the patient as well as for the healthcare professional helps patients to form a therapeutic alliance (Kanerva et al. 2012) and can for example reduce aggressive and violent behavior. Furthermore, issues related to seclusion and restraint (Goetz and Taylor-Trujillo 2012), self-harm, substance-use related harm, suicide and absconding are unique safety concerns in mental health (Brickell and McLean 2011). The difficulty of evaluating patient safety within psychiatric care is that it cannot be considered on its own, given the thin line between the healthcare professionals' safety and patient safety.

The HSOPS was initially selected for use within Belgian acute hospital settings and was translated and validated using the same strategy applied by the AHRQ. Both Dutch (Flemish) and French translations were found to be valid and reliable, without the necessity to change the original structure of safety culture dimensions (Wenqi 2005). Within the federal programme the instrument was also applied within Belgian psychiatric and long-term care hospitals. However, it was found that demographic survey items related to work area and staff position were unsuitable for use within psychiatric hospitals (Vlayen et al. 2012). Therefore, by agreement in several meetings with hospital delegates, these categorical items were adjusted to suit the context of mental care. As patient safety in mental health is also context dependent in relation to the type of psychiatric setting (e.g. emergency department, neuropsychiatry, forensic setting) and the specific therapeutic approach, more refined categories at the level of the hospital unit would increase the relevance of the HSOPS for use within psychiatric hospitals. In relation to work area, new categories were created for (1) mobile teams, (2) admission/observation or crisis units, (3) day- or night-hospitalisation, (4) supporting services (pharmacy, medical-technical services, technical services, administration) and (5) specialised units (including addiction therapy, psychosis care, mood disorders, behaviour disorders, pediatric psychiatry, elder psychiatry, neurology and rehabilitation). No modifications were made to the content items of the survey instrument in order to maintain its original structure.

For both nationwide safety culture measurements, response rates were found to be higher within psychiatric hospitals (72 percent) in comparison with acute hospitals (47.4 percent). This might be an indication of the higher involvement of hospital staff within this usually smaller type of setting. Overall, safety culture perceptions were more favourable for respondents working in psychiatric hospitals, compared with acute hospitals, although the pattern of lower scoring dimensions was similar. As in acute hospitals, the aspect of 'Handoffs and transition' emerged as a major problem area. The positive dimensional score of 36.5 percent indicates that only one third of respondents working in psychiatric hospitals have positive perceptions on the aspect of patient information transmission. Low scores for this dimension comprise negative perceptions of items related to the transfer of patients from one unit to another (only 27.3 percent of positive answers) and the exchange of information across hospital units (only 30 percent of positive answers). On the other hand, 'Teamwork within units' scored relatively high within psychiatric hospitals (positive dimensional score of 70.7 percent) comprising items related to teamwork (76 percent of positive answers), support (79.2 percent of positive answers) and respect (73.3 percent of positive answers).

Based on the survey answers of 8,353 respondents from 46 psychiatric hospitals, the HSOPS was validated for use within Dutch and French speaking psychiatric hospitals. Reliability values of the underlying construct of the questionnaire, commonly measured by the level of Chronbach Alpha, are found to be acceptable at a level of 0.7. Variations in reliability scores might be attributed to linguistic differences when translating the original American questionnaire. Therefore, a level of Chronbach Alpha of 0.60 seems more plausible. In our sample of psychiatric hospitals, Cronbach's Alpha indicated, for most dimensions, an acceptable level of reliability. Similar to the results in the acute hospitals, the dimension of 'Frequency of event reported' showed the highest internal consistency, while 'Staffing' and 'Organisational learning' showed the lowest reliability scores (Table 14.2). Several international studies reported similar reliability estimates (Blegen et al. 2009; Bodur and Filiz 2010; Nieva and Sorra 2003; Sarac et al. 2011; Smits et al. 2008; Waterson et al. 2010).

For both translations, confirmatory factor analysis showed an acceptable fit with the original 12-dimensional model. Although exploratory factor analysis resulted in a 10-dimensional and a 9-dimensional structure for, respectively, the Dutch and French questionnaires, based on the acceptable validity and reliability scores, it was concluded that no modifications were required to the original 12-factor model in order to allow internal and external benchmarking for the psychiatric hospitals.

Table 14.2 **Reliability of the 12 safety culture dimensions for Belgian psychiatric hospitals compared with Belgian (Dutch and French speaking) and American acute hospitals**

Dimensions	Survey Items	Belgian psychiatric hospitals		Belgian acute hospitals		American acute hospitals
		Alpha PH (Dutch)	Alpha PH (French)	Alpha AH (Dutch)	Alpha AH (French)	Alpha AHRQ*
Supervisor/manager expectations and actions promoting safety	b1-b2-b3-b4	0.77	0.74	0.77	0.75	0.75
Organisational learning and continuous improvement	a6-a9-a13	**0.50**	0.58	0.59	0.59	0.76
Teamwork within units	a1-a3-a4-a11	0.65	0.84	0.66	0.82	0.83
Communication openness	c2-c4c-c6	0.66	0.71	0.65	0.72	0.72
Feedback and communication about error	c1-c3-c5	0.76	0.70	0.78	0.76	0.78
Non-punitive response to error	a8-a12-a16	0.70	0.68	0.68	0.64	0.79
Staffing	a2-a5-a7-a14	0.55	**0.50**	**0.57**	**0.52**	**0.63**
Management support for patient safety	f1-f8-f9	0.72	0.79	0.72	0.77	0.83
Team work across units	f2-f4-f6-f10	0.69	0.66	0.66	0.68	0.80
Handoffs and transitions	f3-f5-f7-f11	0.70	0.70	0.71	0.72	0.80
Overall perceptions of safety *(Outcome dimension)*	a10-a15-a17-a18	0.54	0.58	0.58	0.63	0.74
Frequency of events reported *(Outcome dimension)*	d1-d2-d3	**0.85**	**0.84**	**0.85**	**0.87**	**0.84**

Notes: *Results of the American pilot study. PH=psychiatric hospitals; AH=acute hospitals; AHRQ=Agency for Healthcare Research and Quality.

Future Approach

Second 5-Year Plan on Quality and Patient Safety (2013–17)

Results from the first stage (2007–12) of the national quality and patient safety programme confirm that a long-term approach is essential to guarantee and sustain quality improvement in Belgian hospitals. Therefore, a second multi-annual plan for quality and patient safety is being developed for the next five year stage (2013–17). This new programme will focus on specific domains, such as high risk medication, safe surgery, identity-vigilance and integrated care. More generic aspects, such as patient safety management, (clinical) leadership, communication between caregivers and patient and family empowerment will also be addressed within the programme (www.patient-safety.be). For the development of the programme, the federal government relies on the expertise and collaboration of different scientific institutions. Since, currently, many hospitals are developing a hospital-wide accreditation programme, the second federal programme will be aligned with accreditation schemes and will define strategic objectives that will help hospitals to achieve common goals. As with the first plan, participating hospitals will be supported by training and education. Annual reports, surveys and feedback from networks will be used as opportunities to improve the programme and initiatives will be integrated at national and regional levels with the ultimate goal of making quality and patient safety the core business of Belgian hospitals.

Future Research

Future research is needed to further the understanding of safety culture, particularly with an emphasis on theory-driven longitudinal research designs. Recent developments in Belgian hospitals provide a range of new improvement and research opportunities. Also, our safety culture dataset provides an enhanced opportunity for future research in the field of patient safety.

In healthcare limited attention has been given to testing the psychological mechanisms that could mediate the relationship between safety culture and individual safety related behaviour. Future research should examine how person related factors (safety knowledge, safety motivation, job attitudes, personality characteristics, etc.) and situation related factors (management commitment, leadership style) interact in influencing safety behaviour (both compliance and participation) (Clarke 2010). It can be assumed that individual healthcare professionals are motivated to comply with safe working practices and to participate in safety activities if they perceive that there is a positive safety culture overall within the hospital. Healthcare professionals can be trained and supported through positive safety culture to maximise safety motivation and safety knowledge, which in turn leads to safe behaviours (e.g. compliance with a safe surgery checklist, hand washing, reporting incidents, etc.).

Future research should focus on enriching the evidence of effectiveness of strategies aimed at improving patient safety culture (Morello et al. 2012). Improvement strategies should be more selective and flexible and should be adapted to the specific context of the hospital or hospital unit. Since hospitals are organisations with an inherent hierarchical structure and also seem to be built upon subcultures, strategies should be aimed at target groups.

Conclusion

Belgian hospitals proved to be very interested in comparing safety culture scores with other hospitals. Inspired by the American survey approach, Belgian safety culture research proves that large, comparative patient safety databases allow for identification of patterns and trends. Within Belgian hospitals, greater attention should be paid to the transmission of patient information and reporting of (near) incidents. Also, staffing proved to be an area that is less susceptible to improvement. This should be a signal for the federal authorities to invest in higher staffing levels. Positive evolution on the dimension of 'Management support for patient safety' demonstrates the increasing focus of hospital management on patient safety and this is considered as an important precondition for improving safety culture in Belgian hospitals. Our findings on variations in safety culture perceptions between types of hospitals, hospital units and professional groups highlight the need for targeted strategies in order to improve patient safety.

References

Blegen, M.A., Gearhart, S., O'Brien, R., Sehgal, N.L. and Alldredge, B.K. 2009. AHRQ's hospital survey on patient safety culture: Psychometric analyses. *Journal of Patient Safety*, 5(3), 139–44.

Bodur, S. and Filiz, E. 2010. Validity and reliability of Turkish version of 'Hospital Survey on Patient Safety Culture' and perception of patient safety in public hospitals in Turkey. *BMC Health Services Research*, 10, 28.

Brickell, T.A. and McLean, C. 2011. Emerging issues and challenges for improving patient safety in mental health: A qualitative analysis of expert perspectives. *Journal of Patient Safety*, 7(1), 39–44.

Campbell, E.G., Singer, S., Kitch, B.T., Iezzoni, L.I. and Meyer, G.S. 2010. Patient safety climate in hospitals: Act locally on variation across units. *Joint Commission Journal on Quality and Patient Safety*, 36(7), 319–26.

Clarke, S. 2010. An integrative model of safety climate: Linking psychological climate and work attitudes to individual safety outcomes using meta-analysis. *Journal of Occupational and Organizational Psychology*, 83(3), 553–78.

Colla, J.B., Bracken, A.C., Kinney, L.M. and Weeks, W.B. 2005. Measuring patient safety climate: A review of surveys. *Quality and Safety in Health Care*, 14(5), 364–6.

Deilkas, E. and Hofoss, D. 2010. Patient safety culture lives in departments and wards: Multilevel partitioning of variance in patient safety culture. *BMC Health Services Research*, 10, 85.

Flin, R., Burns, C., Mearns, K., Yule, S. and Robertson, E.M. 2006. Measuring safety climate in health care. *Quality and Safety in Health Care*, 15(2),109–15.

Gallego, B., Westbrook, M.T., Dunn, A.G. and Braithwaite, J. 2012. Investigating patient safety culture across a health system: Multilevel modelling of differences associated with service types and staff demographics. *International Journal of Quality in Health Care*, 24(4), 311–20.

Gerkens, S., Farfan Portet, M., Desomer, A., Stordeur, S., De Waroux, M., Van de Voorde, C., Van De Sande, S. and Léonard, C. 2010. Het Belgische Gezondheidssysteem in 2010. *Health Services Research (HSR)*. Brussels: Federaal Kenniscentrum voor de Gezondheidszorg (KCE). KCE Reports 138A.D/2010/10.273/59.

Goetz, S.B. and Taylor-Trujillo, A. 2012. A change in culture: Violence prevention in an acute behavioral health setting. *Journal of American Psychiatric Nurses Association*, 18(2), 96–103.

Hellings, J., Schrooten, W., Klazinga, N. and Vleugels, A. 2007. Challenging patient safety culture: survey results. *International Journal of Quality Assurance*, 20(7), 620–32.

Hellings, J., Schrooten, W., Klazinga, N.S. and Vleugels, A. 2010. Improving patient safety culture. *International Journal of Quality Assurance*, 23(5), 489–506.

Huang, D.T., Clermont, G., Sexton, J.B., Karlo, C.A., Miller, R.G., Weissfeld, L.A., Rowan, K.M. and Angus, D.C. 2007. Perceptions of safety culture vary across the intensive care units of a single institution. *Critical Care Medicine*, 35(1), 165–76.

Jackson, J., Sarac, C. and Flin, R. 2010. Hospital safety climate surveys: Measurement issues. *Current Options in Critical Care*, 16(6), 632–8.

Kaafarani, H.M., Itani, K.M., Rosen, A.K., Zhao, S., Hartmann, C.W. and Gaba, D.M. 2009. How does patient safety culture in the operating room and post-anesthesia care unit compare to the rest of the hospital? *American Journal of Surgery*, 198(1), 70–75.

Kanerva, A., Lammintakanen, J. and Kivinen, T. 2012. Patient safety in psychiatric inpatient care: A literature review. *Journal of Psychiatric Mental Nursing*, 541–8.

Morello, R.T., Lowthian, J.A., Barker, A.L., McGinnes, R., Dunt, D. and Brand, C. 2012. Strategies for improving patient safety culture in hospitals: A systematic review. *BMJ Quality and Safety*, 22(1), 11–18. doi: 10.1136/bmjqs-2011-000582. Epub 2012 July 31.

Nieva, V.F. and Sorra, J. 2003. Safety culture assessment: A tool for improving patient safety in healthcare organizations. *Quality and Safety in Health Care*, 12(Suppl 2), ii17–23.

Sarac, C., Flin, R., Mearns, K. and Jackson, J. 2011. Hospital survey on patient safety culture: Psychometric analysis on a Scottish sample. *BMJ Quality and Safety*, 20(10), 842–8.

Singer, S.J., Gaba, D.M., Falwell, A., Lin, S., Hayes, J. and Baker, L. 2009. Patient safety climate in 92 US hospitals: Differences by work area and discipline. *Medical Care*, 47(1), 23–31.

Smits, M., Christiaans-Dingelhoff, I., Wagner, C., Wal, G. and Groenewegen, P.P. 2008. The psychometric properties of the 'Hospital Survey on Patient Safety Culture' in Dutch hospitals. *BMC Health Services Research*, 8, 230.

Smits, M., Wagner, C., Spreeuwenberg, P., van der Wal, G. and Groenewegen, P.P. 2009. Measuring patient safety culture: An assessment of the clustering of responses at unit level and hospital level. *Quality and Safety in Health Care*, 18(4), 292–6.

Vlayen, A., Hellings, J., Claes, N., Peleman, H. and Schrooten, W. 2012. A nationwide Hospital Survey on Patient Safety Culture in Belgian hospitals: Setting priorities at the launch of a 5-year patient safety plan. *BMJ Quality and Safety*, 21(9), 760–67.

Vlayen, A., Schrooten, W., Hellings, J. and Claes, N. 2012. Benchmark report on Belgian Hospital Survey on Patient Safety Culture. Available at: www.patient-safety.be (last accessed on 22 May 2014).

Vlayen, A., Schrooten, W., Wami, W., Aerts, M., Garcia Barrado, L., Claes, N. and Hellings, J. 2013. Variability of patient safety culture in Belgian acute hospitals. *Journal of Patient Safety*. 27 September 2013 [Epub ahead of print].

Vlayen, A., Hellings, J., Schrooten, W., Garcia Barrado, L., Haelterman, M. and Peleman, H. 2014. Evolution of patient safety culture in Belgian hospitals after implementing a national patient safety plan. *BMJ Quality and Safety*, 23, 346–7.

Waterson, P., Griffiths, P., Stride, C., Murphy, J. and Hignett, S. 2010. Psychometric properties of the Hospital Survey on Patient Safety Culture: Findings from the UK. *Quality and Safety in Health Care*, 19(5), e2.

Wenqi, L. 2005. Validation of a translated version of the hospital survey on patient safety culture. Available at: www.patient-safety.be (last accessed on 25 April 2014).

Chapter 15

Patient Safety Culture in Practice – Experiences and Lessons Learnt by University Hospital Zurich

Francesca Giuliani, Mirjam Meier, Maria Helena Kiss and
Amanda Van Vegten-Schmalzl

Laying the Groundwork for a Patient Safety Culture

Some years ago, in an interview for *The New York Times*, Robert Wachter stated the following:

> Process changes, like a new computer system or the use of a checklist, may help a bit, but if they are not embedded in a system in which the providers are engaged in safety efforts, educated about how to identify safety hazards and fix them, and have a culture of strong communication and teamwork, progress may be painfully slow. (Denise Grady 2010. Study finds no progress in safety at hospitals. *The New York Times*, 24 November 2010, 1)

How, then, can a system such as that outlined by Wachter be established in a hospital? How can a structure and culture be developed in which employees are mindful and in which lessons are learned from incidents on a continuous and systematic basis? Is the design of learning processes the major lever that can bring about a positive patient safety culture? And how can continuous monitoring of patient safety culture contribute to this development?

Quality management and patient safety are important considerations in the daily business of a modern hospital. Over the past few years University Hospital Zurich has implemented a number of tools and procedures as part of a strategic approach to this aspect of hospital management.

2000–2004: Definition and Documentation of Quality and Safety

Towards the end of the 1990s University Hospital Zurich joined forces with many other public, and some private, hospitals in Switzerland to examine the topic of medical quality within the framework of a regulatory project being run by the Canton of Zurich. The project was sponsored by the Zurich Health

Department, health insurers and hospitals. Emphasis was placed on comparisons of the outcome results and in particular on patient-focused comparisons of hospital performance. Measurements performed by hospital staff in accordance with a unitary system drove a continuous improvement in quality in the hospitals and promoted transparency in the services provided. This occurred firstly within the hospitals. However, after a few years of experience, the results were also released to the public.

Regulatory considerations and early discussions on the topic of quality in professional circles prompted many questions, such as 'How can quality be defined, documented and measured both for individual specialties and across an entire hospital?' and 'What type of quality should we be striving for?' These questions prompted many individual activities. At University Hospital Zurich a new department was established in order to coordinate measurements and topics relating to medical quality. Measurements and improvements became routine, until ultimately another question arose, namely 'What level of quality do we have and how can we improve it?' This led to continuous measurement of indicators of treatment quality and service provision and to continuous assessment of patients' accounts of their experiences.

The Focus Shifts from Environmental and Occupational Safety to Patient Safety

At that time hospital departments with responsibilities for quality and safety directed almost all of their attention to environmental and occupational safety. Then, in the late 1990s, the publication of *To Err is Human* startled both the public and professional circles throughout the world and shifted the focus towards patient safety. A small number of individually designed early warning systems for risks were introduced but, for a long time, Switzerland lacked any national patient safety programme. As a result, individual hospitals started by developing activities of their own. In the following years individual activities were mandated, e.g. the Canton of Zurich introduced a reporting system for near misses. In 2012 a number of national programmes were introduced and the Swiss Foundation for Patient Safety was made responsible for their implementation.

2004: A Sentinel Event puts Patient Safety in the Spotlight

In 2004 the occurrence of a sentinel event – an incompatible blood group heart transplant – at University Hospital Zurich abruptly made patient safety a central concern of the hospital. This sentinel event created a major public stir, not least because the patient, who died in the postoperative period, was followed live by a television reporting team. The way in which this sentinel event was dealt with thus became a matter of public interest. Attention was no longer focused just on individual activities relating to quality and safety performed by motivated healthcare workers. Instead, the hospital started to make more concrete efforts to develop an effective safety culture. Firstly the hospital management wanted to know precisely where the University Hospital Zurich stood in terms of its safety culture and how existing activities aimed at improving patient safety could be most effectively supported and coordinated by the new department.

2006: Hospital Survey of Patient Safety Culture

Safety culture, which refers to underlying values, beliefs and behaviour in relation to safety and risk, is scarcely amenable to direct measurement (St. Pierre et al. 2011). The definition of safety culture adopted in the project is borrowed from the nuclear industry and is as follows: 'The safety culture of an organization is the product of individual and group values, attitudes, perceptions, competencies, and patterns of behaviour that determine the commitment to, and the style and proficiency of, an organization's health and safety management' (Health and Safety Commission 1993, cited in Nieva and Sorra 2003: 18). In 2006, therefore, in order to obtain a snapshot of the perceptions of employees' prevailing attitudes, beliefs and perceptions in relation to safety and risk, University Hospital Zurich conducted its first comprehensive, standardised staff survey of patient safety climate. This was the first comprehensive staff survey using a patient safety climate inventory to be conducted in the German-speaking world. The survey was more than just a purely statistical analysis of 'safety relevant issues/culture values' or a means of benchmarking the various specialist units of University Hospital Zurich. As well as establishing the baseline situation, the hospital management aimed to raise its employees' awareness of the importance of safety culture and its various aspects (e.g. team spirit and cooperation) and identify specific areas of action that could improve patient safety. In 2010 and 2011 the staff survey on patient safety climate was repeated in two areas as part of a hospital restructuring process and as a monitoring instrument. These surveys are discussed in the following sections. Further a specific example is used below to illustrate how patients too can be surveyed about patient safety climate and how patients' experiences can furnish important information about patient safety and be used as learning tools. The comprehensive staff survey was conducted in parallel with hospital-wide introduction of an electronic reporting and learning system with accompanying scientific research (van Vegten 2008).

2009: How Effective are Our Activities?

We first asked ourselves 'How should our quality and safety be: hospital-wide and specialty-specific?' and 'How is our quality at present and how can we improve it?' Based on experience obtained over the past few years and findings reported in the literature, we now know that we must also ask another question, namely 'How effective and lasting in impact are our activities?'

Roadmap for a Patient Safety Culture – Monitoring by Means of Surveys

Comprehensive Initial Survey

In 2006 University Hospital Zurich, in collaboration with the Center for Organisational and Occupational Sciences of the Swiss Federal Institute of Technology Zurich, conducted the first German-language survey on safety culture (van Vegten 2008; van Vegten, Pfeiffer et al. 2011). The objectives were firstly to determine the safety culture of the hospital and secondly to raise awareness of the topic. To this end a German adaptation of the Hospital Survey on Patient Safety Culture (Singer et al. 2007) with its various dimensions was used. A validated final questionnaire version is available (Pfeiffer and Manser 2010).

Using a pilot-tested question-and-answer approach, we obtained a response rate of 47 percent (2,897 evaluable questionnaires). Respondents included both clinically active employees and employees with no direct contact with patients. Of the employees who responded, 65 percent were directly involved in patient care, 5 percent worked in research, and 31 percent had no direct contact with patients, while 89 percent had been employed at University Hospital Zurich for at least a year and 74 percent stated that their job entailed no personal responsibility or managerial role. The assessments of the various aspects of safety culture, e.g. teamwork and handling of errors, were positive overall and were similar across all occupational groups. The overall results of the questionnaire suggested the following potential areas for improvement: support from leaders (this was the most important determinant of a positive overall assessment), enhancement of teamwork between departments, improvement of process safety at interfaces and the introduction of an early warning system (incident reporting system).

Responses to the questionnaire created a basis for discussion and for a raising of awareness and a campaign on patient safety issues across all occupational groups and seniority levels within the organisation. The fact that the problematic areas and potential areas for action indicated above had, to a large extent, already been identified was apparent from the fact that the areas identified as problematic in the course of the study – e.g. teamwork and cooperation and the need for incident reporting – coincided with areas of action already being undertaken at University Hospital Zurich.

Six months after the patient safety climate survey was conducted, a written survey on action, taken in response to the results, was conducted among quality officers and senior management in the hospital's various specialist units and institutes. This revealed an uneven picture. In 22 specialist units and institutes survey results were found to corroborate the need to target known areas of action and ongoing projects and were used as a basis for discussion and reflection.

Other specialist units and institutes had put off any discussion of the results for organisational reasons or else showed no willingness to make any changes. In 23 out of 37 specialist units the patient safety climate survey led to a decision to set up an incident reporting system. In addition, targeted projects led to the establishment of safety standards for electronic prescriptions and standards for safe surgery.

Follow-up Survey after Introduction of an Intervention

In a pilot project conducted in the hospital's Plastic Surgery and Hand Surgery Unit in 2009, a 'safe surgery' standard, based on WHO Guidelines for Safe Surgery, was introduced in order to optimise patient safety by avoiding errors related to treating the wrong side of the body or the wrong patient or performing the wrong procedure. To this end employees from all relevant areas were actively involved and trained. Pre-, peri- and post-operative checklists were drawn up for use at joint briefings to check patient and treatment data and work processes. The project included group training sessions on communication and team culture ('human factor training'). The instruments introduced into everyday clinical practice along with the safe surgery standard demonstrably improved interdisciplinary and interprofessional communication and thereby also patient safety. This was shown by repetition of the survey conducted in 2006 using a patient safety climate inventory and by a trigger tool analysis.

The results obtained in the follow-up survey on safety culture showed that the 'safe surgery' standard greatly improved quality and safety. Working procedures were simplified and made safer. Many employees stated that they were glad that the standard had been introduced. They also reported that communication between nursing staff and doctors had improved. The following statements illustrate the main conclusions of the survey:

> We are seen as a professional group with equal rights and are therefore appreciated.

> Less time is wasted, we don't have to make follow-up telephone calls to the doctor all the time.

> The team time-out takes time, but ultimately saves much more time.

> We find out in good time what material and what surgical sets we'll need for the operation.

The effectiveness of the standard was also demonstrated by means of a trigger tool analysis of 100 medical records before and after introduction of the standard. This showed that major harmful events, such as those leading to rehospitalisation and prolongation of hospital stay, were significantly reduced and lasting harm was completely avoided, during the period under consideration. The analysis took account of severity of illness. In this way it was found that adverse events were reduced in the 100 patients whose medical records were analysed, even though the CMI of these patients was higher than that of hospital patients as a whole. Against this backdrop – the project won the 2010 Golden Helix Award for innovative and evident quality improvement projects in Germany, Austria and Switzerland – the project was named 'Safe Surgery University Hospital Zurich' and implemented throughout the hospital. Implementation was accompanied by the use of training material, on-site meetings for the purpose of internal auditing and structured feedback centred on role clarification and communication. The experience gained in this process provided a perfect illustration of how a safety-oriented programme can become fully effective only when perceptions, processes, and decisions are questioned and the expectations, interpretations, intentions and relationship patterns of all those involved are continuously weighed against each other.

Follow-up Survey in an Accredited Organisational Unit

In 2006 the comprehensive patient safety climate survey and the subsequent communication of its results led to a hospital-wide strengthening of the critical incident reporting system (CIRS) (van Vegten 2008). As a result, all of the hospital's specialist units and institutes were progressively included in a unitary, organisation-wide electronic reporting system. This inclusion presupposed an explicit commitment to CIRS on the part of the head of the unit and compliance with certain conditions, namely the holding of an informational session for staff and detailed instruction of local CIRS officers from the various reporting units on the value and use of the CIRS. In 2011 the medical laboratories became one of the last organisational units of the hospital to be linked to the CIRS. With just on 400 employees of various occupational categories, the medical laboratories of University Hospital Zurich are accredited and perform a broad range of analyses and diagnostic tests. Each year about a million laboratory orders are received and processed in the various specialised laboratories.

Following the initial survey in 2006, the medical laboratories were used as a starting point for a cross-sectional survey of the existing patient safety climate in all clinically active occupational categories within this particular organisational unit. The hospital transfusion service 'error reporting and safety culture' instrument was used for this purpose (Sorra et al. 2008). The dimensions used for measuring patient safety climate in this instrument are similar to those of the patient safety climate inventory used in the 2006 survey. However, this instrument puts an additional focus on the views of staff as to how errors can occur in the laboratory. The overall response rate was 41 percent. The survey yielded some interesting findings that proved to be of value for monitoring patient safety culture in this organisational unit and for comparing results with the baseline values obtained in 2006. At the same time the survey identified possible ways of improving working practices in this area. One of the first findings of the survey was that, independently of occupational category, employees with a managerial role had a significantly more positive view of the patient safety climate in this organisational unit than did employees with no managerial role. This finding is in accord with those of the surveys conducted in 2006 and 2010. The more positive view of patient safety climate expressed by managerial staff could be seen simply as a reflection of the attitudes expected of senior staff by the organisation (Benn et al. 2012). Alternatively, it could be that, from their external perspective, managerial staff take a much more positive view of teamwork than do the actual members of a team. The results of the survey were presented at a number of feedback sessions, at which they mostly sparked lively discussion among team members.

When asked how errors can occur, the employees – once again, independently of occupational category – responded most commonly that they are under pressure to supply diagnostic test results rapidly and that, in their opinion, staffing levels are insufficient to enable them to get their work done within deadlines. They also cited interruptions to their work, e.g. telephone calls, equipment alarms and queries from colleagues, as reasons for errors. The results of the survey were also used for planning and implementing specific interventions. In this regard it was found that focusing on a particular occupational group, in this case trained laboratory staff, permitted effective and prompt implementation. In a first step an attempt was made on site to analyse the perception that staffing levels were inadequate. This showed that application of standardised rules for reducing accrued holiday entitlement, extra time worked, and overtime gave senior employees a better perspective and made it easier for them to plan ahead. It was concluded that frequent understaffing of teams due to accrued entitlements must be avoided. The slogan of this intervention was 'Planning creates resources'. Within one calendar year the management team had drawn up, communicated and introduced the new rules. A second intervention aims to tackle the frequent occurrence of interruptions to work and the way in which these lead to errors. It is known from the literature that interruptions cause employees to work more rapidly in order to get their workload back under control (Westbrook et al. 2010). This compensatory strategy can exhaust individual resources and increase individual stress. In a next step, an investigation of what types of interruption occur in medical laboratories and how these interruptions can influence particularly sensitive working activities was initiated.

The surveys of patient safety climate served University Hospital Zurich as a kind of roadmap to the future. Over the period between the comprehensive initial survey, conducted in 2006, and the most recent assessment of patient safety climate in an individual organisational unit, these surveys have become a more or less standardised component of the hospital's quality management system. It has also been found that assessments of patient safety climate can be undertaken both at the level of individual teams in connection with an intervention (e.g. safe surgery) and also as a starting point for projects intended to bring about changes (e.g. introduction of a CIRS in an organisational unit).

So far, however, neither the measurements themselves nor the results of the measurements have been fully incorporated into an integrated approach to the improvement of patient safety. Possible reasons for this include structural changes, changes in managerial staff, and different degrees of willingness to address actively the results of the surveys within individual organisational units.

When deciding when, in which organisational unit or units and in what setting (all occupational categories or within a particular team) a patient safety climate survey was to be conducted, we had to take account of the extent to which the managerial staff concerned were willing to embrace change. Our intention is that, in the future, measurement of patient safety climate will become an integral component of regular measurement programmes conducted in all clinical areas. We see provision of systematic support to our specialist units, resource planning and monitoring of actions taken as keys to the growth of safety culture.

Structuring of Organisational Learning Processes – The Major Lever for Patient Safety Culture?

Structuring of successful learning processes is becoming a matter of increasing importance for hospitals as the setting, e.g. a national healthcare system, in which they operate becomes increasingly complex and fluid. In this context, qualified staff are a crucially important resource for prevention of errors and for safety awareness at work. Faced with a confusing plethora of terms and concepts used in connection with the topic of learning within organisations, clinicians ultimately have to ask themselves how learning processes need to be structured and implemented in everyday clinical practice so as to foster a positive patient safety culture. The following sections give a number of examples of how University Hospital Zurich is attempting to structure learning processes.

Patient Feedback – What Effect Does it Have on Patient Safety Culture?

University Hospital Zurich has been conducting patient surveys for more than 10 years now. In addition, its patient advice service deals directly with complaints and praise from patients and, where appropriate, makes contact with the patients' concerned. The data acquired via the surveys and also via this direct patient feedback are made available to the various specialist units. This feedback makes clinical staff more aware of patients' problems and expectations and also of what procedures and structures are proving to be effective. It can also be used to plug and eliminate deficits in patient care as perceived by patients.

Mrs L., aged 78 years, needs to come to hospital for a routine operation. A letter sent to her by the hospital to prepare her for the hospital stay states that she will be able to go home three days after the operation, on a Thursday. The operation is performed as planned. It goes well and Mrs L. recovers rapidly – so rapidly, in fact, that she is ready for discharge on the Wednesday, a day earlier than anticipated. The nursing staff bring Mrs L. the good news and give her the medicines she needs for the next few days. They also give her precise instructions on how to use the medicines. Mrs L., however, is completely taken aback. She has asked someone to pick her up on Thursday, not on Wednesday. How will she get home? How did they say she was to take the medicines? And is she meant to apply the ointment all by herself? Mrs L. is anxious and has many questions but is afraid to say anything. For its part, the specialist unit concerned does not ask Mrs L. whether she will need any help after leaving hospital. After Mrs L.'s return home her daughter approaches University Hospital Zurich's patient advice service to complain that her mother's discharge from hospital was very poorly managed.

Such complaints, together with feedback given by patients in surveys, are important to the hospital in that they point out safety deficiencies that are not, or are not sufficiently, detectable by other means. Patients provide a valuable reference point for possible ways of improving treatment processes. They are involved as individuals in every step from the beginning to the end of treatment and they possess contextualised knowledge of the treatment steps that have been taken so far (Schwappach and Wernli 2010). They are therefore better informed than anyone else about what is happening to them. They are generally also very good observers and therefore possess expert knowledge about their own treatment processes.

Patients often report events that are not documented in their medical records. There are many examples of this, e.g. references to excessively long waiting times or inadequate communication, either between healthcare professionals and the patient or between the various healthcare professionals involved in the patient's care. For example, many patients get annoyed when they find that the various healthcare professionals who are supposed to be caring for them are not equally well informed about them. Or they feel that they've lost track and no longer know who is responsible for their care. At first glance these problems may appear to be no more than deficiencies in the quality of service. However, on closer examination, they are all seen to be relevant also to patient safety. This is because they often indicate important sources of error that can give rise to a safety hazard in hospitals. Inadequate communication when a patient is handed over from one doctor to another does more than just annoy the patient, it may also be relevant to safety if important information about the treatment given to date or about known risks is lost in the transfer process. Patients who report that they no longer know who is supposed to be looking after them ask fewer questions. Badly informed patients pose safety risks, e.g. if after going home they are unsure what activities they may perform and what activities they should still be avoiding, or if they are still unsure how to take their medicines. Moreover, patients who feel well cared for also feel more secure.

Individual reports such as that of Mrs L. can also be of value to the attending healthcare professionals. Where, as in the case outlined above, an elderly patient states explicitly that questions that are important to her have been left unanswered and that she felt inadequately cared for on discharge, this can serve as a stimulus to the healthcare professionals concerned to pay closer attention to these points in other patients. This is because individual reports of this kind give a face to abstract descriptions of problems and raise awareness of the topic of patient safety among healthcare professionals.

The question arises of whether the views of patients should be used routinely as a source of patient safety topics. For example, as well as being asked neutral questions about their experiences during their hospital stay, patients could perhaps also be asked directly whether they noticed any errors or incidents that they consider to be relevant or prejudicial to patient safety (Wasson et al. 2007).

Learning from Errors and Critical Events –
The CIRS Reporting and Learning System

The reporting and learning system based on a local critical incident reporting system (CIRS) that is used at University Hospital Zurich is an interprofessional and interdisciplinary early warning system for the risk of patient safety-relevant events occurring in all clinical areas of the hospital. The CIRS is operated on established principles for a successful reporting system and by means of a quality assurance procedure (PDCA cycle). Its objective is to make it possible for all employees, independently of the specialist unit in which they work, to report errors, critical events and positive learning experiences related to patient safety anonymously at any time via an easy-to-operate and easily accessible system (van Vegten 2008). Detection and correction of weaknesses in procedures, processes and structures, together with the possibility of learning from the experiences of others, should help to make safe working possible at all times and in all workplaces. All reports are considered and processed at University Hospital Zurich within a prescribed period of time by local case-processing units, known as committees. These committees are made up of a number of people from different occupational groups within an organisational unit. Responsibilities and tasks are determined for all committees. At present 25 local committees process about 1,500 reports per year. Particular attention is paid to assessing serious individual cases and problem clusters and to providing feedback to staff about processed cases. Only if this feedback is prompt will staff remain motivated to submit reports and feel that their commitment (the reports they submit and their involvement in finding solutions to problems) is having a positive impact.

The following CIRS report was received: a two-hour-old neonate is transferred to the neonatal intensive care unit for urgent investigation. The treating doctor then searches in vain in the hospital's clinical information system for the little patient's medical record in order to be able to prescribe medicines electronically.

A medical record for the patient has yet to be created. In order not to waste any more time, the doctor issues handwritten prescriptions and the medicines are dispensed immediately by the nursing staff. Electronic documentation of the medication is supposed to follow later. In similar fashion, a radiographer performs some urgent radiographs. However, because the little patient has no case number the diagnosis is considerably delayed since, without a case number, the radiographs are difficult to locate within the radiology department's records. As it happens, this event has no serious consequences for the neonate, whose condition soon improves. Despite this, the treating team feels helpless in this situation and feels that from then on the management doesn't take it seriously. How can it be that an electronic system requirement such as the existence of a case number can cause a delay in an emergency medical situation?

How Can the Organisation Learn from this Incident?

The first step is to have the report processed by the responsible local CIRS committee. Before being entered in the system, reports that contain names are anonymised so that all reports are anonymous. The question of whether reporting systems of this kind should or should not be anonymous is a topic of debate at University Hospital Zurich. Our experience is that, in general, critical events occurring in treatment teams are not brought to people's attention in the first instance by a local CIRS report. Additionally, the treatment team concerned is generally already aware of what happened and, of course, who was involved in the event. It is therefore difficult to assume that such a CIRS can be anonymous. The fact is that in an electronic system all names and dates are systematically deleted. The important thing, however, is to make members of treatment teams aware that if lessons are to be learned from incidents a culture of confidentiality and safety is essential when CIRS reports are being processed. Our view is that the development of a positive safety culture will ensure that preservation of anonymity does not need to remain a major concern in the long term. This is because, if it is assumed that within a given system errors and successes can occur under the same conditions, the lessons that can be learnt from a report are learnt not by finding a guilty party but by analysing the situation that led the person concerned to act in the way that they did. Every effort should be made to work out why that behaviour made sense to the person in that situation since, if it made sense to that person, it is highly likely that another person in the same situation would act in the same way. For this reason the first insights into the causes of an incident generally result from processing of the case by the local CIRS committee. In the ideal case findings obtained in this way can be used to design and implement measures to improve the situation. In principle, this type of on-site learning by and among staff members should occur in association with the reporting and processing of every incident since, within this small circle of people, motivation not to repeat mistakes is generally high. A disadvantage of this type of learning is that knowledge acquired in this way may not be passed on in full to other relevant people, i.e. it may be of little benefit to the treatment team as a whole. Moreover, the core convictions and attitudes of the organisation are rarely affected. Nevertheless, continuous improvement in the patient safety competence of CIRS committee members certainly occurs and is now observable at University Hospital Zurich. As stated below, the crucial point is that adoption and implementation of measures and provision of adequate resources are the responsibility of the hospital management.

How can the learning process be expanded to include the treatment team and eventually also the organisation? This step often proves difficult in practice. Why is this so? The notion that organisational learning can generally be planned is based on an illusion. Learning processes are impeded not just by the incompleteness of the information available to involved persons in the various organisational units but also by differences between interest groups in terms of degree of motivation, power relationships and a tendency to inertia on the part of organisations. Some of the reasons for this are to be found in organisational routines (Argyris and Schön 1999). The pressure of the working day, lack of management structures and a fixation on 'defensive routines' form obstacles to organisational development in hospitals and need to be taken more carefully into account. Our experience would suggest that there are two important aspects of organisational structure that promote organisational learning and on this basis a horizontal and vertical organisational structure with defined areas of responsibility has been introduced at University Hospital Zurich as a means of ensuring a continuously functioning CIRS.

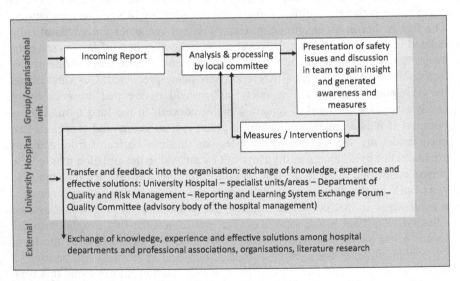

**Figure 15.1   Process model: dialogue-oriented reporting and
               learning system**

Source: van Vegten 2008: 203.

On the horizontal line the local CIRS committees are supported by the Quality Management and Patient Safety Department, which acts as a CIRS manager. It goes without saying, however, that responsibility for patient safety cannot be delegated to individuals with executive positions such as quality officers, whose principal role is to provide expert assistance in the form of process advice (Schein and Bruckmeier 2010). This includes coordination and moderation of case processing via interfaces, as mentioned in the case report at the beginning of this section. The committees are linked firstly by exchange of information via the electronic platform of the CIRS in the form of annotated retransmissions or queries. In addition, once or twice each year, information about critical events is exchanged at a forum. And several times a year, in order to disseminate knowledge more widely within the organisation, CIRS cases of overarching importance are prepared in the form of a 'topical CIRS case' and posted on the hospital intranet. This information format is especially suitable for discussion in team reports. In addition, error analyses based on CIRS cases are presented on site to the treatment teams in accordance with the London Protocol (Taylor-Adams and Vincent 1999). 'Culture workshops' are also held. These focus on group discussion of and compliance with the perceived safety culture of the workshop participants' team or organisational unit. In some cases these workshops reveal perceptible and palpable discrepancies between what is desirable and what is available. Still at an early stage is the use of CIRS cases to create scenarios that can be played out in the hospital's simulator. Structured debriefing by specially trained instructors makes possible an individual and team learning process aimed at promoting safe action in critical situations.

In the vertical structure a Central CIRS Committee has been created. Four times each year this committee processes CIRS cases that are of overarching importance to University Hospital Zurich and can be regarded as serious. The Central CIRS Committee reports its deliberations directly to the Medical Director and the Director of Nursing, who are authorised to take decisions regarding further measures of overriding importance to the hospital as a whole. The local CIRS committees thus have the option of referring difficult or unresolvable cases on to a higher authority. Here again it is important that the knowledge obtained be fed back promptly to the local committees. In this way, equipped with an understanding of how critical situations arise and are discovered and overcome, the organisation can strike new paths and think the previously unthinkable.

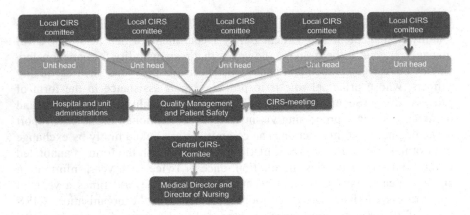

**Figure 15.2   Horizontal and vertical structures of University Hospital
              Zurich CIRS**

The existence of a CIRS does not in itself ensure a positive patient safety
culture. On the other hand, when actively used as an instrument in clinical risk
management, a local CIRS is very useful for promoting organisational learning
and a positive patient safety culture in a university hospital. The knowledge
accrued in hospitals is an important resource in this regard. Systematic availability
of relevant knowledge at the right time and the right place helps the treatment team
to improve patient safety. We also see an urgent need for more ergonomic design
of hospitals and of the health system itself.

Learning from Serious Adverse Events with Consequences

Adverse events with consequences are cases of unintended harm due to treatment
rather than to the patient's underlying disease. The management of such events is
a central element of clinical risk management and promotes the development of a
safety culture in the organisation concerned. When such events are dealt with in
a responsible manner trust in the treatment team can be rapidly restored. Affected
staff members find that they are supported and the institution is able to learn
from the mistakes that have been made. In the ideal case this leads to a deeper
understanding of why such events occur and to system changes that minimise the
likelihood of repeat occurrence and can prevent any increase in the number of
complaints and liability claims. Every avoided accident or harmful event matters.
In order to provide staff with support in the handling of serious adverse events with
consequences, University Hospital Zurich has stipulated procedures for reaction,
communication and provision of on-site assistance in such cases. On the basis of
standard recommendations a range of measures that can contribute to a successful
resolution of the case have been stipulated (Vincent 2010). Along with legally

stipulated investigative measures, the focus here is on communication a) within the team, b) with the affected patients and their carers and c) with the referring doctor. In addition, the event must be analysed promptly and the affected medical staff must be offered support.

It has been found that not only patients and their carers but also medical staff can be emotionally and functionally traumatised by a serious adverse event, especially a serious error (von Laue et al. 2012). Until now, support for affected staff members has been provided by the managerial staff concerned, by the staff member's team or by consultant experts called in for that purpose. These sources of support are now being supplemented by an internal 'coaching' team. This coaching team is made up of experienced and trustworthy clinicians whose own clinical experience has given them a personal appreciation of the consequences of errors. They are to be directly available both to managerial staff and to affected staff members themselves for confidential conversation and support. An open and constructive approach to errors in the organisation, that eschews finger-pointing in favour of an analysis of factors that predispose to errors and provision of support to clinical staff to help them deal with the emotional trauma that follows a harmful event, make it more likely that lessons will be learned from errors while at the same time reinforcing the hospital's patient safety culture (Vincent 2010).

Conclusion and Outlook

How did we bring about, and how can we maintain, an organisational culture that can improve patient safety?

Over the past few years we have developed a programme of continuous monitoring involving targeted interventions aimed at bringing about improvements. Our experience with the individual interventions, namely

- continuous measurement and communication of relevant indicators
- support and further development of the critical incident reporting system (CIRS)
- surveys of patients, referrers and staff
- instruction and training (leadership courses, simulation training, human factors training)
- fostering of individual non-technical skills and also team performance for new models of cooperation such as structured briefing-observation-debriefing

has informed the slogan 'Moving forward together with varied but targeted programmes towards an effective patient safety culture'. By this we mean that because of the organisation's culture, the truly important questions are not off limits and can be addressed. It is important for us to understand how our fellow employees work, how they interact in the team and how they deal with multi and interdisciplinary interfaces. Only an approach that takes account both of the organisation and of the individuals within it, that permits multiple points of view, that fosters a patient safety culture and that aspires to work-system designs that facilitate learning can institute and maintain an effective safety culture. At the same time we must continue to meet the challenge of finding a healthy balance between the meeting of economic demands, learning based on the facts and figures of near-accidents and the development of additional effective tools for improving patient safety culture. In this we aim to focus on new forms of cooperation and learning marked by commitment, competence and trust (van Vegten, Tanner et al. 2011).

Moving Forward Together Towards an Effective Patient Safety Culture

University Hospital Zurich's management acted on the slogan that guides our actions by creating an advisory body in the form of a Quality and Patient Safety Board, made up of representatives of the management of the hospital's various departments and areas. This board focuses on medical treatment (medical diagnosis, therapy and care) and service quality. Some of the most substantial contributions being made to putting the slogan into practice are those of the various medical area and unit heads, the many committed departmental CIRS and quality officers and all the committed specialists who monitor the hospital as a system and design preventive and remedial measures. Together we are working, and will continue to work, for an effective patient safety culture.

References

Argyris, C. and Schön, D.A. 1999. *Die Lernende Organisation. Grundlagen, Methoden, Praxis*. Stuttgart: Klett-Cotta.

Benn, J., Burnett, S., Parand, A., Pinto, A. and Vincent, C. 2012. Factors predicting change in hospital safety climate and capability in a multi-site patient safety collaborative: A longitudinal survey study. *BMJ Quality and Safety*, 21, 559–68.

Grady, D. 2010. Study finds no progress in safety at hospitals. *The New York Times*. 24 November 2010, 1.

Kohn, L., Corrigan, J. and Donaldson, M. 1999. *To Err is Human: Building a Safer Health System*. Committee on Quality of Health Care in America, Institute of Medicine (IOM), Washington, DC: National Academy Press.

Nieva, V.F. and Sorra, J. 2003. Safety culture assessment: A tool for improving patient safety in healthcare organizations. *Quality and Safety in Health Care*, 12(Suppl 2), ii17–23.

Pfeiffer, Y. and Manser, T. 2010. Development of the German version of the Hospital Survey on Patient Safety Culture: Dimensionality and psychometric properties. *Safety Science*, 48, 1452–62.

Schein, E.H. and Bruckmaier, I. 2010. *Prozessberatung für die Organisation der Zukunft*. Bergisch Gladbach: EHP.

Schwappach, D.L.B. and Wernli, M. 2010. Medication errors in chemotherapy: Incidence, types and involvement of patients in prevention: A review of the literature. *European Journal of Cancer Care*, 19, 285–92.

Singer, S., Meterko, M., Baker, L., Gaba, D., Falwell, A. and Rosen, A. 2007. Workforce perceptions of hospital safety culture: Development and validation of the patient safety climate in healthcare organizations survey. *Health Research and Educational Trust*, 42, 1999–2021.

Sorra, J., Nieva, V., Fastman, B.R., Kaplan, H., Schreiber, G. and King, M. 2008. Staff attitudes about event reporting and patient safety culture in hospital transfusion services, *Transfusion*, 48, 1934–42.

St. Pierre, M., Hofinger, G. and Buerschaper, C. 2011. *Crisis Management in Acute Care Settings. Human Factors, Team Psychology, and Patient Safety in a High Stakes Environment*. Berlin: Springer.

Taylor-Adams, S.E., Vincent, C. and Stanhope, N. 1999. Applying human factors methods to the investigation and analysis of clinical adverse events. *Safety Science*, 31, 143–59.

van Vegten, M.A. 2008. *Incident Reporting-Systeme als Möglichkeit zum Organisationalen Lernen (nicht nur) aus Fehlern und kritischen Ereignissen. Chancen, Barrieren und Gestaltungsansätze für Berichts- und Lernsysteme im Krankenhaus*. Dissertation an der ETH Zurich. Available at: http://kobra. bibliothek.uni-kassel.de/bitstream/urn:nbn:de:hebis:34-2009032426765/3/ DissertationVanVegten.pdf (last accessed on 17 January 2013).

van Vegten, M.A., Pfeiffer, Y., Giuliani, F. and Manser, T. 2011. Patient safety culture in hospitals: Experiences in planning, organising and conducting a survey among hospital staff. *Zeitschrift für Evidenz, Fortbildung und Qualität im Gesundheitswesen*, 105, 734–42.

van Vegten, M.A., Tanner, M., Amman, C., Giovanoli, P., Giuliani, Kiss, H. and Noethiger, C. 2011. Check/listenKULTur. *Schweizerische Ärztezeitung*, 92, 1547–50.

Vincent, C. 2010. *Patient Safety*. Chichester: BMJ Books, Wiley-Blackwell.

von Laue, N., Schwappach, D. and Hochreutener, M.A. 2012. "Second victim" – Umgang mit der Krise nach dem Fehler. *Therapeutische Umschau*, 69, 367–70.

Wasson, J.H., MacKenzie, T.A. and Hall, M. 2007. Patients use an internet technology to report when things go wrong. *Quality and Safety in Health Care*, 16, 213–15.

Westbrook, J., Coiera, E., Dunsmuir, W.T.M., Brown, B.M., Kelk, N., Paolini, P. and Tran, C. 2010. The impact of interruptions on clinical task completion. *Quality and Safety in Health Care*, 19, 284–9.

Chapter 16
Leadership Behaviour and Safety Culture in the UK NHS: A Manager's Perspective

Peter Bohan

Introduction

This chapter describes a study (Bohan and Laing 2012) which examined the impact of the leadership behaviour of an executive management team in a UK NHS Acute Foundation Trust on Quality and Safety (Q&S) outcomes. Q&S is a key element in maintaining services and developing a safety culture where change can remove or dislodge the appropriate management style. Leadership continues to play an important role in determining organisational outcomes with strategy, structure and process being key elements of team and organisational effectiveness (Yammarino et al. 2008). Approximately 10 percent of patients admitted to NHS hospitals are subject to a patient safety incident and half of these could have been prevented (NPSA 2004). The Bristol Royal Infirmary Inquiry (Department of Health 2001) concluded that around 25,000 preventable deaths occur in the NHS annually. More recently, the Mid Staffordshire inquiry identified that the Trust did not have its own system for monitoring patient outcomes (Francis 2010, 2013) and chose to ignore the quality indicators signposted by mortality rates. The failure of organisations to meet regulations covering 'effective, safe and appropriate care' was one of three outcomes identified as having the highest level of non-compliance in Care Quality Commission (CQC) inspections during 2010/11 (CQC 2011). The focus on patient safety has been further enhanced recently by the Stepping Hill investigation of three suspicious patient deaths, caused by a yet unknown assailant, and the circumstances in which a saline solution contaminated with insulin was unlawfully administered to a further 17 patients (*The Guardian* 2011).

Recent Changes within the NHS

The NHS Constitution (Department of Health 2010) and the Darzi review (Department of Health 2008) have resulted in raising the expectations of patients regarding their care and their freedom to choose how care is delivered, identifying

High Quality Care for all in the NHS next stage review, which provides a framework to improve Quality through three key areas of safety, efficiency and patient experience. As a result ambitions to strengthen the measurement of quality to drive improvement have never been greater (Boaden et al. 2010).

The UK Coalition Government's vision is to modernise the NHS, which is to be built around patients, led by health professionals and focused on the delivery of world-class healthcare outcomes (DOH 2011). With increased pressure from the CQC (CQC 2011) and DOH to provide higher quality for less cost, it is increasingly difficult for NHS providers to keep pace with demands from service users with complex health needs, which include lifestyle factors such as clinical obesity, diabetes, alcohol and drug use and an aging population. This is compounded by costs of drug treatments that are rising at a much higher rate than inflation. Savings of £20 billion, to be achieved by 2015 will require a 4 percent increase in productivity each year (BBC 2011). The types of leadership behaviour that can influence change and effectively manage Q&S and an emerging safety culture in these circumstances are critical.

Study Background and Aims

The UK Acute Trust described in this chapter has seen significant change over the past twelve months and recently appointed a new Chief Executive. The management team is all relatively new with the longest serving member having been in the organisation for seven years. The culture of the organisation is evolving and group behaviour within teams is already well established. The effectiveness of the organisation's leadership as it struggles to adapt to ever increasing demands forms the basis of this review.

The aim of the study was to provide a better understanding of the complex relationships between leadership behaviour and quality and safety. The process of interviewing executives provided an insight into organisational tensions between rising expectations in terms of targets, to be achieved with lower funding. The key element was to: (1) identify if there is an organisation wide preferred leadership style and behaviour that can influence quality and safety and (2) identify common perceptions of quality and safety and how the process of defining quality and safety can drive change and influence local internal policies, procedures, training programmes and lessons learnt.

Methods of Study and Data Analysis

A purposive sample (Cresswell 2007) using a volunteer strategy targeted the Trust Chief Executive, Chair of the Board and senior executives (nine individuals in total). Each interview lasted between 30 and 55 minutes. Individual, face-to-face, semi-structured interviews took place at the organisation's headquarters and were recorded with the permission of the participants. Data analysis involved using a grounded theory approach (Glaser and Strauss 1967). The validity of the data was verified for accuracy with the participants and cross-referenced by the project sponsor. For the purposes of maintaining anonymity, quotes from individual interviewees are labelled 'Executive A', Executive B' etc. throughout the findings section.

Findings

Leadership Behaviour

The executives talked about the issues that were important to them, which included being honest, inclusive and having a supportive management style that provided individuals with clear accountability and empowered them to take responsibility for their actions with clear direction. Executive E described the individuals within the organisation as 'people [who] liked to be led and it was important to adapt and adopt different styles of management'. This appears to be the general rule applied by most executives. Executive E also said; 'A different approach would be if I was dealing with a cardiac arrest situation, I'd be perfectly happy to be shouted at whereas if we were talking about how we were going to challenge a different department that would be different'.

Executive F described an approach that was very directive, task focused and autocratic at times but recognised, as an individual, that she should be more people focused. The current focus of the executives appears to be that leaders see themselves as transformational, providing integrity, setting clear goals, providing a good example and expecting the best from the teams they lead. The transformational leadership style described in the majority of cases can better rally the support of followers to make changes than can leadership with a transactional style, according to Herold et al. (2008). The problem with the current conflict in management styles between a target-driven transactional approach and the preferred transformational approach is ambiguous. The change process undertaken to achieve targets and to attempt to give clear roles may mask safety culture as it gets lost in the complexity of behaviours, roles and structures of the organisation. The perceptions of employees are particularly important within the executives' workplace, as attempts to promote enduring organisational change are unlikely to succeed if senior management involvement and commitment is not present (Van Dyke 2006).

Involving staff in decision-making and action planning was also seen as important. 'Inclusive', 'supportive', 'engaging' and 'educating' are all words frequently used to describe the leadership approach. Executive A believed 'you start off being open then you get the team behind you. But this may have to change to a more transactional leadership style once the ground rules and team are established (give them enough rope). Individual accountability was also very important. Executive G said 'having to do the right thing not the most popular thing and [being] hugely determined to succeed'.

Elements of autocratic and transactional leadership were required frequently in the achievement of targets, for example waiting times and quality matrices that provide evidence of outcomes for patients (e.g. numbers of hospital acquired infections). Achieving tasks and goals requires an autocratic style, which may be to the detriment of the staff involved. The previous benefits of transactional leadership, including tangible commodities such as pay increases, promotion and recognition are difficult to achieve and the only actual commodity that is generally available is recognition of good work completed (Wallis et al. 2011). This may be problematic with a flat pay structure and limited promotion opportunities, particularly in the current financial climate.

Types of Poor Leadership Behaviour

The most inappropriate style of management for executives was dictatorial bullying. A number of executives had witnessed or been a victim of bullying within the NHS in previous organisations. These types of behaviour were described by all executives. Executive G described a senior staff member ranting and gesticulating at staff, which meant the executive had to initiate support sessions for teams to enable them to feel confident enough to say that they would not accept being spoken to in a derogatory manner. Thirty years ago in another premises, Executive D had witnessed a charge nurse being physically aggressive and seen him assault patients and members of staff. He also described behaviours where there is a clique mentality and you could be excluded if your face doesn't fit – the power base does not always lie with the Ward Manager and a staff nurse or auxiliary may be the leaders, not the person nominally in charge. A number of executives mentioned stress as a significant issue and this could influence behaviour. Working to very autocratic managers was seen as absolutely horrendous by Executive A, who defined harassment and bullying as 'constantly undermining the competence of a competent person'.

Quality and Safety Perspectives

The interview process revealed that a focus on patient safety at the expense of all other Q&S matters could be detrimental as it could mean that issues such as equipment and facilities were overlooked. The concept of patient safety is often described in different terms, with a number of executives expressing the view that the quality of service meant making sure that 'doing what we do is right first time' (Executive E) or 'monitoring the system we put in place' (Executive A) or being 'fit for purpose and acceptable to the user' (Executive H). If expectations are not clear it may be problematic for the workforce thoroughly to understand the vision of the organisation in relation to Q&S, for example if there is a mismatch of clarity between executives' views. Clinical outcomes were also significant and the term 'safety' was often equated with legislative compliance. Patient safety and experience were dominant features in the interviews, with Executive G being focused on patients 'because we are patient driven', but also concluding that staff safety was paramount as one can't be delivered without the other.

The drivers for Q&S and current controls within the organisation include a range of safety matrices described by Executive G as a pro-active approach that seeks out risks, manages stress and learns from others, particularly from organisational failures, patient experience markers, patient trackers, feedback in a variety of dashboards, quality markers, Health & Safety Executive inspections, Environmental Health requirements and CQC outcomes. The Board Assurance Framework or risk register was a common theme concerning the escalation of risks and there was also recognition that data collection was just part of the process. Executive D believed clinical audits were not always robust, raising questions as to their value.. If planned systems are not interlinked and audited against each other, good information may be lost. The external drivers for Q&S included losing Foundation Trust status and the financial penalties from the Commissioners who direct where efforts have to be focused in order to influence policy. All saw external drivers as a major influence on internal policy development and control.

Executive C described the dangers of information overload: 'it's alright if we have 300 dashboards but it is possible to be bombarded by too much and [we] need to prioritise these sorts of approaches'. The data collection system described may not be all encompassing and may not provide all the detail required when triangulated. Finances and doing more for less are constant themes and there is competition to keep Q&S on the agenda when financial pressures drive changes. Raw data may only reveal details of what has been identified as needing to be measured. Attitude and behaviour require different measuring techniques and may not have been evaluated.

Executive G recognised that 'if we had a large increase in H&S incidents [reported to] the Commissioner [they] would be very aggressive about us intervening and stopping this problem. [It is clear that, to improve the culture we will require] more reporting and getting people to raise concerns that have been there a long time; that [is what is required for the organization to move forward]'. We need to move away from a culture where we give most attention to the loudest noise. To embed learning Executive F believed that 'once an incident becomes policy it becomes embedded' but it may be difficult to measure how many people read or follow policy, given the volume of clinical policies in place and an audits system that was described by one Executive as being poor in tracking policy implementation. There was also discussion about how issues get stuck with middle management rather than filtering down to the bottom and how this can result in something catastrophic occurring.

The Ambiguous Nature of Quality and Safety

The Executives had a wide range of views on the subject of quality and safety being perceived as problematic, with some expressing the view that people were beginning to understand the process: 'it's problematic and in the public sector it doesn't come without a price tag, it's difficult to strive to the text book, but everyone knows you can't, you have to find ways to do it with time and imagination'. One of the fundamental conflicts to be resolved is reconciling quality and safety issues with the target driven approach required by Commissioners in terms of getting results and keeping all parties happy. This may become especially problematic if individual Board members pursue their own agendas.

Organisational Culture

One Executive described culture and behaviour as taking many years to change and, if the focus is just on finance, quality and safety improvement issues often move out of focus: 'culturally people only hear the message about finance and targets and they don't hear the message about Q&S and, as you know, CQC are currently working very closely with HSE (Health and Safety Executive) and Monitor and I think even from the inspections we have had in the last week in the Trust where it's kept up there the regulation aspects are being seen as important'. Schein (1992) described culture as the process of complex group learning that is only partly influenced by leadership behaviour. However, if the leader shows poor behaviour this can highlight actions that result in a poor safety culture. Executive H supported this notion 'I think you can easily erode a Q&S Culture very quickly by what you do, by one act of how you lead and your behaviour and its quicker to do that in a minute or two minutes and it takes years and years to build a culture but it can be eroded very very quickly by a few bad examples.'

Outcomes from the Research

A recommendation from the research was to undertake a cultural survey using a set of focus groups in tandem with the MaPSaF tool (National Patient Safety Agency 2004) to capture the attitudes and beliefs of staff within Trust service's areas. The data could also be cross referenced with safety compliance and statistical evidence to identify areas that require improvement. Additional feedback was obtained from a set of in-depth interviews with a range of professionals and managers. This involved all heads of nursing, matrons, ward managers and team leaders (approximately 50 staff). A key finding was that responsibility for managing safety was often loaded onto one individual. Staff believed that safety alerts through e-mails could also be missed on wards, alongside a widespread perception that there are too many layers to get through to get the safety message across.

Ward managers needed to ensure all staff understand their part in the system and pass information up and down communication lines. The change management process was viewed as very complex with information not always going down to local level. It was emphasised that the change process needs regular review and feedback from staff as well as on-going consultation with staff regarding any planned changes.

Two major outcomes came about as a result of the cultural survey. Firstly, a strategy for Quality and Safety within the Trust has been developed and runs in parallel with the Workforce, People and Leadership strategies. The new strategy provides a strong narrative and a 'line of sight' between job roles and the organisation's vision (Macleod 2009). Part of this new safety and quality improvement strategy contains an emphasis on effective employee engagement. Secondly, an executive leadership programme has been developed, the aim of which is to align a vision for safety and quality across all parts of the Trust. Accountability for the leadership programme requires clarity in order to provide more effective feedback to the centre that results in real change in the organisation. Projects need to be linked to hospital management and the implementation process of the overall Quality Improvement Strategy (Duckers et al. 2011). The leadership programme requires effective evaluation of change in management style and behaviours and of how it makes a real difference to the whole organisation. The evaluation of the effectiveness of the leadership programme is currently being reviewed, with particular emphasis on what is required by senior executives in order to promote and influence effective leadership behaviours. Finally, leadership forms part of the development activities during executive away days.

Summary and Conclusions

The complexity of managing quality and safety in Healthcare, alongside the current scale of targets and cuts to services faced by the NHS, can have a profound effect on patient outcomes and staff behaviour. If the drivers for change become too narrow and task focused this will negate innovation in the NHS designed to deliver high quality care for all every time. The people interviewed as part of this study had a view on what is appropriate behaviour. However, if they are overly focused on targets and current financial pressures from DoH and Commissioners, this can increase risks to the culture of the organisation and increase risks to patients.

Leaders are required continually to review organisational norms and how people behave, including self evaluating their own behaviour to ensure it has a positive impact on staff behaviour. Effective leadership is not just about a style or type (e.g. transformational or transactional) as leaders will clearly use a variety of styles in any given situation. Leaders create a culture by casting a shadow that keeps Q&S at the forefront of the mind of all staff and teams, with constant vigilance to tackle norms of poor behaviour. The task of improving healthcare will take time and commitment from all. In healthcare the alignment of vision, leadership, patient safety and culture needs to be a constant and visible process and any diversion must be quickly eliminated to avoid re-occurrence of NHS tragedies like that at Mid Staffs.

It is acknowledged that Commissioners have huge power and influence over where an organisation concentrates its effort to influence policy and this could also create tension with the Q&S strategy. With the competing demands of financial penalties and targets it was often difficult to ensure that Q&S remained high on the agenda for executives; what was really needed was a Q&S champion who wasn't overwhelmed by the financial issues. Q&S has a reputation for being problematic. The identification of conflicts between achieving targets and reducing costs in real terms requires further development across a range of services. The risks attached to increasing efficiency, keeping staff numbers at the right ratio and having all the equipment available were a challenge in looking at the patient safety strategy. The perceptions of not just executives but new commissioners should be examined in order to understand the complexity of new relationships and their effect on Q&S outcomes. Public expectations of the NHS and scrutiny of it in terms of the delivery of appropriate healthcare and robust governance systems is paramount.

References

BBC 2011. *NHS £20bn Savings Progress Questioned*. Available at: http://www. bbc.co.uk/news/health-17541803 (last accessed on 16 July 2013).

Boaden, R., Harvey, G., Moxham, C. and Proudlove, N. 2008. *Quality Improvement: Theory and Practice in Healthcare*. Coventry: NHS Institute for Innovation and Improvement/University of Manchester Business School.

Boaden, R., Harvey, G., Moxham, C. and Proudlove, N. 2010. *Quality Improvement Theory and Practice*. NHS Institute for Innovation and Improvement. Available at: www.institute.nhs.uk (last accessed on 20 November 2011).

Bohan, P. and Laing, M. 2012. Can leadership behaviour affect quality and safety? *British Journal of Healthcare Management*, 18(4), 184–90.

Care and Quality Commission (CQC). 2010. *What Standard of Care Can You Expect from Your Hospital?* Available at: http://www.cqc.org.uk/public/what-are-standards/national-standards (last accessed on 20 November 2011).

Care and Quality Commission (CQC). 2011a. *Quality & Risk Profile*. Available at: http://www.cqc.org.uk/organisations-we-regulate/registered-services/quality-and-risk-profiles-qrps (last accessed on 20 February 2011).

Care and Quality Commission (CQC). 2011b. *Investigation Report Barking, Havering and Redbridge University Hospital King George Hospital*. Available at: http://www.cqc.org.uk/directory/rf4 (last accessed on 11 November 2011).

Cassell, C. and Symon, G. 1999. *Qualitative Methods in Organisational Research: A Practical Guide* (fourth edition). London: Sage.

Cresswell, W. 2007. *Qualitative Inquiry & Research Design: Choosing among Five Approaches*. London: Sage.

Department of Health 2001. *Learning from Bristol. The Report of the Public Inquiry into Children's Heart Surgery at the Bristol Royal Infirmary, 1984–1995*. London: The Stationery Office.

Department of Health 2008. *High Quality Care for All NHS: Next Stage Review*. London: The Stationery Office.

Department of Health 2009. *A Review of Lessons Learnt for Commissioners and Performance Managers following the Healthcare Commission Investigation*. London: The Stationery Office.

Department of Health 2010. *NHS Constitution*, Series no 30063. London: The Stationery Office.

Department of Health 2010a. *White Paper Equity & Excellence Liberating the NHS July 2010*. London: The Stationery Office.

Department of Health 2011. *The Health & Social Care Bill*. London: The Stationery Office.

Duckers, M., Wagner, C., Vos, L. and Groenewegan, P. 2011. Understanding organisational development, sustainability, and diffusion of innovations within hospitals participating in a multilevel quality collaborative. *Implementing Science (Biomed Central)*, 6(18) Available at: www.implemenationscience.com/content/6/1/18 (last accessed on 24 October 2011).

Francis, R. 2010. *Mid Staffordshire NHS Foundation Trust January 2005–March 2009 Independent Inquiry into Care Provided Chaired by Robert Francis QC*. Office of Public Sector Information, Information Policy Team, Kew, Richmond, Surrey.

Francis, R. 2013. *The Mid Staffordshire NHS Foundation Trusts Public Enquiry 6th February 2013 Chaired by Robert Francis QC*. Kew: Public Sector Information, Information Policy Team.

Glaser, B.G. and Strauss, A.L. 1967. *Discovery of Grounded Theory: Strategies for Qualitative Research*. Chicago: Aldine.

The Guardian. 2011. 17 Stepping Hill patients poisoned by insulin in saline solution, police say, 17 October 2011. Available at: http://www.guardian.co.uk/uk/2011/oct/17/16-stepping-hill-patients-saline (last accessed on 16 July 2013).

Herold, D., Fedor, M., Caldwell, S. and Liu, Y. 2008. The effects of transformational and change leadership on employees' commitment to change: A multi-level study. *Journal of Applied Psychology*, 93(2), 346–57.

Macleod, D. and Clarke, N. 2009. *The Macleod Review*. Available at: http://www.nhsemployers.org/employmentpolicyandpractice/staff-engagement/researchandreports/pages/macleodreview.aspx (last accessed on 16 July 2013).

National Patient Safety Agency 2004. *Seven Steps to Patient Safety: An Overview Guide* (second edition). London: NPSA.

Schein, E. 1992. *Organisational Culture and Leadership* (second edition). San Francisco, CA: Jossey Bass.

Van Dyke, D. 2006. *Management Commitment: Cornerstone of Aviation Safety Culture*. Montreal: Royal Aeronautical Society.

Wallis, N., Yammarino, F. and Feyerherm, A. 2011. Individualised leadership: A qualitative study of senior executive leaders. *The Leadership Quarterly*, 22, 182–206.

Yammarino, F.J., Dionne, S.D., Schriesheim, C.A. and Dansereau, F. 2008. Authentic leadership and positive organizational behavior: A meso, multi-level perspective. *Leadership Quarterly*, 19, 693–707.

PART IV
Additional Perspectives and Future Directions

Chapter 17
A View from Elsewhere: Safety Culture in European Air Traffic Management

Barry Kirwan and Steven T. Shorrock

Air Traffic Management – An Ultra-safe Industry?

Safety culture is relevant to all safety-related sectors and, to avoid an insular perspective, it is important to learn from other industries. Healthcare (surgery in particular) has learned several lessons from aviation in terms of the cockpit. A sub-sector of aviation that may provide a different but useful perspective is air traffic management (ATM) which, unlike healthcare, is generally seen as an 'ultra-safe' and 'high-reliability' industry (Amalberti, 1991), especially in Europe. It is ultra-safe not just because of the extremely low number of accidents that involve ATM, given the high volume of air traffic, but because of the ability of the ATM system to succeed under challenging conditions (see Hollnagel 2012).

The difference in safety levels can be traced to many differences between the two systems, including the nature of the *systems* (variety of demand and the nature of flow through the systems), *resources* (variety of equipment and the degree of proceduralisation), the nature of *hazards and defences* (especially variety and controllability and external defences), the nature of *external constraints* (political intervention and individual liability) and the nature of the *organisational and professional cultures* (variety of subcultures, attitude to error and occurrence reporting and dominant cultural influencers, e.g. controllers).

However, there are some commonalities that are relevant to safety culture. Both healthcare and ATM are complex systems with emergent outcomes. Both sectors must balance multiple conflicting goals, particularly safety, cost-efficiency and capacity (with top-down targets increasingly attached to each). Both have very high consequences of failure. And both are human-centred safety-related industries that rely heavily on human intervention and high levels of competency. An air traffic controller in a busy centre or tower may make many decisions an hour to preserve the safety of passengers and crew travelling in aircraft. Only more recently have advanced tools been developed to provide some automated assistance and, even so, this is mostly limited to information automation. Other parts of the ATM/CNS (communication, navigation, surveillance) system are also still highly human-centred. This means that the maintenance or improvement of safety (as improvement is necessary in the face of rising traffic levels) is dependent on not only a well-integrated safety management system, but also a healthy safety culture.

A healthy safety culture (like good design) can seem difficult to see, whereas unhealthy safety culture (like bad design) can seem more apparent, especially from the outside or after a major accident. In Europe, the turn of the twenty-first century was a tumultuous time for ATM safety (Johnson and Shea 2007). In 2001, 114 passengers, crew and ground staff died in the Milan Linate airport ground collision between two aircraft: a MD-87 and a Cessna 525-A taxiing on the runway. In 2002, 71 passengers and crew died in the 2002 Überlingen mid-air collision between a Boeing 757-200 and a Tupolev TU164M. Like several other disasters in other industries (e.g. Challenger, Chernobyl, Piper Alpha, Texas City, Deepwater Horizon) these tragic accidents were ultimately linked to poor safety culture. Such accidents are thankfully rare and, at the time of writing this chapter, there has not been a commercial aviation fatal accident caused by ATM in Europe since these events.

For many years, and particularly since the Überlingen accident, European ATM has been developing and implementing safety management systems. The past decade saw the implementation of a number of EUROCONTROL safety and regulatory requirements in European air navigation service providers (ANSPs: the national providers of air traffic services to airspace users). These involved ensuring ANSPs had a safety management system (SMS), including safety occurrence reporting systems, safety surveys, safety promotion and safety assessments of system changes. Such approaches and legislation have been instrumental in ensuring that ANSPs have the competence, processes and systems to ensure an appropriate focus on operational safety is maintained and the likelihood of mishaps reduced (e.g. see Perrin et al. 2007).

An organisation's safety health is the product of two key elements: the quality of the systems – and processes implemented to deal with risk and safety-related information (the SMS, which may or may not be formalised) – and the safety culture, which includes people's shared values, beliefs and attitudes about safety. These two elements combine to characterise the way that people behave within their organisation, the 'behavioural norms'. The overarching goals are that all personnel (including management) recognise the importance of safety and understand that it is everyone's responsibility and for this to be reflected in everyday decision making and behaviour. Even the best SMS will be ineffective if the safety culture is characterised by counterproductive attitudes and behaviour.

The concept of safety culture considers the critical importance of management action regarding safety as well as operational, engineering and other support personnel. In this way, it reflects the movement (e.g. in safety occurrence investigation) from so-called 'operator error' to systemic issues in recent years (e.g. Reason 1997; Dekker 2006). The concept also does not necessarily have a negative focus; both positive and negative aspects of an organisation's culture are of interest in safety culture work.

Safety culture has been recognised as important in ATM and, following Überlingen, in 2003 a research project commenced to adapt safety culture approaches from nuclear and other industries to ATM (Gordon and Kirwan 2005; Gordon et al. 2007). This was followed from 2006 onwards by a fully-fledged programme of safety culture surveys across more than 20 countries in Europe (Kirwan et al. 2010). This continuing programme of surveys is aimed at assessing and enhancing the safety culture of each ANSP, determining where their safety culture is strong, where it is weak, and where and how improvements could or should be made.

This chapter is in three main parts. First, we outline the context of the EUROCONTROL safety culture programme, including some background on the countries associated with EUROCONTROL and some of the latest major developments that are relevant to safety in European ATM. Second, we describe the development of the approach used by EUROCONTROL. Third, we describe the EUROCONTROL safety culture methodology as it stands today, along with some concluding remarks.

Context of the EUROCONTROL Safety Culture Programme

Apart from the desire to maintain and improve safety culture following such accidents as Überlingen and Linate, there are other forthcoming challenges that reinforce the need to ensure that European ATM as a whole has a strong and healthy safety culture, and maintains its 'ultra-safe' status, protecting passengers, crew, cargo and the airlines who carry them.

Currently there are 40 European Member States associated with EUROCONTROL.[1] The aim of the EUROCONTROL Safety Culture Programme is to help each state to understand its own safety culture, and take action to maintain or improve it, by the end of 2014. The Programme is set in the context of two major changes in the way that European ATM operates. The first major change is the SESAR (Single European Sky ATM Research) Programme, which will upgrade the entire ATM system architecture in Europe. The deployment of SESAR

---

1   These are: Albania, Armenia, Austria, Belgium, Bosnia and Herzegovina, Bulgaria, Croatia, Cyprus, Czech Republic, Denmark, Finland, France, Georgia, Germany, Greece, Hungary, Ireland, Italy, Latvia, Lithuania, Luxembourg, Malta, Moldova, Monaco, Montenegro, the Netherlands, Norway, Poland, Portugal, Romania, Serbia, Slovakia, Slovenia, Spain, Sweden, Switzerland, the former Yugoslav Republic of Macedonia, Turkey, Ukraine and United Kingdom of Great Britain and Northern Ireland. EUROCONTROL itself is defined as 'the organisation for the safety of European air navigation' and works with its member states and other related organisations to harmonise and improve air traffic services across Europe.

improvements may commence as early as 2014, with the major deployment phase starting in 2018.

The second major change is the development of Functional Airspace Blocks (FABs). A FAB is an airspace block based on operational requirements and established regardless of national state boundaries, where the provision of air navigation services and related functions is performance-driven and optimised through enhanced cooperation among ANSPs or, when appropriate, an integrated provider such as the newly formed 'NUAC' (Nordic Upper Airspace Control Centre, a working partnership between the two ANSPs for Sweden and Denmark). Further FABs should come into operation from 2014. The key benefit of FABs is to reduce the fragmentation of airspace in the European Union, which has potential safety, capacity and cost implications. The result would be a reorganisation of the 67 airspace blocks in Europe – all based on national boundaries – into around nine functional airspace blocks. As an example, one of the larger proposed FABs called 'FABEC' (FAB Europe Central) comprises France, Germany, Belgium, the Netherlands, Luxembourg, and Switzerland.

In practice this means that different ANSPs in different countries must work closely together, requiring mergers which will bring together organisational, professional and national cultures. Furthermore, particularly in the larger FABs, the individual ANSPs may have different levels of safety management 'maturity' and different safety cultures and this fusion may lead to the emergence of new cultures. It is critical that such operational mergers do not adversely affect safety culture and operational safety. The ideal outcome is that the ANSPs with less developed safety cultures learn from those with more developed safety cultures, but the possibility that collective safety culture could deteriorate cannot be discounted and this risk must be actively mitigated. This is another reason why it is essential to know one's safety culture strengths and weaknesses *before* embarking on joining a FAB.

For these reasons, the EUROCONTROL Safety Culture Programme aims to survey all EUROCONTROL member states by the end of 2014, prior to significant SESAR or FAB impact. This was endorsed at a workshop held in Luxembourg in November 2012 with 20 ANSPs, where it was concluded by the member states that they needed to understand their own safety cultures first before moving to FABs and FAB-based safety culture assessments.

The current status of safety culture assessment in European ATM is shown in Figure 17.1.

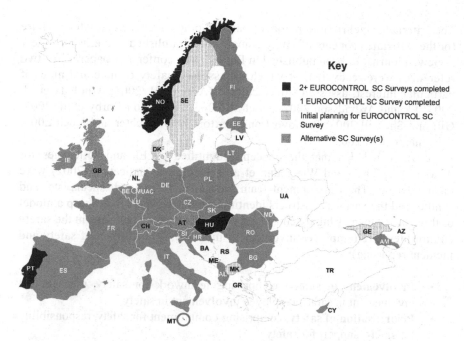

Figure 17.1   Status of European ATM Safety Culture Surveys[2] at the end of 2012

The following sections explain briefly the development of the safety culture approach, how it works in practice and the type of results it generates.

Development of the Approach

The development of the EUROCONTROL safety culture survey approach was an iterative process, involving the following phases (see also Gordon and Kirwan 2005; Gordon et al. 2007; EUROCONTROL 2008; Kirwan et al. 2010, 2012; Mearns et al. 2013; Shorrock et al. 2011).

Phase 1: Literature Review

Work began on the EUROCONTROL safety culture questionnaire in the early 2000s with a review of the safety culture literature. The review identified constructs of potential importance for measurement in ATM. The review focused on safety culture research from 2001–05 (and the review by Guldenmund 2000).

---

2   Sweden's ANSP (LFV) has had several subsequent safety culture surveys prior to the NUAC FAB development.

The criteria for inclusion were that research should have assessed safety culture (or the associated concept of safety climate) been published in English in a peer reviewed journal or been published in English as a conference paper. Fifty-two references were recovered, of which 43 assessed safety climate and nine, of particular interest, assessed safety culture (Adie et al. 2005; Arboleda et al. 2003; Ek and Ardvidsson 2002; Ek et al. 2003; Farrington-Darby et al. 2005; Gill and Shergill 2004; Pronovost and Sexton 2005; Richter and Koch 2004; Wiegmann et al. 2002).

It was concluded that the concepts identified by Ek and colleagues for ATM (2002, 2003) and Wiegmann et al.'s theoretical indicators (2002) were most relevant. The development team worked with industry specialists, and synthesised the various constructs identified in these papers to develop a model of three broad themes highly consistent with over-arching concepts in the safety culture literature (employee involvement in safety, prioritisation of safety and incident reporting):

- **Involvement in safety**, comprising teamwork for safety, management involvement in safety, employee involvement in safety
- **Prioritisation of safety**, comprising commitment for safety, responsibility for safety, support for safety
- **Reporting and learning**, comprising incident reporting, learning and communication on change.

These themes and sub-themes were investigated further through interviews and focus groups in Phase 2, in order to test their face validity as key components of safety culture in ATM.

The literature review (in particular Guldenmund 2000) also pointed out that, whilst safety climate is typically measured through questionnaire instruments, safety culture assessment involves mixed-methods (e.g. qualitative interviews) in order to understand the deeper and more engrained aspects of culture. This approach is supported by research showing the limitations of relying on safety climate questionnaires for assessing organisational safety (O'Connor et al. 2011). While the majority of safety climate research relies solely on questionnaires, qualitative methods, including focus groups and interviews, were in frequent use in safety culture research. From the outset, therefore, it was decided that the resultant methodology would use this mixed-method or triangulation approach in order to go beneath the outer appearance of safety climate to the deeper issues of an ANSP's safety culture.

Phase 2: Interviews, Focus Group and Questionnaire Development

Semi-structured interviews were conducted to investigate the themes identified above and to explore any additional constructs. Two researchers visited four geographically disparate ANSPs (located in the North, South, East and West of

Europe), to ensure that the resultant question set accounted for national culture diversity. The sample of 52 volunteers covered a cross-section of ANSP personnel (i.e. air traffic controllers (ATCOs), maintenance technicians, engineers and managers, including some senior managers).

A total of 80 statements were generated and presented in questionnaire form to a focus group of ANSP stakeholders (European ANSP safety directors and safety managers) at a 2-day workshop convened in Madrid in September 2006. The focus group participants were divided into three groups of 10–12 and each group (A, B and C) was allocated a section of the questionnaire to complete and comment on. Respondents were asked to mark each response category as if they were completing the questionnaire themselves and also to score each statement according to its perceived pertinence. In addition, they were asked to give feedback on the general realism, relevance, sensitivity and usefulness of the questionnaire items.

On the basis of the feedback from these managers a questionnaire consisting of 59 items from the original 80 was developed, with evaluation on a 5-point Likert-type scale ranging from 1 'strongly disagree' to 5 'strongly agree'. In addition, a series of open-ended questions asked for feedback on the content of the questionnaire and ease of completion. It was noted that confidentiality was assured, and results would be reported at group level. The questions were, at this stage, grouped in 5 themes and 19 sub-themes. This marked 'version 1' of the questionnaire.

Phase 3: Validation of the Questionnaire

The questionnaire was piloted with a sample of air traffic controllers, engineers and managers from four ANSPs (different from those already canvassed) from across Europe in 2007 and 2008. In order to test the validity of the survey instrument, its 'construct validity' (to ensure it was measuring safety culture and not something else), both exploratory factor analysis (EFA) and confirmatory factor analysis (CFA) statistical techniques were applied (Gerbing and Hamilton 1996).

As suspected, it was difficult to find a 'one-size-fits-all-cultures' model, with respondents from the different countries and different occupational groups responding to the items in different ways. In addition the data was possibly compromised by language, complexity of the items and/or genuine underlying cultural differences between the different national groups.

Since there was little evidence for the hypothesised structure of five main themes with 19 sub-themes in the initial modelling attempts, it was decided to return to the three higher order themes identified from the literature review and attempt to find sub-themes under these higher order factors. Based on the initial modelling process, a set of 32 general items as well as specific items for air traffic controllers, technical staff and management could be identified. Data were collected in three additional ANSPs in order to test the underlying factorial structure of the 32 identified general items (specific statements for occupation groups were excluded as the sample size was insufficient). Exploratory factor analysis was conducted on the three datasets. Across countries a six-factor solution, including 29 general items, was subjected to a confirmatory factor analysis to determine how well the data fit the theoretical model for the ANSPs (total n = 883). This confirmed a six factor model for the 'general' section, with 'Management commitment/involvement'; 'Teaming for safety'; 'Incident reporting'; 'Communication about change'; 'Team commitment' and 'Support for safety' loading onto the three higher order factors: 'How we are involved in safety'; 'How we learn' and 'How we prioritise safety'.

Since the questionnaire was still in a formative stage of development and factor analysis work had focussed only on the general section, it was not possible to determine a final model from the factor analyses but eight safety culture 'elements' were proposed to reflect the results of the factor analyses and the themes derived from the literature and interviews. The eight elements were: Commitment, Involvement, Reporting and Learning, Teamwork, Communication, Risk Awareness, Trust and Responsibility. The first five of these elements had strong support from the factor analyses, while the last three had support from the literature and seemed to fit items from the occupation-specific sections of the questionnaire.

The questionnaire was, at this stage, applied to several more ANSPs which provided valuable information on how the items were being interpreted or understood and on the distribution of responses for items. For a more comprehensive review of the questionnaire development and validation, see Mearns et al. (2013).

Phase 4: ANSP Workshops

The difficulty of finding a factor structure that is consistent between diverse countries and occupational groups, coupled with the complexity and depth of the safety culture concept (DeJoy 2005; Guldenmund 2000; Schein 2004), reinforced the need for a multi-methods approach. The safety culture literature highlighted the need for complementary methods for understanding safety culture (Battles and Lilford 2003; Cox and Cheyne 2000), as survey measures do not explain the reasoning behind responses (Guldenmund 2000; Mearns et al. 2001). Qualitative methods are often best suited to explain why perceptions of safety culture are positive or negative (Guldenmund 2007). In combination, mixed methods can reveal 'complementary data rather than alternatives' for understanding organisational culture (Cox and Cheyne 2000: 111). The EUROCONTROL Safety Culture Programme therefore uses a mixed methods approach and follow-up workshops are an integral part of this.

In 2007 interviews were utilised in several ANSPs, as well as data gathering on safety performance and events at the ANSP. However, interviews yielded contradictory information – e.g. one person or group might 'blame' another for certain conditions affecting safety, leaving the surveyors in an impasse. Workshops were then implemented, involving 6–10 participants led by a team of three facilitators (a Human factors specialist, an operational controller and a note-taker) and, for example, including air traffic controllers from different teams along with engineers in the same workshop groups. These workshops resulted in a far more satisfactory outcome, since different opinions could be aired and explored, resulting in consensus and a deeper understanding of the rationales of the different 'sides'. Workshops quickly became a major component of the safety culture survey process. Furthermore, the workshops were (and still are) profoundly valued by the ANSP participants, including management, and often led to prompt localised actions to improve safety culture issues. Whilst the questionnaire characterises safety climate and points the survey team in the right direction, it is the workshops that enable a deeper insight into the safety culture of the organisation.

Phase 5: Continuous Review and Development of the
EUROCONTROL Safety Culture Questionnaire

In 2010, Following the application of the EUROCONTROL safety culture questionnaire to more than a dozen ANSPs, a major review was undertaken to determine the validity of the items and the underlying model of safety culture elements (Kirwan et al. 2012; Shorrock et al. 2011). By this stage there had been significant feedback and learning from application to ANSPs but a degree of questionnaire 'creep' had occurred, with subtle changes to each subsequent survey based on feedback from the previous one. This began to bring into question the stability of the questionnaire, which was important given that some ANSPs had already begun their second survey and, naturally, wished to see if they had improved – which is difficult if some of the questions have changed. The aim of the review was therefore to refine the questionnaire to a more definitive and stable set for wider and repeatable distribution across the whole of Europe.

Each questionnaire item was considered, for example, in terms of

- Concept/Model fit – the degree to which the item appears to fit the concept and model of safety culture
- Distribution (skewedness) – the degree to which item responses are evenly distributed. This was based on the distributions across several applications
- Loading (cross-loading) – the degree to which item responses load clearly on one factor. This was based on factor analyses
- Wording – the degree to which respondents understand the item, independently of their cultures. This was based on their feedback
- Quantifiability – the degree to which agreement or disagreement with a statement can be deemed positive or negative for safety culture.

The following are examples of some questions that were removed from the questionnaire, on the basis of this review:

- (General) Safety is a responsibility shared throughout the organisation

- (General) Each member of staff has a responsibility to keep up with changes to procedures
- (Operational) I consider error reporting to be an important professional responsibility
- (Engineering) We ensure that Operations are aware of any system degradation that may occur due to maintenance or engineering works.

Some items seemed to have conceptual problems. One Operational item 'I trust the ATC equipment that I use in my job' did not fit the concept of safety culture well, since distrust of equipment could indicate good safety awareness, even though it may, for instance, also indicate – indirectly – poor design or maintenance. Similarly, a management item 'The Regulator makes life difficult for us', while trying to tap into the attitude to and relationship with the Regulator, does not fit well the concept or model of the safety culture within an organisation. Indeed, it was posited in workshops that the regulator should not make life easy.

Some items seemed to reflect more the respondents' preferred self-image, e.g. responses to the statement 'I clearly show that safety is one of my core personal values' were skewed – few managers disagreed with the statement – even though it was often difficult for them to articulate in workshops how they showed this.

Some questions cross-loaded onto more than one factor in factor analysis (e.g. Teamwork and Communications), or loaded on different factors in different factor analyses, suggesting that the item may not fit consistently into a coherent model of safety culture. Many of these concerned issues that were better covered by other items statistical analysis was used to determine the best – most precise – questions. An engineering-related example was 'I sometimes have to do "workarounds" to compensate for lack of resources (equipment, manpower or time)'. Wording was another reason for removal of certain questionnaire items. For instance, controllers pointed out that they could not be sure what would constitute sufficient system checks by maintenance staff when asked whether 'Maintenance staff perform sufficient system checks'.

Some items did not appear to transfer well across different national cultures, the issue of 'workarounds' being a notable example. While this is an important safety culture issue, the concept or word, even when translated, does not always transfer across cultures and languages. The connotations and transferability of concepts or words across cultures was one reason for modification. The concept of 'concern' can have different connotations between languages (and within the same language). These items were intended to have negative connotations but it became apparent in workshops that they could be seen as positive. As one example, there was at one stage an item 'Everyone knows about an accident just waiting to happen.' In certain cultures, agreement with this statement indicates a poor safety culture, because it is widely believed that an accident is imminent but nobody does anything to prevent it. But in other cultures where this statement was rated with strong agreement, subsequent workshops revealed that the respondents associated this with heightened risk awareness – a positive characteristic.

These statements needed to be reworded. Such surprises in cultural interpretations can still occur and they reinforce the need for workshops where facilitators can check participants' understanding of the questions and correct any misunderstandings.

The review, plus subsequent reviews following further large-scale applications of the questionnaire, has led to a stable set of questions. This set of questions appears to be working; there are very few questions left unanswered or queried by participants in each new survey.

Having described the methodology's derivation and evolution, it is next described in terms of how it works today.

The EUROCONTROL Safety Culture Methodology

The evolved methodology includes questionnaires and workshops and may include other methods (such as interviews) as necessary. The key stages of the process are described in the following sections.

Pre-meeting

A pre-meeting is usually the first step in the safety culture survey process. This establishes a working relationship with the Safety Manager (or whoever is leading the work within the ANSP) with respect to the application; it allows a discussion about the feasibility of the survey, in terms of timing. The pre-meeting also provisionally ascertains the scope of the survey, i.e. whether the whole organisation will be surveyed or the operational and managerial part only, excluding non-operational support services – the ideal is the whole organisation. The pre-meeting is also used to determine the level of commitment to conducting and acting on the survey.

Kick-off Meeting

Following the pre-meeting, a formal kick-off meeting establishes a wider relationship with the senior management team at the ANSP and establishes formal commitment to conduct and act on the survey. This includes the CEO and Directors (or senior management team) and, ideally (sometimes separately), the relevant union bodies. The aim is to set realistic expectations (there will always be some areas for improvement recommended) and address any questions and concerns that the senior management team may have, including timing, confidentiality and requirements for staff involvement in the workshops. Following a presentation and discussion, a formal decision is made by the management team regarding whether to continue with the survey and what its scope (full or part-organisation participation) should be. The kick-off meeting usually concludes with detailed survey planning, including discussing questionnaire translation (multi-language) and presentation (paper-based or digital versions) options, a visit to various operational areas (e.g. area control room, approach control room, tower) and collection of any relevant and available safety documentation, such as safety strategy reports, safety newsletters and safety policies. Demographic details are also collected (e.g. size and demographic breakdown of the organisation, number of air traffic centres and towers, organisational directorates and departments, etc.). It is also pertinent to find out if any other surveys are planned at the ANSP during the same period, e.g. staff satisfaction or other surveys, as this may affect response rate due to 'survey fatigue'.

Launch

A launch of the survey is organised to communicate the aims and process of the survey and to encourage staff engagement. Prior to the launch, the ANSP communicates about the survey aims and process internally. This activity varies depending on the organisation but may include a letter or email from the CEO, articles in company magazines and on the intranet, briefings from managers and posters. Following this internal communication, the EUROCONTROL safety culture survey team lead visits to ANSP sites to inform staff and promote the survey. Experience has shown that the level of engagement with the survey process is largely dependent on this internal communication and the launch process. A typical launch would comprise presentations of the survey rationale and approaches to the management and all available staff. This takes place at sites agreed at the kick-off meeting and typically focuses on major air traffic control centres and towers. Multiple presentations are often run over several consecutive days, starting with headquarters, to allow as many staff as possible to attend. On occasions it has been necessary to translate some or all of the slides to the native language(s) to aid comprehension by the audience.

Questionnaire

The EUROCONTROL questionnaire is distributed during or immediately following the launch. The development of the questionnaire has been described above. The questionnaire currently comprises around 50 items, with two sections. The safety culture questionnaire is structured in the following way.

- Cover sheet – background information, instructions and important notes
- Section A: demographics – department, job category and location
- Section B: general – items for all staff in the organisation
- Section C: operational and technical – items for operational staff in four areas: air traffic services (ATS), aeronautical information services (AIS), meteorology (MET) and technical staff in communication, navigation, surveillance (CNS). Sometimes, these four staff groups are licensed staff and, typically, all can report safety occurrences
- Section D: feedback – for respondent feedback on the questionnaire and any additional comments respondents wish to make about safety culture.

The questionnaire is administered, either in paper or electronic form, with the paper questionnaire distributed by the ANSP to all targeted staff (which is usually all staff in the ANSP, excluding external contractors). An electronic version of the questionnaire has also been developed. This is more efficient in the analysis stage but participation depends on the availability of computers for operational staff. The questionnaires are returned directly to an academic partner, currently the London School of Economics (until 2011 the University of Aberdeen provided this service), for analysis. The academic partner ensures that data analysis remains independent and thus provides additional assurance of confidentiality to ANSP staff, as neither ANSP nor EUROCONTROL personnel see the raw data or filled-in questionnaires.

Data Analysis

Around one month is provided for staff to respond to the questionnaire. The questionnaires are processed to provide descriptive statistics and a summary of the questionnaire results for all demographic sub-groups.

Workshops

Following the submission of questionnaires, a series of workshops is held with managers and staff. These are an integral part of the process and a 'survey' therefore comprises both questionnaires and workshops. There are several cases where conclusions based on the questionnaire results alone could have led to a misunderstanding of safety culture, due to the issues being deeper and more complex than a questionnaire can reveal. Issues captured by questionnaire can also be more transient, and in some cases may be less pertinent by the time of the workshops. This is a conclusion being reached in some other areas, e.g. medical safety culture (Waterson 2012) where there are calls for more 'methodological pluralism' beyond simple questionnaire use, mentioning workshops in particular. The purpose of the workshops is to:

- Discuss the results of the survey to determine the possible reasons for the responses
- Gather and verify specific examples of the findings of the survey within the work environment
- Provide an opportunity for staff and management to comment on the findings and suggest a way to address any problems.

Each workshop normally comprises 6–10 staff from the ANSP and will focus on issues raised by the questionnaire results. An example is shown in Figure 17.2, where a bar graph shows results from a questionnaire survey.

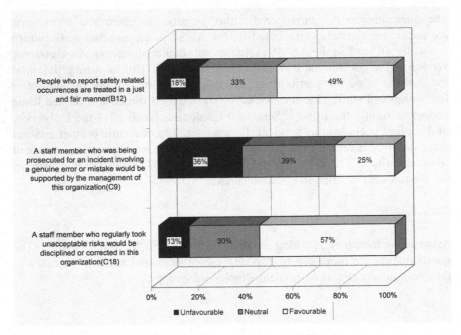

Figure 17.2   Example of results discussed in workshops

The EUROCONTROL team comprises a facilitator, recorder and operational specialist. The facilitator and recorder are Human Factors and/or Safety specialists, while the operational specialist is a current or former air traffic controller (from a different ANSP or EUROCONTROL). Each workshop lasts 2–3 hours. Staff and managers engage with separate workshops to reduce power-distance effects. A typical set of workshops would comprise the following:

- Senior management – e.g. CEO and board of directors
- Middle managers – e.g. middle managers in operations, engineering, purchasing, training, quality, etc.
- Supervisors – e.g. engineering supervisors, watch managers/shift supervisors
- Operational and technical staff – controllers, assistants, meteo staff, aeronautical information services staff, engineers (no line managers are permitted in these workshops.)
- Support and specialist staff – HR, purchasing, legal, contracts, communications, etc.
- Safety team workshop – safety team (e.g. safety risk, safety investigation, safety auditing).

The workshops explore the findings and possible factors associated with them, instantiate the findings with examples and suggest ways to improve. Experience has shown that it is important to focus on both positive and negative findings in a balanced way.

A recorder makes detailed notes and the discussion of the workshop constitutes the majority of the report. Following the workshops, the survey team reviews the data and records key recommendations.

Reporting

Two reports are prepared, detailing the rationale for and purpose of the survey, the methodology, and the findings, conclusions and recommendations. One report is a full report containing the questionnaire and workshop findings in detail. The other is a short report of key findings and recommendations.

Experience has shown that a long list of detailed prescriptive recommendations is less effective than a shorter list of goal-oriented recommendations with suggested means of implementation. Typically, some are short term (months) and others are longer term (years). These means of implementation usually arise in the workshops and from good practice noted in comparable environments, either within the ANSP or in other ANSPs. The recommendations are not mandatory, since the safety culture surveys are not a mandatory or regulatory exercise.

Feedback

Following review and finalisation of the report, the next phase is to give feedback to management and staff. The purpose of this activity is to

- Feed back the results and key recommendations of the survey
- Foster understanding of the results and recommendations
- Encourage commitment to implementing the recommendations.

The first activity is to have the report reviewed by the ANSP point of contact (e.g. safety manager or director) in the ANSP. As the report is independent, actual findings and recommendations cannot be altered significantly, since these derive from the data but, sometimes, an inside view can help in phrasing the recommendations to make them more palatable to the organisation. The review is however largely restricted to ensuring that the facts (rather than the perceptions) presented are correct. A management feedback session provides the ANSP leadership with an overview of the findings and recommendations and allows them to comment on them before the information is shared with staff. It is important that the senior management team is informed and is aware of the issues raised.

For the management feedback, the results need to be balanced, avoiding the tendency or temptation to focus on negative safety culture findings and rather giving a balanced picture of the results (which are usually more positive than negative). Usually, a subset of key issues and recommendations is presented to focus attention.

A few days or weeks after the management feedback, feedback is provided to staff (at multiple sites for medium and large ANSPs). This includes references to agreed actions, which help to overcome any cynicism about management commitment to acting on the results.

Action Planning

Following reporting and feedback, the key focus of activity returns to the ANSP. A key danger is that momentum is lost. A safety culture survey can generate hope and expectation among staff, though many who have been in the same organisation for a long time may be cynical that anything will really change. It is therefore vital that the recommendations from the safety culture survey are translated into an action plan to help guide the change process. The action plan, in itself, achieves nothing without the commitment of management and staff to provide resources and work within the organisation. Occasionally, there is a long delay before any action happens and sometimes nothing happens, or else the actions become diluted or diverted or fail. If this happens, then subsequent attempts to improve safety culture are much less likely to succeed.

The process of considering the report findings and recommendations and writing an action plan, typically takes up to three months and is always at risk of being sidelined due to other changes or priorities within the ANSP.

After-care

Since 2010 the EUROCONTROL safety culture team has offered an 'after care' service. In practice this means visiting the ANSP some 12–24 months after the survey to see how it is proceeding, how the actions are being implemented and whether they are working. An external visit can sometimes help remind the organisation of its plans and get them back on track. Additionally, if certain actions are not working or are encountering resistance, the survey team may be able to propose alternative solutions that have worked for other ANSPs.

Finally, it is worth noting that a number of lessons have been learned along the way, from both successes and failures. These are noted in the summary box below, and may help those working in patient safety.

Assessing and Improving Safety Culture: Some Key Lessons Learned

1. Promotion and acceptance: the idea of safety culture may not be accepted initially, especially where safety management is seen primarily through a technocratic lens. Promoting safety culture is both Europe-wide and for each survey. We have found that having a European map showing the status of safety culture assessment is a useful tool for promoting the acceptance of surveys by each national service provider. Within a service provider, each survey is launched in person, with multiple presentations to management and staff to provide information, and opportunities for questions. The promise of similar face-to face feedback is also important. Acceptance from unions and associations is particularly useful.

2. The 'science': a scientifically underpinned methodology and theory separates safety culture from fads or other seemingly similar interventions, which staff may well have seen before. It also helps anchor the questionnaire tool, as otherwise there is a tendency to adapt and change questions after each implementation, thus affecting the possibility of learning across interventions, or between successive surveys for a single organisation.

3. Independence: it is important, and sometimes critical for many staff, that a survey or study is independent and impartial. This is helped by having organisational independence (as the EUROCONTROL organisation) and further firewalls of confidentiality (by working with a university for questionnaire data collection). Working with a university psychology department helps, for example, as they have strict codes of ethics concerning confidential data-handling and are seen as being impartial.

4. Credibility: the credibility of the survey team will affect acceptance of the process and its outcomes. A blend of human factors (or applied psychology), safety and operational experience in the team seems to work best. The credibility of the internal coordinator is also important. A good 'safety champion' within an ANSP is priceless. A key predictor in their effectiveness in this role relates to the quality of their connections with both management and the operational workforce.

5. Principles and ethics: a set of principles for conducting a survey needs to be developed and adhered to. This might outline issues such as: anonymity and confidentiality, independence and impartiality, respect for opinions, robustness of methodology, continuous improvement, lesson learning, breadth of perspective, sustainable improvement, feedback to management and staff and follow-up support.

6. Sense-making: it has been clear from the EUROCONTROL programme that a questionnaire is useful, but of questionable value if used alone. It is necessary to have a way to make sense of the data. Workshops are often the most useful method for this and individual workshops usually need to be separated along vertical, hierarchical lines to avoid power-distance effects. Interviews can be useful, especially for senior managers, but can also pose a challenge in making sense of contrasting views.

7. Improvement and momentum: recommendations, which typically emerge from workshops, are easier to make than to implement. Recommendations need to be carefully constructed, practically-oriented and only as many as necessary – having too many recommendations is counterproductive. Aftercare sessions with managers and staff (interviews are useful for this) help to track progress and other informal methods help to maintain momentum, such as the EUROCONTROL safety culture discussion cards (Shorrock 2012; SKYbrary 2013).

Concluding Comments

The EUROCONTROL Safety Culture Programme was developed following two fatal European ATM accidents. After an initial three year development period it was tested with a number of ANSPs spread around Europe and then rolled out as a pan-European programme. Since 2010 in particular the process has stabilised and has been accepted by most of the European Member States.

It is important to point out that there is no regulatory requirement to do a survey although, more recently, having carried out a survey gains an ANSP 'credit' on their benchmarked Safety Maturity rating. Despite such enticements, most ANSPs take the process seriously and use it to gain a more comprehensive picture of their safety risks and vulnerabilities. Many of these states recognise that, whilst they may have mature safety management systems and processes in place, they may still have safety culture vulnerabilities and these can contribute significantly to risk.

To date, several ANSPs have carried out more than one survey (two ANSPs have carried out three surveys), which allows them to see if they are progressing or slipping backwards in certain areas. Additionally, the ANSPs come together once a year to discuss safety culture issues at the annual 'regional workshop'. The first of these comprised only three ANSPs but the most recent one, the fifth, had 20 ANSPs present, including several who had not done the survey yet and came along to discover what it is all about.

The purpose of the programme however is not only organisational self-awareness but improvement. To this end there is now more focus on what works and what does not and an attempt to learn across Europe, for example considering the 'top 10' pan-European safety culture issues. Ultimately such initiatives are aimed at taking safety culture to the 'macro' (cross-ANSP) level, encouraging ANSPs to work together to tackle common issues. This is important, because one single accident will affect all the ANSPs and the industry. Pan-European partnerships to tackle major safety culture issues are in everyone's interests, from the ANSPs to the passengers they aim to protect.

Acknowledgements

The two authors would like to acknowledge the work of other members of the EUROCONTROL Safety Culture team over the years. The authors would also like to thank the ANSPs for their hospitality through the years and their openness in sharing information for the good of safety.

References

Adie, W., Cairns, J., Macdiarmid, J., Ross, J., Watt, S., Taylor, C.L. and Osman, L.M. 2005. Safety culture and accident risk control: Perceptions of professional divers and offshore workers. *Safety Science*, 43, 131–45.

Amalberti, R. 1991. The paradoxes of almost totally safe transportation systems. *Safety Science*, 27, 109–26.

Arboleda, A., Morrow, P.C., Crum, M.R., Shelley, I. and Mack, C. 2003. Management practices as antecedents of safety culture within the trucking industry: Similarities and differences by hierarchical level. *Journal of Safety Research*, 34, 189–97.

Battles, J.R. and Lilford, R.J. 2003. Organizing patient safety research to identify risks and hazards. *Quality Safety Health Care*, 12(II), ii2–ii7.

Cox, S. and Cheyne, A. 2000. Assessing safety culture in offshore environments. *Safety Science*, 34, 111–29.

DeJoy, D. 2005. Behavior change versus culture change: Divergent approaches to managing workplace safety. *Safety Science*, 43, 105–29.

Dekker, S. 2006. *The Field Guide to Understanding Human Error*. Aldershot: Ashgate.

Ek, A. and Arvidsson, M. 2002. Safety culture in the Swedish air navigation services. Paper Presented at the *European Academy of Occupational Health Psychology*, Vienna, 4–6 December.

Ek, A., Arvidsson, M., Akselsson, R., Johansson, C. and Josefsson, B. 2003. *Safety Culture in Air Traffic Management: Air Traffic Control*. Paper Presented to the 5th USA/Europe ATM 2003 R&D Seminar, Budapest, Hungary, 23–7 June.

EUROCONTROL. 2008. *Safety Culture in Air Traffic Management: A White Paper*. December 2008. Brussels: EUROCONTROL. Available at: http://www.skybrary.aero/bookshelf/books/564.pdf (last accessed on 24 April 2014).

Farrington-Darby, T., Pickup, L. and Wilson, J.R. 2005. Safety culture in railway maintenance. *Safety Science*, 43, 39–60.

Gerbing, D.W. and Hamilton, J.G. 1996. Viability of exploratory factor analysis as a precursor to confirmatory factor analysis. *Structural Equation Modelling: A Multidisciplinary Journal*, 3, 62–72.

Gill, G.K. and Shergill, G.S. 2004. Perceptions of safety management and safety culture in the aviation industry in New Zealand. *Journal of Air Transport Management*, 10, 231–7.

Gordon, R. and Kirwan, B. 2005. Developing a safety culture in a research and development environment, in D. de Waard, K.A. Brookhuis, R. van Egmond and T. Boersema (eds), *Human Factors in Design, Safety and Management*, Maastricht: Shaker Publishing, 493–504.

Gordon, R., Kirwan, B., Mearns, K., Kennedy, R. and Jensen, C.L. 2007. *A Safety Culture Questionnaire for European Air Traffic Management*. Paper to ESREL 2007, Stavanger, June, 2007.

Guldenmund, F.W. 2000. The nature of safety culture: A review of theory and research. *Safety Science*, 34(1–3), 215–57.

Guldenmund, F.W. 2007. The use of questionnaires in safety culture research. *Safety Science*, 43, 723–43.

Hollnagel, E. 2012. *A Tale of Two Safeties*. Available at: http://www.resilienthealthcare.net/ (last accessed on 6 June 2013).

Johnson, C.W. and Shea, C. 2007. *A Comparison of the Role of Degraded Models of Operation in the Causes of Accidents in Rail and Air Traffic Management*. Paper to the second Institution of Engineering and Technology International Conference on System Safety, 22–4 October 2007, 89–94.

Kirwan, B., Mearns, K., Jackson, J., Reader, T., Leone, M., Kilner, A., Wennenberg, A. and Grace-Kelly, E. 2010. *Measuring Safety Culture in European Air Traffic Management*. Paper to PSAM 10, Seattle, Washington, USA, 7–11 June 2010.

Kirwan, B., Mearns, K. and Shorrock, S. 2012. *The EUROCONTROL Safety Culture Questionnaire: Lessons from Application*. Paper to PSAM 11, Helsinki, Finland, 25–9 June 2012.

Mearns, K., Flin, R., Gordon, R. and Fleming, M. 2001. Human and organizational factors in offshore safety. *Work & Stress*, 15(2), 144–60.

Mearns, K., Kirwan, B., Reader, T.W., Jackson, J., Kennedy, R. and Gordon, R. 2013. Development of a methodology for understanding and enhancing safety culture in Air Traffic Management. *Safety Science*, 53, 123–33.

O'Connor, P., Buttrey, S.E., O'Dea, A. and Kennedy, Q. 2011. Identifying and addressing the limitations of safety climate surveys. *Journal of Safety Research*, 42, 259–65.

Perrin, E., Stroup, R. and Kirwan, B. 2007. *Future Considerations in ATM Safety RandD. A Summary of the Safety Gap Analysis Report by FAA/Eurocontrol Action Plan 15, Safety Research and Development*. Paper to International System Safety Conference, Baltimore, August. Available at: http://bit.ly/10VsxZh (last accessed on 6 June 2013).

Provonost, P. and Sexton, B. 2005. Assessing safety culture: Guidelines and recommendations. *Quality Safety Health Care*, 14, 231–3.

Reason J. 1997. *Managing the Risks of Organisational Accidents*. London: Ashgate.

Richter, A. and Koch, C. 2004. Integration, differentiation and ambiguity in safety cultures. *Safety Science*, 42, 703–22.

Schein, E. 2004. *Organisational Culture and Leadership* (third edition). San Francisco: Jossey Bass.

Shorrock, S.T. 2012. Safety culture in your hands: Discussion cards for understanding and improving safety culture. In M. Anderson (ed.), *Contemporary Ergonomics and Human Factors 2012*. London: Taylor and Francis, 321–8.

Shorrock, S.T, Mearns, K., Laing, C. and Kirwan, B. 2011. Developing a safety culture questionnaire for European air traffic management: Learning from experience. In M. Anderson (ed.), *Contemporary Ergonomics and Human Factors 2011*. London: Taylor and Francis, 56–63.

SKYbrary. 2013. *Safety Culture Discussion Cards*. Available at: http://bit.ly/safetycards (last accessed on 6 June 2013).

Uttal, B. 1983. The corporate culture vultures. *Fortune*, 17, 66–72.

Waterson, P.E. 2012. Measuring patient safety culture: How far have we come and where do we need to go? In M. Anderson (ed.), *Contemporary Ergonomics and Human Factors 2012*. London: Taylor and Francis, 19–26.

Wiegmann, D.A., Zhang, H., Thaden, T., Sharma, G. and Mitchell, A. 2002. *A Synthesis of Safety Culture and Safety Climate Research*. University of Illinois Aviation Research Lab Technical, Report ARL-02-03/FAA-02-2.

Chapter 18
The Prospects for Patient Safety Culture

Patrick Waterson

This final chapter aims to summarise a number of the themes which regularly crop up within the book. A second aim of this chapter is to offer some speculations regarding the likely directions the field of Patient Safety Culture (PSC) will take in the coming years, as well as potential areas for future exploration. A key argument is that the field of PSC needs to move beyond current concerns and expand its theoretical and methodological horizons. Part of this will involve improving the way in which we develop and test PSC surveys, tools and instruments. A second element is the need to take a closer look at some of the key lessons from other industries and work towards an improved application of a wider systems approach to future PSC research and practice (Wilson 2014).

Recurrent Themes and Future Challenges for PSC

The Controversy Surrounding Safety Culture

As many of the chapters have pointed out, the term safety culture tends to elicit very strong opinions – everything from full acceptance of its existence and its validity, through to outright rejection and, in some cases, contempt (e.g. *The Independent* 2012). Many of these debates might be characterised as shedding 'more heat than light.' However, they also reflect the tensions amongst both academics and practitioners involved in safety improvement. As Guldenmund describes in Chapter 2, it still remains difficult to define what we actually mean by 'safety culture', even in domains where the term is well established (e.g. the nuclear industry). Some of the difficulties are reflected in the perennial discussions centred on the so-called 'climate vs. culture' debate (Mearns and Flin 1999), others reflect more fundamental conceptual, theoretical and practical issues (e.g. the relationship between safety culture and the adoption of a systemic approach to human error – Reiman and Rollenhagen 2013). Safety culture remains controversial. Despite this it is a widely respected approach to understanding and improving safety. Within healthcare and other domains, these debates are likely to continue and, in most cases, are important in terms of challenging researchers and practitioners to think deeply about the role played by culture within safety and how safety 'unfolds' and is enacted over time within healthcare organisations (Vogus et al. 2010).

The Challenge of Measuring Medical Error

Part of the problem with attempting to measure error and safety culture relates to the characteristics of healthcare environments. Clinical work tasks do not always follow a clear 'linear' path – rather healthcare is a complex sociotechnical system and many of the parts of healthcare delivery are messy and non-linear (Karsh et al. 2010). Even processes which appear at first glance to be simple (e.g. hip replacement surgery), involve a large range of clinical roles, technologies (e.g. electronic record systems) and locations (Eason et al. 2012). PSC will vary a great deal according to the characteristics of the work tasks, locations, people involved etc. There is huge variability in the way healthcare systems operate. In our previous work on PSC data from the UK we found that culture on one hospital ward may be radically different to another only a floor above or below it (Waterson et al. 2010). Different types of roles and specialisms may have different attitudes towards safety which make the measurement of PSC difficult. Pfeiffer and Manser (2010) for example, found that nurses are more likely to rate PSC as high when staffing and workload levels are adequate, whereas physicians are more likely to be influenced by the quality of teamworking in their work environment. More attention should be paid to the contextual influences which are likely to 'shape' attitudes towards safety in healthcare environments. These span a range of dimensions including types of work tasks, technological characteristics, healthcare roles and job specialisms (see for example, the SEIPS model as described in Chapter 1 and in Chapter 5 by Phipps and Ashcroft).

Conceptual Challenges – Widening Perspectives

A number of chapters in the book (e.g. Hammer and Manser; McDonald and Waring) discuss the various different ways in which safety culture has been conceptualised. These include functionalist approaches which assume that the prime function of safety culture is to support management strategies and systems. The measurement of safety culture as seen from this point of view can be reduced to a relatively simple model of prediction and control (Glendon and Stanton 2000). By contrast, interpretive approaches emphasise the importance of regarding safety culture as an emergent complex phenomenon where issues such as identity, beliefs and behaviour need to be taken into account. Silbey (2009) similarly discusses the influence of three different perspectives on safety culture which have dominated research since the 1990s: (1) 'culture as causal attitude' – here culture is seen as formed on individual attitudes and organisational behaviour which can be decomposed into values and attitudes; (2) 'culture as engineered organisation' – here the emphasis is on how culture leads to outcomes such as reliability and efficiency (e.g. as reflected in research on 'high-reliability organisations' and models of accidents which draw on control theory – Eisenhardt 1993; Weick 1987; Leveson 2012) and (3) 'culture as emergent and indeterminate' – culture from this point of view is seen as an 'indissoluble dialectic of system and practice' which 'cannot be engineered and only probabilistically predicted with high variation from certainty (Silbey 2009: 356).

What is perhaps most noticeable about these approaches is that, within the field of PSC, the majority of studies so far adopt a functionalist, 'culture as causal attitude' approach towards theory and measurement. Most studies involve the use of quantitative surveys; very few studies in the mainstream medical literature are based on observation or the use of ethnographic studies. Theory still remains poorly defined in most studies (Halligan and Zecevic 2011), whilst other areas (e.g. contemporary work on 'risk' – e.g. Aven et al. 2011; Ben-Ari and Or-Chen 2009) are much more developed. The field of PSC needs to move beyond an exclusive focus on the adoption of a 'functionalist' approach to culture and safety. Part of this will involve the involvement of other disciplines, as well as the use of other types of methods (e.g. ethnography, qualitative studies), whether in isolation or in a mixed method format.

Whose Safety Culture are We Measuring?

Many studies which report findings based on PSC surveys focus on a limited set of participants, roles and healthcare specialists. A closer look at the characteristics of respondents reveals that, often, the largest percentage of them are nurses or healthcare assistants, with a much smaller response rate from physicians and other medical practitioners (Waterson et al. in preparation). Managers and senior administrators are also under-represented (see Hammer et al. 2011 for a PSC survey targeted at these groups). Sample characteristics such as these raise a number of issues, not least, how far we can generalise the findings from these types of studies, based as they are on a limited sample of healthcare staff.

A related issue is the inclusion of groups of people who are often ignored or excluded from studies of patient safety, namely patients and their carers. Very little work of this kind has been so far reported. One recent exception is the study by Cox et al. (2013) which used an adapted version of the AHRQ HSPSC survey (Sorra and Battles – Chapter 12) to assess the perceptions of parents of safety on the hospital units where their children were being cared for. Front-line healthcare staff (in particular nursing staff) appear to make up the bulk of respondents to PSC surveys. Accordingly, there needs to be a degree of caution in interpreting findings from these studies. Both researchers and healthcare practitioners are encouraged to ask questions about study sample characteristics and respondent profiles. Part of this improvement might involve the development of guidelines or standards covering the reporting of PSC studies (e.g. making sample characteristics more explicit).

Methodological Gaps and Opportunities

Currently, the field of PSC is dominated by the use of a very limited set of methods and tools. The European Network for Patient Safety Project (2010), for example, identified 19 different PSC instruments in use throughout the EU member states. The most frequently used of the 19 identified PSC instruments were the Hospital Survey on Patient Safety Culture (used in 12 EU states), the Manchester Patient Safety Framework (used in 3 EU member states) and the Safety Attitudes Questionnaire (used in 4 member states). In contrast to other industries, only a small subset of the types of methods which could be used to assess PSC have been taken up by researchers and healthcare practitioners. LeMaster and Wears (2012), for example, argue that direct observation and ethnography should be used more to assess actual or simulated behaviour (e.g. adverse events, near misses). They argue that there is a need to unpack a wider range of system factors involved in patient safety and to move beyond the limited range covered within culture/climate surveys. Future work should try to adopt a more eclectic approach towards alternative methodologies for PSC. There is a need to move beyond an exclusive focus on surveys and to examine how other tools, methods, interventions and theory might be linked to PSC. Much could be learnt from other more established industries where the use of mixed methods is common (see for example Mearns et al. 2013).

'Culture in the Large'

The question of how well PSC instruments developed in one nation translate to another has been raised by a number of studies and chapters in the book (e.g., Hammer and Manser – Chapter 11). Many instruments (e.g. HSPSC, SAQ) are derived from a US context and analysis of their psychometric properties has shown that, in some cases (e.g. measures of staffing and organisational learning), there are difficulties in using them in other national contexts. A variety of reasons might be given why these and other problems occur: the larger health system is very different and its operational functioning may vary a great deal across nations (e.g. private vs. publicly funded); staff may have different attitudes towards safety and this may vary greatly across roles and national contexts. These sorts of considerations underline the need to be cautious of attempts to simply translate and use PSC surveys without paying attention to national and wider health system characteristics. Surveys need to be tailored and modified in line with national and other cultural dimensions which have been shown to be important (e.g. power distance, individualism – Hofstede 1993). Pre-testing in the form of the use of focus groups and cognitive interviews is highly recommended.

The Relationship between PSC and Outcomes for Staff and Patients

Research establishing the link between PSC and a range of clinical (e.g. incident and near miss reports) and organisational outcomes (citizenship and organisational commitment; job and work design variables – Talati and Griffin – Chapter 3; Phipps and Ashcroft – Chapter 5) is still very much underdeveloped. This criticism is not new (Flin 2007; Flin et al. 2006). However, it represents a large gap in our understanding about the relationship between PSC and other phenomena and healthcare organisational dynamics. Similarly, criticisms can be made of the lack of studies relating PSC to measures derived from staff and patients (e.g. workload, satisfaction with care, hospital re-admission rates). The link between PSC and outcomes for staff and patients is still, as yet, relatively unexplored. By contrast, other areas outside of PSC have been more successful in demonstrating relationships between human resource management practices and healthcare productivity (e.g. Patterson et al. 2010). Much could be learned from such examples.

Improving the 'Science' Underpinning Patient Safety Culture

The reporting of psychometric data has improved a good deal since Flin et al.'s (2006) review; most studies now report data covering the reliabilities of survey dimensions for example. However, a number of problems still remain. Many studies appear not to be aware of established procedures for assessing the acceptability of survey item validity or reliability. It is common to read a study which concludes that a specific instrument is acceptable for use within a nation whilst at the same time reporting poor levels of reliability for specific dimensions. Likewise, many studies fail to report the results of exploratory or confirmatory factor analyses. There is also a large degree of variation when comparing the results of studies (e.g. factor structures) which have used an identical instrument, perhaps in modified form (e.g. the HSPSC – Hammer and Manser, Chapter 11). In combination, these problems indicate that there is still some way to go before we can be confident in assessing the degree to which PSC is being measured by specific instruments and how this compares across multiple levels of analysis, including nations, healthcare systems and locations (Karsh et al. 2014). There are a number of areas for future work which might help to improve the scientific basis of PSC as well facilitating comparison between studies. These include: (1) the development of a set of procedures covering survey development and evaluation (including psychometric criteria); (2) developing similar procedures or guidelines covering the reporting of PSC data and findings and (3) constructing a comparative database of worldwide studies which have used specific tools and instruments (e.g. HSPSC, SAQ).

Learning from Other Industries

As Chapter 17 (Kirwan and Shorrock) demonstrates, there is a huge amount that could be learnt from other safety industries and sectors. Some of the methods which have been tried and tested within the nuclear and oil and gas sectors are only just starting to be applied within healthcare (e.g. safety cases and human reliability assessment – Health Foundation 2012). Others such as the concept of 'safety intelligence' (discussed in Chapter 17; Fruhen et al. 2014) offer the potential for future work, particularly as it relates to senior managers within healthcare. Even a cursory glance at the accounts and recollections of well-known researchers in the field of safety science (e.g. Kletz 1993; Reason 2013) is enough to underline the dangers of not learning from other domains. Interdisciplinarity means not only looking at what other scientific research has been done outside 'traditional' studies in healthcare and patient safety. It also means paying closer attention to possible lessons from practitioners working in other industries. The field of PSC needs to pay more attention to past work within the safety-critical industries – some of this may well not translate to healthcare but in other cases there may be potential which is as yet unrealised.

Moving beyond 'What Went Wrong' – from 'Safety I' to 'Safety II'

A final area which has recently grown in importance is studies which have looked at how to improve the resilience of healthcare organisations. The emphasis here is not only on errors or failures to deliver care but also on identifying successes and other examples where routine practice has resulted in safety and reliability in care delivery. The focus on identifying error and the 'latent conditions and pathways' (Reason 1990) which contribute to mistakes and accidents is sometimes characterised as 'safety I' (Hollnagel et al. 2013; Rowley and Waring 2011). By contrast, 'safety II' concentrates on everyday performance and how this acts as a necessary resource for system flexibility and resilience. The safety management principle is continuously to anticipate developments and events (Eurocontrol 2013). Work on PSC has so far focused on 'safety I.' Future work could be carried out on developing tools and methods for identifying and assessing the conditions where performance variability can become difficult or impossible to monitor and control.

Summary and Conclusions

It is clear that over the course of a brief history, spanning less than 20 years, the field of PSC has made some real progress. Needless to say there is a long way to go.

> Measurement issues cut across many of the themes identified in this chapter and are an important priority for the future. Recent work by the Health Foundation in the UK following the publication of the Francis report (Francis 2013) has underlined the importance of accurate diagnosis and assessment of quality and safety in healthcare: 'our ability to measure and assess quality of care is improving, [*however*] … there are still many aspects of care, and care services, for which routinely available information on quality is inadequate or non-existent'. (Health Foundation 2013: 6)

Similar conclusions have been reached by a number of prominent authors in the field of patient safety (e.g. Provonost et al. 2011; Wachter 2010). Part of the 'solution' might be to develop further a core part of current work in PSC and patient safety, namely the application of the systems approach (Carayon et al. 2014; Reiman and Rollenhagen 2013; Waterson 2009; Wilson 2014). The approach remains under-exploited, particularly in terms of unpacking the multi-level properties of safety culture as well as understanding external systemic influences on safety (e.g. the influence of regulators and political and wider societal factors). Perhaps the most important aspect of the systems approach within PSC is that it might facilitate a shift towards a focus not only on senior managers in healthcare organisations but also other stakeholders, including patients and the wider public. As this book has hopefully demonstrated, the study and practical application of patient safety culture faces considerable challenges in the future. The chapters in this book give some indication of the vibrancy and extent of interest in PSC. Much remains to be done and to be improved however, in order to reduce unacceptably high rates of accidents, errors and near misses which occur across the world every year in hospitals and other healthcare environments.

References

Aven, T., Renn, O. and Rosa, E. 2011. The ontological status of the concept of risk. *Safety Science*, 49, 1074–9.

Ben-Ari, A. and Or-Chen, K. 2009. Integrating competing conceptions of risk: A call for future direction of research. *Journal of Risk Research*, 12, 865–77.

Carayon, P., Wetterneck, T.B., Rivera-Rodriguez, A.J., Schoofs Hundt, A., Hoonakker, P., Holden, R. and Gurses, A.P. 2014. Human factors systems approach to healthcare quality and patient safety. *Applied Ergonomics*, 45, 14–25.

Cox, E.D., Carayon, P., Hanson, K.W., Rajamanickam, V.P., Brown, R.L., Rathouz, P.J., DuBenske, L.L., Kelly, M.M. and Buel, L.A. 2013. Patient perceptions of children's hospital safety climate. *BMJ: Quality and Safety in Health Care*, 22, 664–71.

Eason, K., Dent, M., Waterson, P., Tutt, D., Hurd, P. and Thornett, A. 2012. *Getting the Benefit from Electronic Patient Information that Crosses Organisational Boundaries. Final Report*. NIHR Service Delivery and Organisation programme. Available at: http://www.nets.nihr.ac.uk/projects/hsdr/081803226 (last accessed on 23 April 2014).

Eisenhardt, K. 1993. High reliability organisations meet high velocity environments: Common dilemmas in nuclear power plants, aircraft carriers, and microcomputer forms. In K.H. Roberts (ed.), *New Challenges Understanding Organizations*. New York: Macmillan.

Eurocontrol 2013. *From Safety I to Safety II: A White Paper*. Available at: http://www.eurocontrol.int/sites/default/files/content/documents/nm/safety/safety_whitepaper_sept_2013-web.pdf (last accessed on 23 April 2014).

European Society for Quality in Healthcare 2010. Use of patient safety culture instruments and recommendations. *EUNetPas Project Report*, Aarhus, Denmark. Available at: http://ns208606.ovh.net/~extranet/images/EUNetPaS_ Publications/eunetpas-report-use-of-psci-and-recommandations-april-8-2010. pdf (last accessed on 22 April 2014).

Flin, R. 2007. Measuring safety culture in healthcare: A case for accurate diagnosis. *Safety Science*, 45(6), 653–67.

Flin, R., Burns C., Mearns, K., Yule, S. and Robertson, E.M. 2006. Measuring safety climate in health care. *Quality and Safety in Health Care*, 15, 109–15.

Francis, R. 2013. *The Mid Staffordshire NHS Foundation Trust Public Inquiry*. Available at: http://www.midstaffspublicinquiry.com/ (last accessed on 23 April 2014).

Fruhen, L.S., Mearns, K.J., Flin, R.H. and Kirwan, B. 2014. Safety Intelligence: An exploration of senior managers' characteristics. *Applied Ergonomics*, 45, 967–75.

Glendon, A.I. and Stanton, N.A. 2000. Perspectives on safety culture. *Safety Science*, 34, 193–214.

Halligan, M. and Zecevic, A. 2011. Safety culture in healthcare: A review of concepts, dimensions, measures and progress. *BMJ: Quality and Safety*, 20, 338–43.

Hammer, A., Ernstmann, N., Ommen, O., Wirtz, M., Manser, T., Pfeiffer, Y. and Pfaff, H. 2011. Psychometric properties of the Hospital Survey on Patient Safety Culture for hospital management (HSOPS_M). *BMC Health Services Research*, 11, 165.

Health Foundation 2012. *Using Safety Cases in Industry and Healthcare*. Available at: http://www.health.org.uk/public/cms/75/76/313/3847/Using%20safety%20 cases%20in%20industry%20and%20healthcare.pdf?realName=09HlEo.pdf (last accessed on 23 April 2014).

Health Foundation 2013. *Is the Quality of Care in England Getting Better?* Available at: http://www.qualitywatch.org.uk/sites/files/qualitywatch/field/ field_document/131010_QualityWatch_Annual-Statement-2013_Summary. pdf (last accessed on 23 April 2014).

Hofstede, G.H. 1993. Cultures and organizations: Software of the mind. *Administrative Science Quarterly*, 38(1), 132–4.

Hollnagel, E., Braithwaite, J. and Wears, R.L. (eds) 2013. *Resilient Health Care*. Farnham: Ashgate.

The Independent 2012. David Cameron: I will kill off safety culture. 5 January 2012. Available at: http://www.independent.co.uk/news/uk/politics/david-cameron-i- will-kill-off-safety-culture-6285238.html (last accessed on 23 April 2014).

Karsh, B.-T., Waterson, P.E. and Holden, R. 2014. Crossing levels in systems ergonomics: A framework to support 'mesoergonomic' inquiry. *Applied Ergonomics*, 45, 45–54.

Karsh, B.-T., Weinger, M.B., Abbott, P.A. and Wears, R.L. 2010. Health information technology: Fallacies and sober realities. *Journal of the American Medical Informatics Association*, (JAMIA) 17, 617e623.

Kletz, T. 1993. *Lessons from Disaster: How Organisations Have No Memory and Accidents Recur*. London: Institute of Chemical Engineers.

LeMaster, C.H. and Wears, R.L. 2012. Stepping back: Why patient safety is in need of a broader view than the safety climate survey provides. *Annals of Emergency Medicine*, 60(5), 564–6.

Leveson, N. 2012. *Engineering a Safer World*. Cambridge, Mass.: MIT Press.

Mearns, K. 2010. Safety culture and safety leadership – Do they matter? Keynote presentation at *Working on Safety 2010 Conference*. Available at: http://www.wos2010.no/assets/presentations/0-Key-note-Mearns.pdf (last accessed on 22 April 2014).

Mearns, K.J. and Flin, R. 1999. Assessing the state of organizational safety – Culture or climate? *Current Psychology*, 18(1), 5–17.

Mearns, K., Kirwan, B., Reader, T.W., Jackson, J., Kennedy, R. and Gordon, R. 2013. Development of a methodology for understanding and enhancing safety culture in Air Traffic Management. *Safety Science*, 53, 123–33.

Patterson, M., Rick, J., Wood, S., Carroll, C., Balain, S. and Booth, A. 2010. Systematic review of the links between human resource management practices and performance. *Health Technology Assessment*, 14(51), 1–iv.

Pfeiffer, Y. and Manser, T. 2010. Development of the German version of the Hospital Survey on Patient Safety Culture: Dimensionality and psychometric properties. *Safety Science*, 48, 1452–62.

Provonost, P.J., Berenholtz, S.M. and Morlock, L.L. 2011. Is quality of care improving in the UK? Yes, but we do not know why. *BMJ*, 342, c6646.

Reason, J. 1990. *Human Error*. Cambridge: Cambridge University Press.

Reason, J. 2013. *A Life in Error*. Farnham: Ashgate.

Reiman, T. and Rollenhagen, C. 2013. Does the concept of safety culture help or hinder systems thinking in safety? *Accident Analysis and Prevention*, 68, 5–15.

Rowley, E. and Waring, J. (eds) 2011. *A Socio-Cultural Perspective on Patient Safety*. Farnham: Ashgate.

Silbey, S.S. 2009. Taming Prometheus: Talking about safety and culture. *Annual Review of Sociology*, 35, 341–69.

Vogus, T.J., Sutcliffe, K.M. and Weick, K.E. 2010. Doing no harm: Enabling, enacting, and elaborating a culture of safety in health care. *Academy of Management Perspectives*, 24, 60–77.

Wachter, R.M. 2010. Patient safety at ten: Unmistakable progress, troubling gaps. *Health Affairs*, 29(1), 165–73.

Waterson, P.E. 2009. A critical review of the systems approach within patient safety research. *Ergonomics*, 52(10), 1185–95.

Waterson, P.E., Griffiths, P., Stride, C., Murphy, J. and Hignett, S. 2010. Psychometric properties of the Hospital Survey on Patient Safety: Findings from the UK. *BMJ: Quality and Safety in Health Care*, 19, 1–5.

Waterson, P.E., Jackson, J., Stride, C., Hutchinson, A., Hammer, A. and Manser, T. in preparation. *The Hospital Survey of Patient Safety Culture (HSPSC): A Cross-national Review of Study Characteristics and Psychometric Properties.* Unpublished manuscript.

Weick, K.H. 1987. Organizational culture as a source of reliability. *California Management Review*, 29, 112–27.

Wilson, J.R. 2014. Fundamentals of systems ergonomics/human factors. *Applied Ergonomics*, 45, 5–13.

Index

hypothetical measure of 81
improving 109–10, 125, 189, 291, 371
individual 314
managing 185, 342
monitoring 189
operational 350, 352
overall perceptions of 239, 244–7, 249
potential 352
prioritising of 48
surgical 147
units 286–7
valuing and prioritising 122
Safety Attitude Questionnaire, *see* SAQ
safety awareness 72–7, 85–7, 326
good 358
higher 87
staff's 80
safety climate 30–31, 44–6, 48, 52, 67–70,
88–91, 104–6, 108, 110, 113, 139,
141–2, 229–30, 253, 265, 277,
285–7, 354
assessed 354
assessment of 253
culture of 44
differences 238
dimensions 232, 253
environment 286
improvement 294
influence patient 105
measures 91
perceived 53
poor 90
positive 53, 140
questionnaires 140, 354, 357
regarding patient 253
research 354
scales 108
scores 108, 141, 287–8
strong 108–9
studies 104, 140
survey data 141
survey results 140
surveys 139–40, 287, 292
comprehensive patient 324
cross-sectional 106
in healthcare 140
tools 141, 149
safety climate surveys

Hospital Safety Culture Survey 141
safety climate surveys *Safety Attitudes
Questionnaire* 141
safety communication, as a proxy measure
of safety 89
safety compliance 209, 211, 213, 342
safety culture 1–4, 28–30, 44–6, 67–74,
119–23, 191–3, 207–9, 220–23,
229–30, 263–5, 276–9, 289–95,
301–3, 309–15, 321–4, 349–52,
356–8, 371–3
acceptable 75
American survey approach 315
approaches to 30
assessing and measuring of 3, 139,
210, 264, 292, 302, 339, 354
assessment 28, 33–4, 68, 72, 91–2,
104, 222, 279, 352, 354, 365
assessment process 68, 223
in aviation 54
bureaucratic 150
collective 352
comprehensive 291
concept of 28, 119, 356
core elements and dimensions of 71–2,
141, 276
correlates with safety outcomes 68
criteria 192–4
current 34
dataset 314
defined 113
developed 352
developing of mature 142, 150
development of 291
diagnosing of 67
dimensions 67–92, 159, 207–23, 230,
246, 250–51, 303, 308, 311, 313
early work on 2
effective 321, 334
emerging 338
in European Air Traffic Management
349–66
evaluations 201
factors 72–7, 85–8, 209, 214
generative 56
good patient 45, 183, 200, 271
growth of 326
healthy 350–51